EXPERIMENTAL AND COMPUTATIONAL TECHNIQUES IN SOFT CONDENSED MATTER PHYSICS

Soft condensed matter physics relies on a fundamental understanding of the interfaces between physics, chemistry, biology, and engineering for a host of materials and circumstances that are related to, but outside, the traditional definition of condensed matter physics. Featuring contributions from leading researchers in the field, this book uniquely discusses both the contemporary experimental and computational manifestations of soft condensed matter systems.

From particle tracking and image analysis, novel materials and computational methods, to confocal microscopy and bacterial assays, this book will equip the reader for collaborative and interdisciplinary research efforts relating to a range of modern problems in nonlinear and nonequilibrium systems. It will enable both graduate students and experienced researchers to supplement a more traditional understanding of thermodynamics and statistical systems with knowledge of the techniques used in contemporary investigations. Color versions of a selection of the figures are available at www.cambridge.org/9780521115902.

JEFFREY OLAFSEN is an Associate Professor of Physics at Baylor University in Waco, Texas. His research focuses on experimental nonequilibrium and nonlinear dynamics, including granular systems, biomechanics, and imaging algorithms.

EXPERIMENTAL AND COMPUTATIONAL TECHNIQUES IN SOFT CONDENSED MATTER PHYSICS

Edited by

JEFFREY OLAFSEN
Baylor University

CAMBRIDGE
UNIVERSITY PRESS

University Printing House, Cambridge CB2 8BS, United Kingdom

One Liberty Plaza, 20th Floor, New York, NY 10006, USA

477 Williamstown Road, Port Melbourne, VIC 3207, Australia

314-321, 3rd Floor, Plot 3, Splendor Forum, Jasola District Centre, New Delhi - 110025, India

79 Anson Road, #06-04/06, Singapore 079906

Cambridge University Press is part of the University of Cambridge.

It furthers the University's mission by disseminating knowledge in the pursuit of education, learning and research at the highest international levels of excellence.

www.cambridge.org
Information on this title: www.cambridge.org/9780521115902

First published 2010

A catalogue record for this publication is available from the British Library

Library of Congress Cataloging in Publication data
Experimental and computational techniques in soft condensed matter physics /
edited by Jeffrey S. Olafsen.
p. cm.
Includes index.
ISBN 978-0-521-11590-2 (hardback)
1. Soft condensed matter – Mathematics. 2. Experimental mathematics.
I. Olafsen, Jeffrey S., 1970– II. Title. III. Series.
QC173.458.S62E97 2010
530.4′1 – dc22 2010016136

ISBN 978-0-521-11590-2 Hardback

Additional resources for this publication at www.cambridge.org/9780521115902

For all those who succeed us, and
in deep appreciation of those who preceded us.

Contents

Contributors

Dr. George W. Baxter
School of Science
Penn State Erie
The Behrend College
4205 College Drive
Erie, PA 16563-0203

Dr. Nicholas C. Darnton
Visiting Asst. Prof. Physics
118 Merrill Science Center
PO Box: AC# 2244
Amherst College
Amherst MA 01002-5000

Dr. John de Bruyn
Department of Physics and Astronomy
The University of Western Ontario
1151 Richmond Street
London, Ontario, Canada, N6A 3K7

Dr. Michael Dennin
University of California, Irvine
4129 Frederick Reines Hall
Irvine, CA 92697-4575

Dr. Anthony D. Dinsmore
Associate Professor of Physics
University of Massachusetts Amherst
Hasbrouck Lab 411
666 North Pleasant Street
Amherst, MA 01003

Dr. Corey S. O'Hern
Associate Professor of Mechanical Engineering and Physics
Department of Mechanical Engineering
Yale University
P. O. Box 208286
New Haven, CT 06520-8286

Dr. Jeffrey S. Olafsen
Department of Physics
Baylor University
One Bear Place #97316
Waco, TX 76708-7316

Felix K. Oppong
(1) Department of Physics and Astronomy
The University of Western Ontario
1151 Richmond Street
London, Ontario, Canada N6A 3K7
(2) Unilever R & D
Colworth Science Park
Sharnbrook, Bedfordshire MK44 1LQ
UK

Dr. Nicholas T. Ouellette
Assistant Professor of Mechanical Engineering
Department of Mechanical Engineering
Yale University
P.O. Box 208286
New Haven, CT 06520-8286

Carl F. Schreck
Department of Physics
Yale University
New Haven,
CT 06520-8120

Dr. Leonardo E. Silbert
Assistant Professor
Department of Physics
Southern Illinois University
Carbondale, IL 62901

Dr. Brian Utter
MSC 4502
Dept. of Physics & Astronomy
James Madison University
Harrisonburg, VA 22807

Dr. Eric R. Weeks
Emory University
Physics Department
mail stop 1131/002/1AB
400 Dowman Drive
Atlanta, GA 30322

1

Microscopy of soft materials

ERIC R. WEEKS

1.1 Introduction

"Soft materials" is a loose term that applies to a wide variety of systems we encounter in our everyday experience, including:

- Colloids, which are microscopic solid particles in a liquid. Examples include toothpaste, paint, and ink.
- Emulsions, which are liquid droplets in another immiscible liquid, for example milk and mayonnaise. Typically a surfactant (soap) molecule or protein is added to prevent the droplets from coalescing.
- Foams, which are air bubbles in a liquid. Shaving cream is a common example.
- Sand, composed of large solid particles in vacuum, air, or a liquid; examples of the latter include quicksand and saturated wet sand at the beach.
- Gels are cross-linked polymers such as gelatin, or sticky colloidal particles. Usually the components of a gel (the polymers or particles) are at low concentration, but the gel still is elastic-like due to strong attractive forces between the gel components.

One common feature to all of these materials is that they are all comprised of objects of size 10 nm–1 mm; that is, objects much larger than atoms. In fact, it is these length scales that gives them their softness, as a typical elastic modulus characterizing these sorts of materials is $k_B T/a^3$, where k_B is Boltzmann's constant, T is the absolute temperature, and a is the size of the objects the material is made from [1]. For example, a could be the radius of a colloidal particle or of a sand grain, or in a conventional crystalline solid a would be the lattice spacing. For soft materials such as those listed above, a is much larger than the lattice spacing of a crystalline solid, resulting in a much reduced elastic modulus. This then is why soft materials are "soft."

Experimental and Computational Techniques in Soft Condensed Matter Physics, ed. Jeffrey Olafsen.
Published by Cambridge University Press.

While grains of sand are large enough to be seen with the naked eye, smaller objects, such as micron-diameter colloidal particles or emulsion droplets, are sufficiently large enough that they are easily viewed with optical microscopy. For this reason, microscopy has become an important tool for studying the structure of these types of samples, for example how the bubbles in a foam are arranged.

Along with the spatial scales, the temporal scales of these soft systems are often compatible with conventional video microscopy. For example, consider food products such as mayonnaise or peanut butter. These are somewhat solid-like, in that one can imagine a glop of peanut butter but not a puddle of peanut butter; however, they are also fairly easy to spread with a knife. The speed of spreading with a knife is set by human time scales. For example, a knife moved with velocity $v = 10$ cm/s spreading peanut butter with thickness $h = 4$ mm results in a strain rate given by $\dot{\gamma} = v/h = 25\ \text{s}^{-1}$. Thus, to understand what happens to the peanut butter microscopically when it is deformed by the knife, one might wish to take 25 images per second, which is easily achieved with inexpensive video cameras that are straightforward to connect to a microscope.

Another relevant time scale is set by diffusion. Very small particles undergo Brownian motion due to thermal fluctuations. The diffusion constant D for a sphere of radius a is given by the Stokes–Einstein–Sutherland formula:

$$D = k_B T / 6\pi\eta a \tag{1.1}$$

where η is the liquid viscosity [2, 3]. D then can be used to quantify the particle motion as follows. The motion in the x direction is equally likely to be left or right, so the mean displacement $\langle \Delta x \rangle = 0$, where the angle brackets indicate an average over many different particles. However, the mean square displacement will not be zero, but instead will be proportional to the time Δt over which the displacements are measured:

$$\langle \Delta x^2 \rangle = 2D\Delta t; \tag{1.2}$$

thus a larger value for D results in faster motion. $\langle \Delta x^2 \rangle$ is often called the variance. If you can imagine injecting a tiny blob of dye molecules at a single point, then the expanding cloud of dye has a characteristic size $\sqrt{\langle \Delta x^2 \rangle}$.

Using the mean square displacement, we can estimate the Brownian time scale τ_B as the time a typical particle takes to diffuse its own size as:

$$\tau_B \equiv a^2/2D = 3\pi\eta a^3 / k_B T. \tag{1.3}$$

For a particle of diameter $2a = 1\ \mu$m in water ($\eta = 1$ mPa·s) at room temperature, $\tau_B \approx 1$ s. Again, this motion is easy to study with a conventional video camera and a microscope. Of course, particles that are ten times smaller move 1000 times faster, by Equation (1.3); nonetheless, many systems of interest have micron-sized components.

This chapter will discuss several types of optical microscopy, although it will not provide a complete survey of the variety of microscopy techniques; the interested reader should consult reference [4]. Likewise, several representative applications of microscopy will be presented, although this will not be an exhaustive review; for more comprehensive review of articles of the applicability of microscopy to soft matter experiments, see references [5–8].

1.2 Video microscopy

Conventional optical microscopes are powerful, highly sophisticated instruments, which are nonetheless straightforward to use [4]. They are common in biology and biochemistry laboratories, and therefore it is easy to borrow time on one if you do not wish to purchase a microscope. For data acquisition, it is simple to attach a camera to the microscope, whether it be a conventional "snapshot" camera or a CCD video camera. The latter is now more common, and well-suited for studying samples which move or change. The output of the CCD video camera is usually attached to a frame grabber card in a computer, so that data are saved digitally, although a conventional VCR (video cassette recorder) can also be used.

There are several types of optical microscopy, and frequently the same microscope can be used in different ways by making slight modifications. The simplest technique is termed *brightfield microscopy*. Here, the light source is focused by a lens onto the sample, and the objective lens on the other side of the sample collects the light, allowing the user to see an image of the sample, for example the left image in Figure 1.1. Note that microscopes can be either "upright," where the objective lens is above the sample and the light source below, or "inverted," where the situation is reversed. Upright configurations are good for samples that "float," such as Langmuir–Blodgett films, which are layers of surfactant molecules on the surface of a water bath [9, 10] (see also Chapter 4). Inverted configurations are useful for samples that "sink," such as suspensions of dense particles. In some cases, inverted microscopes are also useful as the light source can be moved far above the sample, and other instrumentation then placed above the objective.

In brightfield microscopy, the image contrast can be due to components in the sample which absorb light (dyes, for example) or variations in the index of refraction of the sample. Variations in the refractive index are common, such as between oil and water in an emulsion; both oil and water are transparent, but the differences in the index of refraction allow the oil droplets to be seen. A good example of this effect is milk, which is white not because it contains white components, but because the variations in the index of refraction between the water and the milk fat scatter light randomly, resulting in a white appearance. A similar argument explains why snow is white, despite the transparency of ice; it is the reflection and refraction at the ice/air interfaces within the snow.

A second common method is *fluorescence microscopy*. Here, short-wavelength light is sent in, typically through the objective. Fluorescent molecules in the sample absorb this light, and radiate slightly longer wavelength (lower energy) light, which is collected through the objective. Special filters and mirrors are used to direct the light appropriately from the light source to the sample, and from the sample to the camera. In particular, a "dichromatic" (or "dichroic") mirror reflects the excitation light onto the sample, but allows the emitted light to pass through to the camera or microscope eyepiece. The advantage of fluorescence microscopy is that the dye can be placed in specific parts of the sample, such as in the solid particles of a colloidal suspension, or even in the surfactant molecules stabilizing emulsion droplets. This makes it easier to distinguish different sample components.

Fluorescence microscopy has one significant limitation: photobleaching. After dye molecules absorb the excitation light, but before they emit light, they can chemically react with oxygen present in the sample to form a non-fluorescent molecule. This only happens when they are excited, so photobleaching happens in direct proportion to the illumination light (with the limitation that once all the dye molecules are excited, they have reached saturation). Thus a sample can sit stably for a long time in the dark, but the photobleaching starts precisely when the sample is observed, that is when the excitation light is turned on (although, this only affects the local region illuminated by the microscope). Photobleaching manifests itself as the image becoming gradually darker. In some cases, photobleaching can be delayed by adding chemicals such as propyl gallate to the sample, which will bind to the free oxygen in the sample. This trick is limited to samples which are compatible with such chemicals, generally restricting the use of these chemicals to nonbiological samples. In other cases, photobleaching can be useful for studying local diffusion in samples, a technique known as "fluorescent recovery after photobleaching" [11]. Intense light is used to photobleach a region of the sample, and then low-intensity light is used to monitor the recovery of fluorescence as nonbleached dye molecules diffuse back into the region. With this method, the diffusivity of the dye molecules can be measured.

There are other common types of optical microscopy such as darkfield microscopy, phase contrast microscopy, and differential interference contrast microscopy (see the right image in Figure 1.1). These are modifications of bright-field microscopy and are able to enhance the contrast from very slight differences in the index of refraction. These techniques are more often used in biology, where one might wish to study a cell, filled with water and biopolymers, immersed in a cell culture medium that also is primarily water. These are less often used for soft condensed matter; for more details, see reference [4].

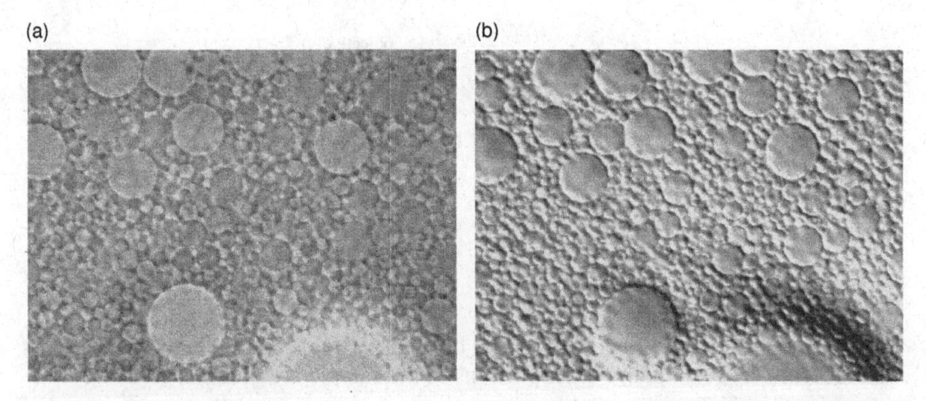

Figure 1.1 (a) Image of an oil-in-water emulsion taken using brightfield microscopy. The picture is ~60 μm wide. (b) Image of the same field of view taken using differential interference contrast microscopy. This form of microscopy produces a fictitious three-dimensional appearance; in reality, these are the cross-sections of spherical droplets. *Note:* the contrast in both pictures has been enhanced for better appearance.

1.3 Confocal microscopy

A confocal microscope is a laser scanned optical microscope. This is a fluorescent technique; the laser light is used to excite fluorescence in dye added to a sample. Typically, the laser beam is reflected off two scanning mirrors that raster the beam in the x and y directions on the sample. Any resulting fluorescent light is sent back through the microscope, and becomes descanned by the same mirrors. A dichroic mirror is used to then direct the fluorescent light onto a detector, usually a photomultiplier tube.

One additional modification is necessary to make a confocal microscope: before reaching the detector, the fluorescent light is focused onto a screen with a pinhole. All of the light from the focal point of the microscope passes through the pinhole, while any out-of-focus fluorescent light is blocked by this screen. This is crucial for viewing samples which may be full of fluorescent objects. Figure 1.2 shows excitation light being focused through a sample, and clearly the highest intensity of the excitation light is at the focus of the lens. However, the weaker out-of-focus light still excites fluorescence in other layers of the sample. While this emitted light is much weaker (in proportion to the excitation light), there is a large volume where this out-of-focus emitted light emanates. The pinhole filters out most of this out-of-focus light, allowing a strong and clean signal to come from the in-focus region. The pinhole is *conjugate* to the *focal* point of the lens, meaning that a point in focus at the focal point of the objective lens is re-imaged onto the pinhole, and this is the origin of the term "confocal." Intriguingly, the confocal microscope was

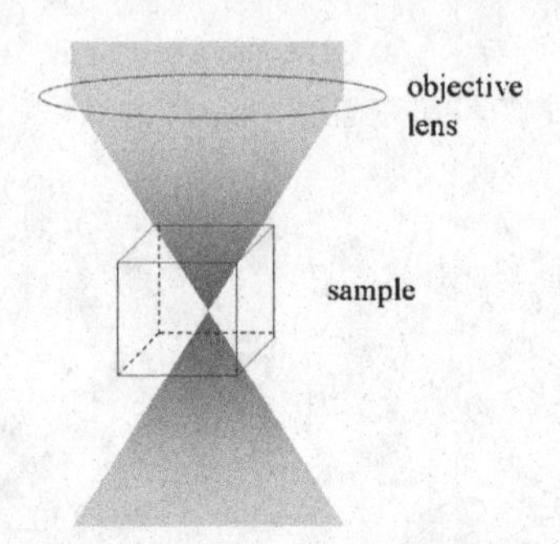

Figure 1.2 Sketch of light being focused by an objective lens. While the intensity is highest at the focal point, other portions of the sample are illuminated as well.

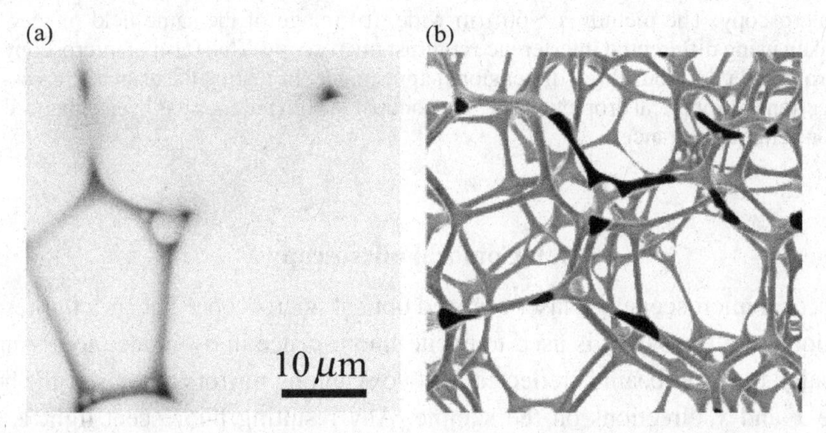

Figure 1.3 Images from confocal microscopy. (a) 2D image of a foam. Because of the narrow depth of focus of confocal microscopy, only foam channels that are close to parallel to the field of view are visible. At the upper right, a channel perpendicular to the field of view can be seen as a small black triangle. The depth of focus of this image is approximately 0.5 μm. (b) 3D image of a foam, 100 μm wide. (Images from Doug Wise and Eric R. Weeks.)

invented by Marvin Minsky in 1955, who is much better known for his work in artificial intelligence [12, 13].

This ability to reject out-of-focus fluorescent light directly results in the main strength of confocal microscopy, the ability to take three-dimensional pictures of samples. By rejecting out-of-focus light, a crisp two-dimensional image can be obtained, as shown in Figure 1.3(a). The sample (or objective lens) can be moved so as to focus at a different height z within the sample, and a new 2D image obtained. By collecting a stack of 2D images at different heights z, a 3D image is built up, as shown in Figure 1.3(b).

The time to scan one 2D image can range from 10 ms to several seconds, depending on the details of the confocal microscope and the desired image size and quality. For faster rates one typically substitutes an Acoustic-Optical Deflector

("AOD") for one of the scanning mirrors. This uses a radio-frequency sound wave to set up a standing density wave pattern in a crystal. The standing wave acts as a diffraction grating and steers the laser light depending on the wavelength of the standing wave, which is controlled by the sound wave. Because there are no moving parts, the AOD is faster than a scanning mirror. However, because the diffraction grating behavior is dependent on the wavelength of the diffracted light, the fluorescent light (being a longer wavelength than the excitation light, and not monochromatic) cannot be descanned by the AOD. Thus, AOD-based confocal microscopes replace the confocal pinhole with a confocal slit, with some slight loss of optical performance.

Another high-speed confocal microscopy technique is the Nipkow disk confocal microscope. This uses a spinning disk with many pinholes in it, such that some fraction of the field of view is illuminated at any given moment. As the disk spins, the entire field of view is scanned. The collected light is imaged by a standard video camera rather than a photomultiplier tube, and thus by using different cameras, the technique can be adapted to different conditions (low light levels, higher speed, etc.) Given that there are no moving mirrors, the scanning speed can be quite fast, 100 images/s or more depending on the camera choice. The drawback is that the resolution is not as good, as more out-of-focus light returns through the pinholes.

1.4 Strengths, weaknesses, and tradeoffs

1.4.1 Strengths of optical microscopy

Like the other methods described in this book, optical microscopy has strengths and weaknesses. One strength is the ability to visualize the heterogeneous structure. Often, complex fluids are spatially heterogeneous (for example, colloidal gels such as the one shown in Figure 1.4), and it may be desirable to understand this structure. A second strength is the ability to distinguish different features of the material, for example by using fluorescence techniques and adding different dyes to different regions. A third strength is that the data provided by video microscopy are fairly straightforward to understand; directly imaging what is in a sample can often be easier to interpret than more indirect measurements. A fourth strength is the ability to understand local properties. An example of this is studying how granular particles pack next to a wall, and comparing that with the packing in the interior. Other examples will be given in Section 1.6.

1.4.2 Weaknesses of optical microscopy

A weakness of video microscopy is the necessity to prepare the sample to be compatible with microscopy. Often this means matching the index of refraction of the components. For example, consider a dry sample of small glass spheres,

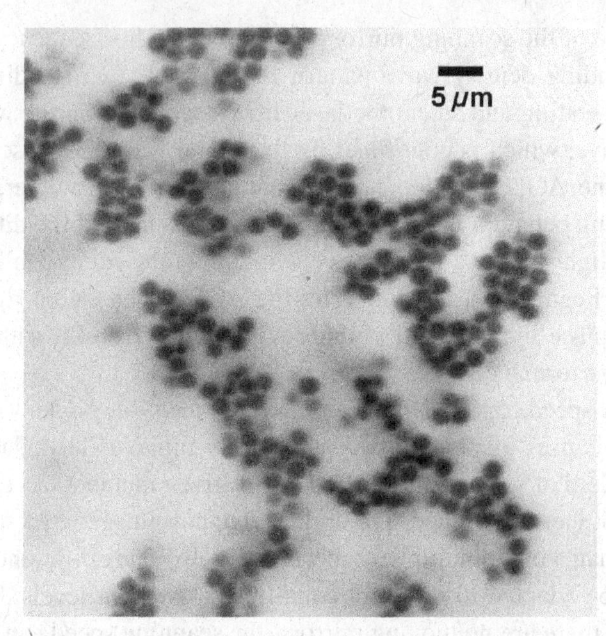

Figure 1.4 Image of a colloidal gel, taken with confocal microscopy. The particles are 2 μm diameter and aggregate due to the depletion force which is caused by adding small polymers to the solvent [14]. Image from Gary Hunter and Eric R. Weeks. See Chapter 3 for more information about colloidal gels.

perhaps each 50 μm diameter. These are of the right length scale for microscopy, but each sphere acts like a small lens and scatters light. Thus, this sample will look like white powder, for the same reason as the milk and snow examples discussed above. Microscopy would only be able to see the first layer of particles clearly. To better observe this sample, you would need to add an index-matching fluid. However, then you would be studying a wet sample, and the properties might be much different [15–19].

A second weakness of video microscopy is the lack of averaging. This shortcoming is complementary to the strength of visualizing spatially heterogeneous structures: while a good picture is gained of the local structure, it might be necessary to examine a large number of different regions to get a true picture of the average structure. For example, consider the case of colloidal crystallization. A sample of monodisperse hard sphere-like colloidal particles can form crystals, usually hexagonal-close-packed, as shown in Figure 1.5. This occurs at volume fractions $\phi > 0.494$, for reasons due to maximizing entropy by improving the local packing; for a fuller description, see references [1, 20]. Microscopy can be used to study the nucleation of these crystals, but is limited to the crystals that happen to nucleate within the field of view [20, 21]. Direct observation with microscopy can thus determine the specific shapes of a crystal nucleus. On the other hand,

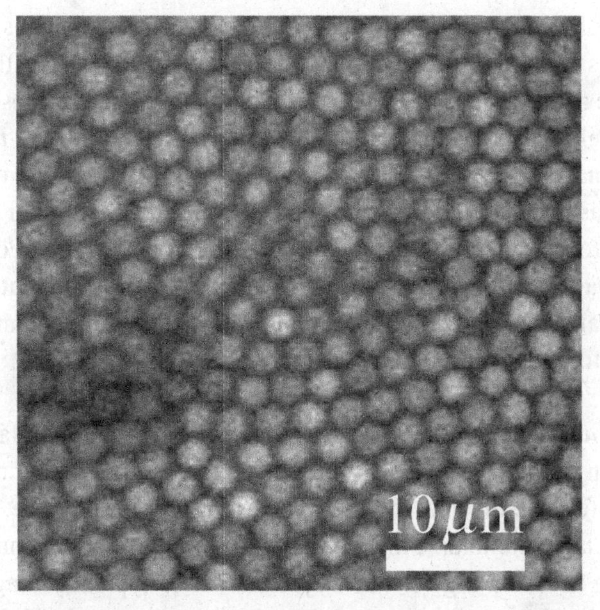

Figure 1.5 Confocal microscope image of a colloidal crystal. The black patch at the lower left is due to out-of-focus particles, likely because of a crystalline defect. A grain boundary is seen running through the middle of the image. The particles have a radius $a = 1.18$ μm and the sample is otherwise similar to those described in reference [20].

a spatially averaging technique like light scattering is better able to measure the average nucleation rate, but is unable to determine nuclei shapes [22].

1.4.3 Tradeoffs when doing optical microscopy

The speed of image acquisition can be either a strength or weakness, depending on the experiment. Video cameras typically acquire images at 30 frames per second (fps). Interlaced video cameras acquire an image alternately from the odd and even rows of pixels, and thus acquire half-images at 60 half-frames per second, which can be useful in some cases [23]. Confocal microscopes, as noted above, have speeds ranging from 1 fps to 30 fps, depending on the hardware. It is possible to use faster cameras, but the tradeoff is then the need to have an illumination level sufficient for the camera. Higher illumination levels (required for faster imaging) will often result in substantial heating of the sample. For fluorescent samples, higher illumination levels cause faster photobleaching.

For confocal microscopy, the main way to achieve faster speeds is to decrease the field of view and number of pixels in the image, so that the scanning optics have shorter distances to cover. One can then maintain the same light levels and the same photobleaching rate as with the slower scanning speed over a larger field of view. It is, of course, harder to take three-dimensional images very quickly; the limitation is how fast the objective can be moved to scan in the z direction. Scanning in the z direction is often achieved by attaching the microscope objective to a fast piezo-electric transducer, which is in turn attached to the microscope. In this way, 3D images of reasonable size can be acquired at speeds of more than 1 image/s, with faster rates possible for thinner image stacks (thinner in z).

One straightforward tradeoff is the optical resolution as compared to the field of view. Higher magnification lenses have better optical resolution, but at the price of looking at a smaller region within the sample. However, to understand this tradeoff one first needs to understand the relevant terms:

- **Magnification**: Technically this is defined as the apparent angular extent of the image as seen by the eye, compared to the actual angular extent of the object if it was at a reference distance of 25 cm from the eye. In practice, magnification is not a crucial parameter. Any image can be magnified as much as desired by projecting it onto a big screen. Objective lenses typically range from $5\times$ to $100\times$ magnification, and the rest of the microscope optics typically provide an additional factor of $10\times$.
- **Field of view**: More directly useful than the magnification, this is the region within the sample that is viewed. For the highest magnification lenses, the field of view can be as small as $50 \times 50\ \mu\text{m}^2$. Field of view also relates to the camera and the optics attaching the camera to the microscope. Typically the field of view as seen by the camera is a quarter of the area seen through the microscope eyepieces.
- **Size of image in pixels**: This depends on the camera. Having more pixels over the same field of view is often helpful, although it requires more room to store the data on a computer. Note, however, that switching to a camera with more pixels does not increase the field of view, which is set by optics. However, the physical size of the CCD chip within a camera can impact the field of view.
- **Resolution**: The optical resolution is set by diffraction effects, quantified by Rayleigh's criterion, and is tied with the quality of the optics and the wavelength of light used to view the sample [4]. The best (smallest) resolution of optical microscopes is about 200 nm. The optical resolution quantifies the ability to distinguish two closely spaced objects. For example, a typical test pattern is a grid of lines, and if the grid spacing is smaller than the resolution, no lines will be seen. Another way to think about resolution is to consider the fluorescence microscopy image of a small fluorescent molecule. The size of the molecule is much smaller than the resolution, but the light emitted from the molecule can still be seen. However, rather than appearing as an extremely sharp point of light, the molecule appears as a fuzzy round spot with a diameter equal to the optical

resolution. The best way to think of optical resolution is that all images will be blurry on the scale of the resolution.

- **Ability to resolve features**: Consider the same fluorescent molecule, which appears as a diffraction limited spot. As just mentioned, the presence of this molecule can be detected by the emitted light, and the true size of the molecule is irrelevant (although the intensity of the emitted light may be highly relevant for the ability of the camera to see it). How accurately can the position of the center of this molecule be determined? Perhaps surprisingly, it is easy to do much better than the optical resolution. Suppose the image of the fluorescent molecule spans a region N pixels wide in the image, and each pixel has size δ in microns. Then, typically the position of the molecule can be determined to a precision of $\approx \delta/N$. This argument also holds true for other optical microscopy techniques, not just fluorescence. The real limitation of the resolution is that two such molecules that are closer together than the resolution will blend together into one diffraction limited spot, and their separation will be nearly impossible to measure. The fact that there are two molecules rather than one might still be measurable by the quantity of emitted light, although this ability applies only to fluorescent techniques.

Overall, as with many experimental techniques, these abilities can be tweaked by spending more money. Higher magnification lenses tend to be more expensive. At a given magnification, lenses with higher resolution are more expensive. In general, the resolution of a lens is reasonably well-suited to the field of view. That is, a high-resolution lens with low magnification would not be useful, as then the limitations would be set by the limited pixel size of the camera rather than optical considerations. It is best to work with a camera with pixel sizes at least as small as the resolution of the microscope; that is, an image of an object the size of δ (the resolution limit) should be at least one pixel wide. Without this, you do not get the best abilities from your microscope; one does not want a cheap camera limiting the capability of an expensive microscope.

1.5 Particle tracking

Particle tracking is a powerful and common technique used to study soft materials [23]. The general idea is to embed tracer particles in our sample that are imaged using video microscopy. By taking a movie of these particles moving, a computer can post-process the images to determine where the particles are at each moment. These particle positions are then used for the tracking step. If the particles do not move large distances between subsequent images, then they can be easily tracked, as shown in Figure 1.6(a). Specifically, they need to move less between images than their typical inter-particle spacing [23].

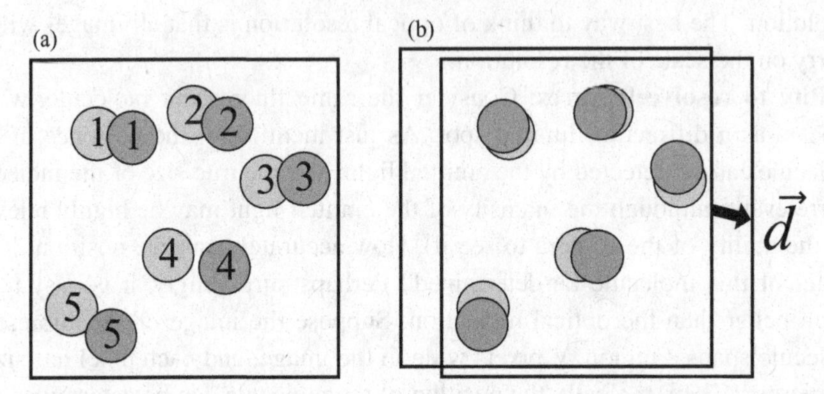

Figure 1.6 (a) Schematic of particle positions at one time (light gray) and a later time (dark gray). If the particles have not moved much between the two times, particle tracking can assign each particle a unique identity number at each time. (b) In particle image velocimetry, two regions within an image are compared at two different times separated by Δt, and the displacement vector necessary to shift one region onto the other provides the velocity as $\vec{v} = \vec{d}/\Delta t$. This method neither requires identifying individual particles, nor requires that the regions are exact matches, as indicated in this sketch.

In some cases, the particles themselves may be a key part of the sample, for example with colloids or granular materials. With confocal microscopy, particles can be tracked in three dimensions, allowing the structure of the materials to be determined or three-dimensional flow fields to be measured [7, 24]. Specific examples will be discussed in Section 1.6.

A complementary method is *particle image velocimetry* (PIV). In this method, again a sequence of images is taken. PIV uses the fact that if there is motion within the sample, then small regions within the image will move roughly uniformly between consecutive frames. By identifying how these regions move with time, the coarse-grained displacement can be measured. The key assumption is that the spatial variation of the velocity field is not too great; that is, within a small region the velocity is fairly uniform. Consider two sequentially taken images that are slightly different from each other, and examine a small region of the first image, say 10×10 pixel^2. To determine the velocity vector, compare this region with nearby 10×10 pixel^2 regions of the second image. The region from the second image that most closely resembles the region from the first image then indicates that whatever you are looking at has moved to the position of the second region. (If the region in the second image has the same location as the original region, then nothing has moved.) Thus, this measures a displacement vector for the position of the original region, as shown in Figure 1.6(b). By checking regions everywhere in your picture, you get a displacement vector field. The advantage of this method is that individual particles do not need to be identified. Typically the size of the compared regions are large enough that there are several particles within each region, and so there is enough information within that area to have a sensible determination of where those particles have moved.

One advantage of PIV (compared to particle tracking) is that the particles are never identified as individual particles, and so PIV easily identifies the flow field even for noisy images. This technique works well for samples with good contrast, and is well suited for movies where particles move large distances between images. The advantage of particle tracking is that information is obtained about each and every individual particle, which can be useful in many applications where flow fields are not the primary interest. Furthermore, with careful techniques that combine PIV and particle tracking, it is possible to track particles even in quickly moving flows where particles move large distances [7]. See Chapter 7 for more information about using particles to measure flow fields.

1.6 Specific applications

This section discusses a few examples of the utility of microscopy for studying soft materials. As noted earlier, these are only representative examples rather than an exhaustive review.

1.6.1 Structure of colloidal glasses

A major application of microscopy is to visualize the structure of soft materials. In 1995, van Blaaderen and Wiltzius used a confocal microscope to study the structure of dense colloidal glasses; this study was one of the very first major applications of confocal microscopy to soft materials [25]. To understand their experiment, a few comments about the glass transition should be made.

When some materials are rapidly cooled, they form an amorphous solid known as a glass. The transition from a liquid to a disordered solid is called the glass transition [26, 27]. As the temperature of a molecular glass-forming material is decreased, the viscosity rises smoothly, but rapidly, with little change in the microscopic structure [25, 28, 29]. Colloidal suspensions are a useful model system for studying the glass transition. In particular, colloidal particles can often be considered as hard spheres with an interaction potential due only to excluded volume effects [30, 31]. Hard spheres only interact when their surfaces touch, and then they are strongly repulsive (imagine frictionless marbles, for example). Because of their simplicity, hard spheres are used to simulate properties of crystals [32, 33], liquids [34–36], and glasses [37–39]. Clearly, attractive interactions between atoms and molecules are responsible for dense phases of matter, but given dense states of matter, repulsive interactions play the dominant role in determining the structure [40, 41].

The control parameter for hard sphere systems is the concentration, expressed as the volume fraction ϕ. For $\phi > \phi_g \approx 0.58$, the system acts like a glass. The transition is the point where particles no longer diffuse through the sample; for $\phi < \phi_g$, spheres do diffuse at long times, although the asymptotic diffusion coefficient D_∞ decreases sharply as the concentration increases [28, 39]. The transition at ϕ_g occurs even though the spheres are not completely packed together; in fact, the volume fraction must be increased to $\phi_{\text{RCP}} \approx 0.64$ (for "random-close-packed" spheres [34, 42–45]) before the spheres are motionless. See Chapters 3 and 5 for more details of hard sphere systems.

The work of van Blaaderen and Wiltzius studied the structure of colloidal glasses. A key question about glass transitions is whether there is a structural change that explains the increase in viscosity as the glassy state is approached. Indirect measurements suggest there is no change [28, 29]. In their experiment, van Blaaderen and Wiltzius used confocal microscopy to directly observe particle positions in a colloidal glass. Even with full information about particle positions, they did not find any structural change in the glassy state [25]. In particular, they could find no structural correlation length scale, and therefore no direct structural link to the growing viscosity. Several quantities related to structure *could* be measured in their experiments, however, and agreed quite well with simulations of random close-packed spheres [25].

This is merely one example of microscopy being used to study the structure of soft materials. Microscopy has also been used to study the structure of colloidal gels [14, 46], colloidal crystals [20, 47, 48], and emulsions [49, 50].

1.6.2 Dynamics of colloidal glasses

At the time of the earlier work of van Blaaderen and Wiltzius (1995), confocal microscopy was still a relatively slow technique, taking one second to scan a 2D image and correspondingly longer for a 3D stack of images [46]. The study of colloidal glasses, with their naturally slow dynamics, was thus well-suited to confocal microscopy. Faster microscopy techniques allow the study of faster motion, and, in particular, particle tracking techniques enable investigation of how particle motion changes in liquids as the glass transition is approached, rather than only motion within the slower glassy state.

Light scattering techniques allow the measurement of the average behavior of samples. Thus, early light scattering experiments that studied dense colloidal samples found results analogous to the average mean square displacement of particles, as shown in Figure 1.7. At time scales less than 100 s, particles are "caged" by their neighbors. Because they cannot move past their neighboring particles, the mean square displacement shows a plateau, with the plateau height related to the size of the "cage" formed by the neighbors. At long time scales, particles are able again

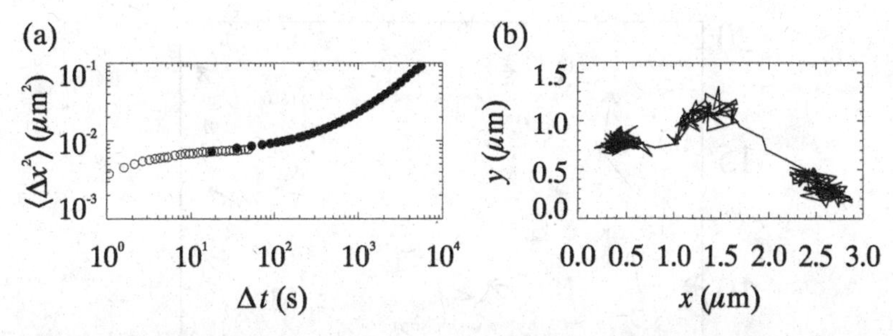

Figure 1.7 (a) Mean square displacement, measured by confocal microscopy, on a colloidal sample with volume fraction $\phi = 0.52$. Open symbols are from two-dimensional images taken within the 3D sample, and solid symbols are from three-dimensional images of the same sample. The particles in the sample are monodisperse with a radius $a = 1.18$ μm. (b) Particle trajectory from the same sample, showing 120 min of data. The particle started at the left, and changed position at $t \approx 51$ min and $t \approx 84$ min. In addition to moving in x and y, the particle also moved $\delta z = 0.3$ μm during the first jump, and $\delta z = 0.2$ μm during the second jump.

to diffuse through the sample when the cages rearrange. That is, the neighboring particles somehow rearrange so that the caged particle can move to a new location. Of course, every particle is caged; in other words, not only is a given particle caged, but it also forms part of the cages for its neighbors.

This general picture has been known for a while, and microscopy allows us to directly visualize the motion responsible for cage rearrangements. Figure 1.7(a) suggests a smooth evolution from diffusion, to cage trapping, to cage rearrangement; perhaps cage rearrangements could be a slow wandering of the position of the particle. Figure 1.7(b) shows that this is not the case; rather, cage rearrangements can be rather abrupt. The jumps in particle position shown take approximately 1–2 minutes to occur. This is slow compared to a freely diffusing particle, which would diffuse approximately 1 μm in 10 s, but fast compared to the time scales between rearrangements in this dense sample. Averaging over thousands of particles, which undergo many cage rearrangements, results in the smooth curve shown in Figure 1.7(a).

Furthermore, with microscopy we can answer the question about the motion of the adjacent particles when the cage rearrangement occurs; that is, what does a cage rearrangement look like? Figure 1.8 shows the motion of several particles in one region. Neighboring particles with large displacements typically move in similar directions, although occasionally cases of neighboring particles with different directions can also be seen [51]. Notably, in general the most mobile particles are clustered together, and also there are regions where particles move relatively little [52, 53]. Overall this is termed "dynamical heterogeneity" – the motion of particles is spatially and temporally heterogeneous. Many experiments provided indirect evidence of dynamical heterogeneity in glassy materials [54],

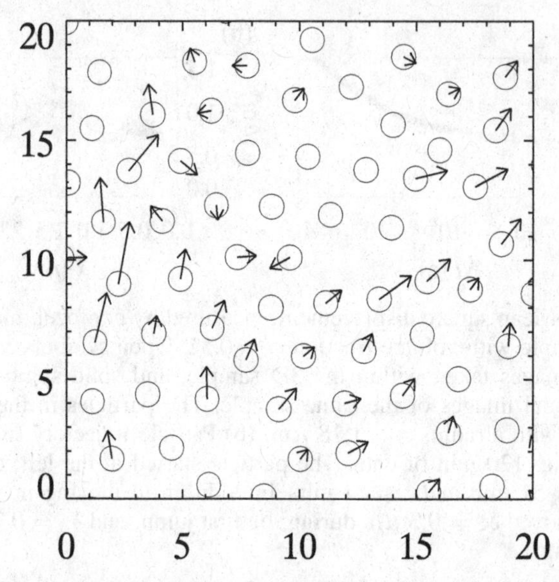

Figure 1.8 Displacement vectors for particles from the sample shown in Figure 1.7. The axes are labeled in microns, and the particle radius is $a = 1.18\ \mu$m; the circles indicating particles are not drawn to scale. The time scale for the displacement is 10 min, and the displacement vectors have been stretched by a factor of 5 to make the cooperative motions more apparent. Only displacements longer than 0.1 μm are shown. The particles are taken from a 2 μm thick slice, and because some have different z positions, they may be closer in x and y than their diameters.

and microscopy allows us to directly see it. While the data shown here are from confocal microscopy, similar results were obtained earlier by video microscopy [55,56].

1.6.3 *Microrheology: determining macroscopic properties from microscopic measurements*

Another widespread application of microscopy is *microrheology*. For example, consider adding a few tracer particles to a Newtonian liquid such as water. Using particle tracking methods, their diffusive motion can be observed and a mean square displacement $\langle \Delta x^2 \rangle(\Delta t)$ calculated. The diffusion constant D can be calculated using Equation (1.2). If the particle size a and temperature T are known, then the viscosity η of the solvent can be determined from Equation (1.1). This is the simplest example of microrheology, using the motion of tracer particles to measure the viscosity.

Likewise, consider a particle trapped in a purely elastic medium like gelatin. It too undergoes thermal vibrations, but now the particle is trapped in place. The amplitude of the vibrations is related to $k_B T$ and the elastic modulus of the material. By measuring $\langle \Delta x^2 \rangle (\Delta t \to \infty)$, the elastic modulus can be determined. More generally, for a viscoelastic material, the Δt dependence of $\langle \Delta x^2 \rangle$ can be used to determine the frequency-dependent viscoelastic moduli [57–60].

This method is complementary to macroscopic rheology. In a macroscopic rheometer, a few milliliters of sample are placed in the instrument, for example between two circular plates. In a "controlled strain" rheometer, the top plate is oscillated at a given frequency, and the torque required to maintain that rotation is measured. In a "controlled stress" rheometer, this is reversed: the torque is controlled and the plate displacement is measured. In either case, the relationship between the applied strain and required torque determines the viscoelastic moduli [1, 61]. (In general, there is a phase angle between the strain and torque, and the sample has both viscous and elastic moduli. The physical meaning of the phase angle is that the torque and the strain, while both oscillating sinusoidally, do not do so in phase.) A large advantage of microrheology is that the same rheological properties can be measured with *microliters* of sample rather than milliliters. For materials such as biopolymers, purified samples may only be available in tiny quantities. Another advantage is that microrheology can sometimes cover a larger frequency range. Essentially, the camera speed of the microscope sets the highest frequency obtainable, and the duration of the measurement sets the lowest frequency. Using a fast camera, higher frequencies can be studied (>1000 Hz); macroscopic rheometers typically are limited to about 100 Hz at the fast end. Microrheology can also be done using light scattering techniques, which can get to still higher frequencies [57].

A further advantage of using microscopy for microrheology is that the measurement is local. If the sample is spatially heterogeneous, then this can be observed with microscopy and the local rheology determined. In some cases, microrheology can be used to diagnose spatial heterogeneity. The most straightforward method for this is termed *two-particle microrheology* [62, 63]. This method uses the cross-correlation of the motion of pairs of particles to infer the rheological properties. The cross-correlation can be examined as a function of the separation R of particles. For a homogeneous viscoelastic medium, the correlation should decay as $1/R$. Deviations from this can be used to infer spatial structure of the sample [64]. This method is very robust, and does not depend on the tracer size, shape, or interactions between the tracers and the medium [62]. For example, imagine trying to study microrheology of a polymer network with added tracer particles. The single-particle mean square displacement would be different if particles stick to the polymers, or if the polymers avoid the particles. However, two-particle microrheology will give the correct results in either case, as it only depends on the long-wavelength behavior of the material. The local details which depend on the specific particle/system interactions are uncorrelated over the longer distances, and thus drop out of the analysis.

A limitation of microrheology is that for very stiff or very viscous materials, the particle motion may be hard to observe. For purely viscous materials, $\langle \Delta x^2 \rangle \sim \Delta t / \eta a$, and for purely elastic materials, $\langle \Delta x^2 \rangle \sim 1/Ga$ (using the elastic modulus G). Thus, for very stiff or very viscous materials, either the particle size a needs to be decreased or else one must be able to measure very small displacements Δx. A limitation of the two-particle method is that, in very spatially heterogeneous samples, one must study the correlated motion of particles separated by distances larger than any spatial heterogeneity, and, at those separations, the amount of correlated motion may be very hard to distinguish from the uncorrelated Brownian motion. (Note that macroscopic rheometers avoid this problem by having a macroscopically large gap between the two plates. Samples which are heterogeneous on the scale of a macroscopic rheometer will still have problems; for example, rheology of granular materials is tricky [61].)

See Chapter 6 for more details about rheology and microrheology.

1.6.4 Flow fields

A final application of microscopy is to determine the flow fields within microscopic samples and microfluidic devices. As mentioned in Section 1.5, this can be done either with particle tracking or particle image velocimetry.

In microscopic systems, the Reynolds number characterizing flow is typically very small. The Reynolds number, given by:

$$\mathrm{Re} = UL\rho/\eta, \tag{1.4}$$

is based on a characteristic velocity U, characteristic length L, viscosity η, and fluid density ρ. So, for water flowing with a maximum velocity of 1000 μm/s in a cylindrical tube of radius 100 μm, Re = 0.1. The Reynolds number indicates the relative importance of inertial and viscous forces. Viscous forces are much more significant in situations with Re $\ll 1$. For a cylindrical tube, one expects the velocity profile to be parabolic with zero velocity at the walls (due to the no-slip boundary condition) and maximal velocity along the long axis of the tube. This simple flow field is termed "pipe Poiseuille flow" [65]. Given that microscopes study things at small length scales corresponding to low Reynolds numbers, and flow fields at low Reynolds numbers are simple, why would one want to measure flow fields in microscopic devices? Here we consider three examples where measurements are indeed useful.

Experiments by Frank *et al.* examined the flow of particle-laden suspensions flowing through rectangular capillary tubes [66]. Indeed, to first approximation, the flow was parabolic in accordance with pipe Poiseuille flow. However, due to shear forces between particles, the particle concentration evolves such that more particles are in the center of the tube compared to near the walls. This is because of the parabolic velocity profile which ensures that particles move at different velocities. A particle will overtake slower particles which are closer to the wall, and be overtaken by faster particles which are closer to the center of the tube. When two particles moving at different velocities approach each other, they interact via the fluid, which jostles them into new positions. There is an effective diffusivity for this particle which depends on the shear rate, that is the slope of the velocity gradient. Particles in the center of the tube see a much smaller velocity gradient (since they are at the local maximum of the flow) and tend to stay in the center of the tube. The overall result is that particles move out of regions of high shear (near the walls) and into regions of low shear (in the center) (although note that the full explanation of this phenomenon is more detailed as discussed in reference [66]). Ultimately for small colloidal particles, this is balanced by diffusion which attempts to homogenize the particle distribution [66]. The heterogeneous particle distribution modifies the velocity profile. Microscopy allows simultaneous measurements of the concentration and velocity profile, and in fact the deviations from a pure parabolic velocity profile agree well with theoretical predictions [66].

The second example relates to mixing in microfluidic devices. In macroscopic applications, turbulence is often used to provide rapid mixing. In microfluidic applications, $Re \ll 1$, so turbulence is suppressed. Experiments by Gao *et al.* studied mixing in microfluidic flow chambers designed to enhance mixing [67]. The microchannels had a rectangular cross-section, with occasional raised ridges on the bottom of the channel. The ridges had a "herringbone" pattern, shaped like a "V" with the pointed end of the V pointing downstream. Using confocal microscopy, they studied the flow of moderately concentrated colloidal suspensions through these microchannels. Their observations enabled them to determine the best geometries to optimize mixing.

The third example is that of water draining through a foam. Between pairs of bubbles in a foam, one has a foam face. Where three bubbles meet, there is a liquid channel in between; these are visible in Figure 1.3. Between four mutually adjacent bubbles, four liquid channels come together and meet in a node. Thus the structure of channels and nodes form a network, somewhat like connected pipes, through which liquid can drain out of a foam. Figure 1.9 shows such a draining foam, and clearly the channels are thicker at the bottom where the liquid has not yet completely drained. Several questions then arise: Are the walls of the bubbles slippery or rigid? Does water flow through the channels like pipe Poiseuille flow, or do the surfactants at the boundaries of the channels prevent a no-slip boundary condition?

Using confocal microscopy, Koehler *et al.* studied the flow of water through a draining foam [68, 69]. They added water with a few tracer particles at the

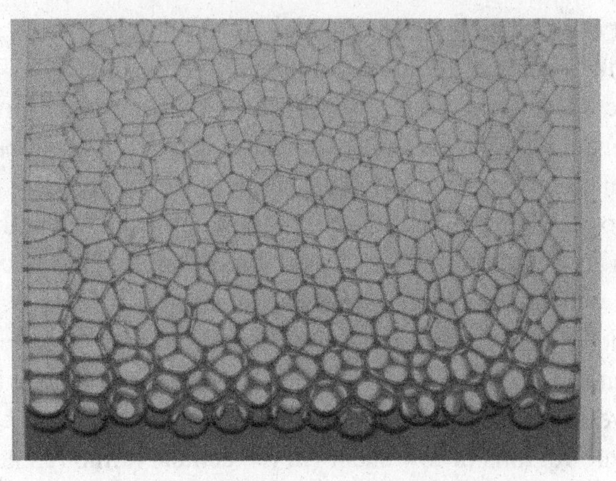

Figure 1.9 Image of a draining foam (taken by Gianguido C. Cianci and Eric R. Weeks).

top of a foam, and, by tilting the microscope sideways, were able to study the flow of the water through the foam channels. Foams made with a small-molecule surfactant resulted in plug-like flow, indicating that the channel walls were indeed slippery. For plug-like flow, the velocity is nearly constant across the width of the channel (similar to how toothpaste comes out of the tube). Using a larger protein as a surfactant, they found parabolic flow, indicating that the no-slip boundary condition was recovered. Thus, changing the surfactant can dramatically change the character of foam drainage (and thus the rate at which foams drain).

1.7 Summary

Microscopy as a scientific technique dates back many centuries, for example to the important work of Antonie van Leeuwenhoek in the late 1600s. This chapter has only touched on suggestive applications of microscopic techniques, and only a few varieties of optical microscopy. Overall, microscopes coupled with video cameras, which in turn can be directly connected to a computer, have greatly expanded the applications of microscopy. Within the limitations of Section 1.4, microscopy is a powerful tool for studying soft condensed matter systems. Given that you get to see the samples, it is also a fun technique, as hopefully the images in this chapter have suggested.

Acknowledgments

I thank J. C. Crocker for helpful discussions on microscopy and particle tracking over many years, and D. Chen, K. W. Desmond, L. Golick, and G. L. Hunter for useful comments on this chapter. This work was supported by the National Science Foundation under Grant No. DMR-0804174.

References

[1] R. A. L. Jones, *Soft Condensed Matter* (Oxford University Press, 2002).
[2] A. Einstein, "On the movement of small particles suspended in a stationary liquid demanded by the molecular-kinetic theory of heat," *Annalen der Physik (Leipzig)* **17**, 549–60 (1905).
[3] W. Sutherland, "A dynamical theory of diffusion for non-electrolytes and the molecular mass of albumin," *Phil. Mag.* **9**, 781–5 (1905).
[4] S. Inoué and K. R. Spring, *Video Microscopy: The Fundamentals* (The Language of Science), 2nd edition (Springer, 1997).
[5] P. Habdas and E. R. Weeks, "Video microscopy of colloidal suspensions and colloidal crystals," *Current Opinion in Colloid and Interface Science* **7**, 196–203 (2002).
[6] V. Prasad, D. Semwogerere, and E. R. Weeks, "Confocal microscopy of colloids," *J. Phys.: Cond. Matt.* **19**, 113102 (2007).
[7] R. Besseling, L. Isa, E. R. Weeks, and W. C. K. Poon, "Quantitative imaging of colloidal flows," *Advances in Colloid and Interface Science* **146**, 1–17 (2009).
[8] V. J. Anderson and H. N. Lekkerkerker, "Insights into phase transition kinetics from colloid science," *Nature* **416**, 811–15 (2002).
[9] M. Sickert, F. Rondelez, and H. A. Stone, "Single-particle Brownian dynamics for characterizing the rheology of fluid Langmuir monolayers," *Europhys. Lett.* **79**, 66005 (2007).
[10] R. Walder, C. F. Schmidt, and M. Dennin, "Combined macro- and microrheometer for use with Langmuir monolayers," *Rev. Sci. Inst.* **79**, 063905 (2008).
[11] D. Axelrod, D. Koppel, J. Schlessinger, E. Elson, and W. Webb, "Mobility measurement by analysis of fluorescence photobleaching recovery kinetics," *Biophys. J.* **16**, 1055–69 (1976).
[12] M. Minsky, "Memoir on inventing the confocal scanning microscope," *Scanning* **10**, 128–38 (1988).
[13] M. Minsky, "Memoir on inventing the confocal scanning microscope," http://web.media.mit.edu/~minsky/papers/ConfocalMemoir.html (1988).
[14] A. D. Dinsmore, V. Prasad, I. Y. Wong, and D. A. Weitz, "Microscopic structure and elasticity of weakly aggregated colloidal gels," *Phys. Rev. Lett.* **96**, 185502 (2006).
[15] R. A. Bagnold, "Experiments on a gravity-free dispersion of large solid spheres in a Newtonian fluid under shear," *Proc. Roy. Soc. London. Series A.* **225**, 49–63 (1954).
[16] A. Samadani and A. Kudrolli, "Segregation transitions in wet granular matter," *Phys. Rev. Lett.* **85**, 5102–5 (2000).
[17] N. Jain, D. V. Khakhar, R. M. Lueptow, and J. M. Ottino, "Self-organization in granular slurries," *Phys. Rev. Lett.* **86**, 3771–4 (2001).
[18] J. C. Géminard, W. Losert, and J. P. Gollub, "Frictional mechanics of wet granular material," *Phys. Rev. E* **59**, 5881–90 (1999).
[19] H. A. Barnes, "Shear-thickening ('dilatancy') in suspensions of nonaggregating solid particles dispersed in Newtonian liquids," *J. Rheo.* **33**, 329–66 (1989).
[20] U. Gasser, E. R. Weeks, A. Schofield, P. N. Pusey, and D. A. Weitz, "Real-space imaging of nucleation and growth in colloidal crystallization," *Science* **292**, 258–62 (2001).

[21] P. Wette, H. J. Schöpe, and T. Palberg, "Microscopic investigations of homogeneous nucleation in charged sphere suspensions," *J. Chem. Phys.* **123**, 174902 (2005).
[22] P. Wette, H. J. Schöpe, and T. Palberg, "Crystallization in charged two-component suspensions," *J. Chem. Phys.* **122**, 144901 (2005).
[23] J. C. Crocker and D. G. Grier, "Methods of digital video microscopy for colloidal studies," *J. Colloid Interf. Sci.* **179**, 298–310 (1996).
[24] A. D. Dinsmore, E. R. Weeks, V. Prasad, A. C. Levitt, and D. A. Weitz, "Three-dimensional confocal microscopy of colloids," *App. Optics* **40**, 4152–9 (2001).
[25] A. van Blaaderen and P. Wiltzius, "Real-space structure of colloidal hard-sphere glasses," *Science* **270**, 1177–9 (1995).
[26] M. D. Ediger, C. A. Angell, and S. R. Nagel, "Supercooled liquids and glasses," *J. Phys. Chem.* **100**, 13200–12 (1996).
[27] C. A. Angell, K. L. Ngai, G. B. McKenna, P. F. McMillan, and S. W. Martin, "Relaxation in glassforming liquids and amorphous solids," *J. App. Phys.* **88**, 3113–57 (2000).
[28] W. van Megen and P. N. Pusey, "Dynamic light-scattering study of the glass transition in a colloidal suspension," *Phys. Rev. A* **43**, 5429–41 (1991).
[29] R. M. Ernst, S. R. Nagel, and G. S. Grest, "Search for a correlation length in a simulation of the glass transition," *Phys. Rev. B* **43**, 8070–80 (1991).
[30] P. N. Pusey and W. van Megen, "Phase behaviour of concentrated suspensions of nearly hard colloidal spheres," *Nature* **320**, 340–2 (1986).
[31] I. Snook, W. van Megen, and P. Pusey, "Structure of colloidal glasses calculated by the molecular-dynamics method and measured by light scattering," *Phys. Rev. A* **43**, 6900–7 (1991).
[32] R. L. Davidchack and B. B. Laird, "Simulation of the hard-sphere crystal–melt interface," *J. Chem. Phys.* **108**, 9452–62 (1998).
[33] S. Auer and D. Frenkel, "Prediction of absolute crystal-nucleation rate in hard-sphere colloids," *Nature* **409**, 1020–3 (2001).
[34] J. D. Bernal, "The Bakerian lecture, 1962: the structure of liquids," *Proc. Roy. Soc. London. Series A* **280**, 299–322 (1964).
[35] B. J. Alder and T. E. Wainwright, "Decay of the velocity autocorrelation function," *Phys. Rev. A* **1**, 18–21 (1970).
[36] J. Mittal, J. R. Errington, and T. M. Truskett, "Does confining the hard-sphere fluid between hard walls change its average properties?" *J. Chem. Phys.* **126**, 244708 (2007).
[37] Z. T. Németh and H. Löwen, "Freezing and glass transition of hard spheres in cavities," *Phys. Rev. E* **59**, 6824–9 (1999).
[38] B. Doliwa and A. Heuer, "Cage effect, local anisotropies, and dynamic heterogeneities at the glass transition: a computer study of hard spheres," *Phys. Rev. Lett.* **80**, 4915–18 (1998).
[39] R. J. Speedy, "The hard sphere glass transition," *Molecular Physics* **95**, 169–78 (1998).
[40] J. D. Weeks, D. Chandler, and H. C. Andersen, "Role of repulsive forces in determining the equilibrium structure of simple liquids," *J. Chem. Phys.* **54**, 5237–47 (1971).
[41] P. M. Chaikin, "Thermodynamics and hydrodynamics of hard spheres: the role of gravity, in M. E. Cates and M. R. Evans (eds.), *Soft and Fragile Matter, Nonequilibrium Dynamics, Metastability and Flow*," pp. 315–48 (Institute of Physics, London, 2000).

[42] S. Torquato, T. M. Truskett, and P. G. Debenedetti, "Is random close packing of spheres well defined?" *Phys. Rev. Lett.* **84**, 2064–7 (2000).
[43] C. S. O'Hern, L. E. Silbert, A. J. Liu, and S. R. Nagel, "Jamming at zero temperature and zero applied stress: the epitome of disorder," *Phys. Rev. E* **68**, 011306 (2003).
[44] A. Donev, S. Torquato, F. H. Stillinger, and R. Connelly, "Comment on 'jamming at zero temperature and zero applied stress: the epitome of disorder'," *Phys. Rev. E* **70**, 043301 (2004).
[45] C. S. O'Hern, L. E. Silbert, A. J. Liu, and S. R. Nagel, "Reply to 'Comment on "Jamming at zero temperature and zero applied stress: the epitome of disorder"'," *Phys. Rev. E* **70**, 043302 (2004).
[46] A. Van Blaaderen, A. Imhof, W. Hage, and A. Vrij, "Three-dimensional imaging of submicrometer colloidal particles in concentrated suspensions using confocal scanning laser microscopy," *Langmuir* **8**, 1514–17 (1992).
[47] H. Yoshida, K. Ito, and N. Ise, "Localized ordered structure in polymer latex suspensions as studied by a confocal laser scanning microscope," *Phys. Rev. B* **44**, 435–8 (1991).
[48] M. Elliot, B. Bristol, and W. Poon, "Direct measurement of stacking disorder in hard-sphere colloidal crystals," *Physica A* **235**, 216–23 (1997).
[49] J. Brujic, C. Song, P. Wang, C. Briscoe, G. Marty, and H. A. Makse, "Measuring the coordination number and entropy of a 3d jammed emulsion packing by confocal microscopy," *Phys. Rev. Lett.* **98**, 248001 (2007).
[50] J. Zhou, S. Long, Q. Wang, and A. D. Dinsmore, "Measurement of forces inside a three-dimensional pile of frictionless droplets," *Science* **312**, 1631–3 (2006).
[51] E. R. Weeks and D. A. Weitz, "Properties of cage rearrangements observed near the colloidal glass transition," *Phys. Rev. Lett.* **89**, 095704 (2002).
[52] E. R. Weeks, J. C. Crocker, A. C. Levitt, A. Schofield, and D. A. Weitz, "Three-dimensional direct imaging of structural relaxation near the colloidal glass transition," *Science* **287**, 627–31 (2000).
[53] W. K. Kegel and A. van Blaaderen, "Direct observation of dynamical heterogeneities in colloidal hard-sphere suspensions," *Science* **287**, 290–3 (2000).
[54] M. D. Ediger, "Spatially heterogeneous dynamics in supercooled liquids," *Annu. Rev. Phys. Chem.* **51**, 99–128 (2000).
[55] A. Kasper, E. Bartsch, and H. Sillescu, "Self-diffusion in concentrated colloid suspensions studied by digital video microscopy of core-shell tracer particles," *Langmuir* **14**, 5004–10 (1998).
[56] A. H. Marcus, J. Schofield, and S. A. Rice, "Experimental observations of non-Gaussian behavior and stringlike cooperative dynamics in concentrated quasi-two-dimensional colloidal liquids," *Phys. Rev. E* **60**, 5725–36 (1999).
[57] T. G. Mason and D. A. Weitz, "Optical measurements of frequency-dependent linear viscoelastic moduli of complex fluids," *Phys. Rev. Lett.* **74**, 1250–3 (1995).
[58] T. G. Mason, K. Ganesan, J. H. van Zanten, D. Wirtz, and S. C. Kuo, "Particle tracking microrheology of complex fluids," *Phys. Rev. Lett.* **79**, 3282–5 (1997).
[59] T. A. Waigh, "Microrheology of complex fluids," *Rep. Prog. Phys.* **68**, 685–742 (2005).
[60] V. Breedveld and D. J. Pine, "Microrheology as a tool for high-throughput screening," *Journal of Materials Science* **38**, 4461–70 (2003).
[61] R. G. Larson, *The Structure and Rheology of Complex Fluids* (Oxford University Press, 1998).

[62] J. C. Crocker, M. T. Valentine, E. R. Weeks, T. Gisler, P. D. Kaplan, A. G. Yodh, and D. A. Weitz, "Two-point microrheology of inhomogeneous soft materials," *Phys. Rev. Lett.* **85**, 888–91 (2000).

[63] A. J. Levine and T. C. Lubensky, "Two-point microrheology and the electrostatic analogy," *Phys. Rev. E* **65**, 011501 (2001).

[64] D. T. Chen, E. R. Weeks, J. C. Crocker, M. F. Islam, R. Verma, J. Gruber, A. J. Levine, T. C. Lubensky, and A. G. Yodh, "Rheological microscopy: local mechanical properties from microrheology," *Phys. Rev. Lett.* **90**, 108301 (2003).

[65] D. J. Tritton, *Physical Fluid Dynamics*, 2nd edition (Oxford Science Publications, Oxford University Press, USA, 1988).

[66] M. Frank, D. Anderson, E. R. Weeks, and J. F. Morris, "Particle migration in pressure-driven flow of a Brownian suspension," *J. Fluid Mech* **493**, 363–78 (2003).

[67] C. Gao, B. Xu, and J. F. Gilchrist, "Mixing and segregation of microspheres in microchannel flows of mono- and bidispersed suspensions," *Phys. Rev. E* **79**, 036311 (2009).

[68] S. A. Koehler, S. Hilgenfeldt, E. R. Weeks, and H. A. Stone, "Drainage of single plateau borders: direct observation of rigid and mobile interfaces," *Phys. Rev. E* **66**, 040601 (2002).

[69] S. A. Koehler, S. Hilgenfeldt, E. R. Weeks, and H. A. Stone, "Foam drainage on the microscale II: imaging flow through single plateau borders," *J. Colloid Interf. Sci.* **276**, 439–49 (2004).

2

Computational methods to study jammed systems

CARL F. SCHRECK AND COREY S. O'HERN

2.1 Introduction

Jammed materials are ubiquitous in nature and share several defining characteristics. They are disordered, yet solid-like with a nonzero static shear modulus. Jammed systems typically exist in metastable states with structural and mechanical properties that depend on the procedure used to create them. There are a number of different routes to the jammed state, including compressing systems to densities near random close packing [1], lowering the applied shear stress below the yield stress [2], and quenching temperature below the glass transition for the material [3]. Examples of jammed and glassy particulate systems include dense colloidal suspensions [4], attractive glasses and gels [5], static packings of granular materials [6], and quiescent foams [7] and emulsions [8]. Due to space constraints, we will limit our discussion to athermal jammed systems in which thermal energy at room temperature is unable to induce local rearrangements of particles. We note though that there are deep connections [9] between athermal jammed systems and thermal, glassy systems [10]. An important open problem in the field of jammed materials is identifying universal features that are not sensitive to the particular path in parameter space taken to create them.

In this contribution, we will review the computational techniques used to generate athermal jammed systems and characterize their structural and mechanical properties. We will focus on frictionless model systems that interact via soft, pairwise, and purely repulsive potentials. (Computational studies of frictional granular materials will be the focus of Chapter 5.) The methods for generating jammed particle packings discussed here are quite general and can be employed to study both two- and three-dimensional systems; both monodisperse and polydisperse systems; a spectrum of particle shapes, including spheres, ellipsoids, and rods; and a variety of boundary conditions and applied stress.

Experimental and Computational Techniques in Soft Condensed Matter Physics, ed. Jeffrey Olafsen.
Published by Cambridge University Press.

The remainder of the chapter will be organized as follows. In Section 2.2, we will review computational methods for creating static particle packings. Many prior studies of jamming have focused on spherical particles under isotropic compression. We will therefore describe how packing-generation methods can be generalized, with a particular emphasis on grains with nonspherical shapes. This review will be restricted to completely jammed, *mechanically stable* (MS) packings, and thus in Section 2.3, we will define mechanical stability of particle packings in terms of the normal modes of the dynamical matrix [11]. In Section 2.4, we will describe many of the computational measurements that can be performed to characterize the structural and mechanical properties of jammed packings, including the pair distribution function, structure factor, translational and orientational order parameters, correlation functions, spectra of vibrational modes, and elastic moduli.

2.2 Methods to generate static-particle packings

Computational methods to generate jammed particle packings fall into two general categories: 'hard' [12] and 'soft' [13–15] particle methods. Hard particle methods strictly enforce the constraint that particles cannot overlap; these include the Lubachevsky–Stillinger algorithm of binary collisions between elastic particles coupled with compression [16], single-particle and collective Monte-Carlo moves with successive compressions, and various geometrical techniques [17–19].

In this contribution, we will review two soft-particle packing-generation techniques: (1) the *isotropic compression* method in which we successively compress or decompress the system followed by energy relaxation until all particles are just touching and (2) the *applied shear* method in which we generate just-touching particle packings at arbitrary values of shear strain. We focus on soft-particle packing-generation methods for several reasons. First, the structural and mechanical properties can be studied as a function of overcompression $\phi - \phi_J$, where ϕ_J is the jamming packing fraction, which allows numerical results to be compared to experiments on foams, emulsions, and granular materials. Second, features of hard-particle MS packings can be recovered from soft-sphere packings in the just-touching limit. Finally, many hard-particle methods yield locally jammed packings, which are not mechanically stable. In contrast, soft-particle methods, which satisfy the constraints of force and torque balance on all grains, reliably produce mechanically stable packings.

2.2.1 Isotropic compression

The isotropic compression method for soft particles consists of initializing the system in a dilute, fluid-like configuration and successively compressing/

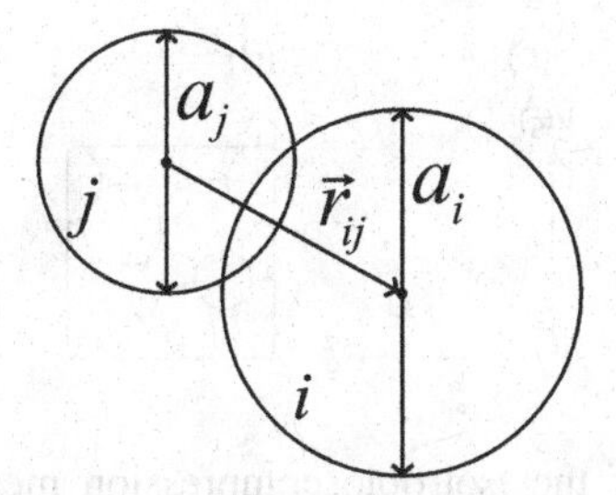

Figure 2.1 Schematic showing the definition of the contact distance σ_{ij} for two disks i and j, with diameters a_i and a_j and center-to-center separation r_{ij}. The two particles overlap when $r_{ij} < \sigma_{ij}$, where $\sigma_{ij} = (a_i + a_j)/2$ is the average diameter.

decompressing the system followed by energy relaxation to the nearest local minimum. The method terminates at packing fraction ϕ_J when all particles (except floater particles that are not locally stable) achieve force and torque balance with infinitesimal particle overlaps. In these studies, we will employ pairwise, short-range repulsive interactions between particles. The pair potential $V(r_{ij}/\sigma_{ij})$, where r_{ij} is the center-to-center separation and σ_{ij} is the contact distance between grains i and j, is positive if particles are overlapped ($r_{ij} < \sigma_{ij}$) and zero otherwise. Thus, the packings possess nonzero, but infinitesimal pressure and potential energy at jamming.

We will focus on two forms for the interaction potential: the repulsive spring potential V_s in (2.1) and the repulsive Lennard–Jones potential in (2.2):

$$V_s(r_{ij}) = \frac{\epsilon}{\alpha}\left(1 - \frac{r_{ij}}{\sigma_{ij}}\right)^{\alpha} \Theta\left(\frac{\sigma_{ij}}{r_{ij}} - 1\right), \tag{2.1}$$

$$V_{\rm RLJ}(r_{ij}) = \epsilon\left[\left(\frac{\sigma_{ij}}{r_{ij}}\right)^{12} - 2\left(\frac{\sigma_{ij}}{r_{ij}}\right)^{6} + 1\right]\Theta\left(\frac{\sigma_{ij}}{r_{ij}} - 1\right), \tag{2.2}$$

where $\alpha = 2$ ($\alpha = 5/2$) correspond to the linear (Hertzian) repulsive spring potential, $\Theta(x)$ is the Heaviside step function, which ensures that particles only interact when they overlap. For spherical particles, the contact distance is simply the average of their diameters $\sigma_{ij} = (a_i + a_j)/2$, as shown in Figure 2.1. The contact distance σ_{ij} for nonspherical grains, which depends on $\hat{r}_{ij}$ and the orientations of grains i and j, will be discussed in detail in Section 2.2.3.2. The total potential energy per particle for the system is given by $V = N^{-1} \sum_{i,j} V(r_{ij})$.

The isotropic compression method can be viewed in terms of the potential energy landscape $V(\vec{\xi})$ of the system, as shown in Figure 2.2. All possible configurations of N particles, each denoted by $\vec{\xi} = (\vec{r}_1, \vec{r}_2, \ldots, \vec{r}_N)$, yield a value of the potential energy per particle $V(\vec{\xi})$. Static, force- and torque-balanced packings correspond to local minima in the potential energy landscape. Jammed packings

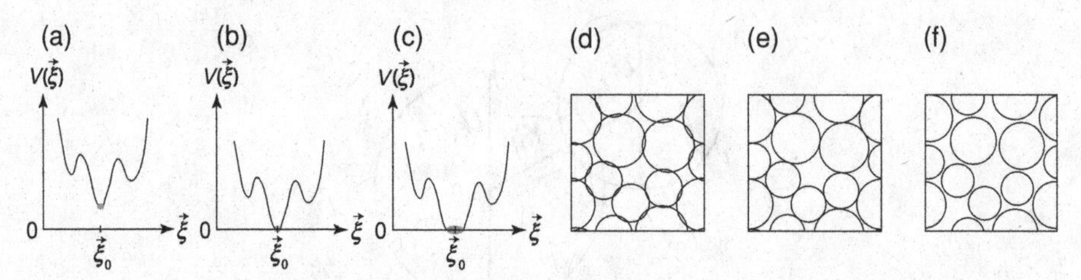

Figure 2.2 Schematic of the isotropic compression method to create jammed packings. In panels (a)–(c), we show the potential energy landscape $V(\vec{\xi})$ in the vicinity of the static granular packings (at point $\vec{\xi}_0$ in configuration space) in panels (d)–(f). If the system exists in a nonoverlapped configuration (panel (f)) with gaps between particles and $V = 0$, it will be compressed followed by energy minimization. If the system exists in an overlapped configuration (panel (d)) at a local potential energy minimum with $V > 0$, it will be decompressed followed by energy minimization. When the system switches between the cases displayed in panels (d) and (f), the compression/decompression increment is decreased. The process stops when the system exists in a static packing at a local potential energy minimum that is infinitesimally above zero.

correspond to minima with $V(\vec{\xi}) \gtrsim 0$. During the packing-generation process, if there are significant particle overlaps (Figure 2.2(d)) and the energy of the system at a local minimum is nonzero (Figure 2.2(a)), the system will be subsequently decompressed. If the potential energy of the system is zero (Figure 2.2(c)) and gaps exist between particles (Figure 2.2(f)), the system will be compressed at the next step. The increment by which the packing fraction ϕ is changed at each compression or decompression step is gradually decreased. After a sufficiently large number of steps, a jammed packing with infinitesimal overlaps (Figure 2.2(e)) and potential energy $V \gtrsim 0$ (Figure 2.2(b)) is obtained.

The isotropic compression method for generating jammed particle packings consists of four basic steps: (1) initialization of particle positions and orientations, (2) compression or decompression of the system, (3) relaxation of the potential energy to the nearest local minimum, and (4) repetition of steps (2) and (3) until a jammed packing with infinitesimal overlaps is obtained. We will describe the implementation of these steps in detail here:

Step (1) Choose an initial configuration for N particles in the simulation cell. This is typically accomplished by assigning each particle a random position and orientation, in the case of nonspherical particles, in the simulation cell. For the repulsive Lennard–Jones interaction potential, random initial configurations without significant particle overlaps must be used. If the initial packing fraction, ϕ_0, is well below the mean jammed packing fraction, $\langle \phi_J \rangle$, the initial conditions will not bias the final set of packings. The results obtained for $\phi_0 > \langle \phi_J \rangle$ and for $\phi_0 < \langle \phi_J \rangle$ only show small differences for frictionless grains, but these differences are important and should be studied in more detail in future studies. When choosing orientations, directors can be selected randomly from a uniform distribution on the unit disk (sphere) in 2D (3D).

Step (2) Compress the system (or increase particle sizes uniformly) if it is below the jamming point ($V < V_{\mathrm{tol}}$) or decompress the system (or decrease particle sizes uniformly) if it is above the jamming point ($V > 2V_{\mathrm{tol}}$). V_{tol} is the potential energy threshold that dictates how close the final packing is to the point at which all particles are just touching. We typically consider $V_{\mathrm{tol}} = 10^{-16}$ (where V is normalized by the energy-scale parameter ϵ) based on numerical precision. We must choose the initial packing fraction increment $\Delta\phi_0$ to be sufficiently small so that it does not influence the jammed packing fraction, but also large enough to efficiently generate jammed packings. Previous studies have used $\Delta\phi_0 = 10^{-4}$ [14, 15]. The packing fraction increment is successively decreased to locate ϕ_J. If the potential energy of the system at successive compressions or decompressions $n-1$ and n satisfy $V_{n-1} > 2V_{\mathrm{tol}}$ and $V_n < V_{\mathrm{tol}}$ or $V_{n-1} < V_{\mathrm{tol}}$ and $V_n > 2V_{\mathrm{tol}}$, the step size is halved. Otherwise it remains at the current value of $\Delta\phi$.

Step (3) Minimize the total potential energy of the system after each compression or decompression step to find a local potential energy minimum. The energy minimization can be performed in several ways, including (a) numerical energy minimization procedures, such as the conjugate-gradient technique [20] and (b) molecular dynamics (MD) simulations with dissipative forces proportional to velocity. The conjugate-gradient method is a numerical scheme that begins at a given point in configuration space and moves the system to the nearest local potential energy minimum without traversing any energy barriers [20]. In contrast, molecular dynamics with finite damping is not guaranteed to find the nearest local potential energy minimum since kinetic energy is removed from the system at a finite rate. The system can thus surmount a sufficiently low energy barrier. A comparison of these two methods provides important geometric information about the width of basins and heights of energy barriers separating the basins in the energy landscape.

In the molecular-dynamics method, each particle i obeys Newton's equations of motion:

$$m\vec{a}_i = \sum_{j\neq i}\left[-\frac{dV(r_{ij})}{dr_{ij}} - b\vec{v}_{ij}\cdot\hat{r}_{ij}\right]\hat{r}_{ij}, \tag{2.3}$$

where $\vec{a}_i$ is the acceleration of particle i, $\vec{v}_{ij}$ is the relative velocity of particles i and j, $\hat{r}_{ij}$ is the unit vector connecting the centers of these particles, and b is the damping coefficient. In the infinite-dissipation limit, $b \to \infty$, the potential energy cannot increase during a molecular-dynamics relaxation, and thus the molecular-dynamics and conjugate-gradient methods should give very similar

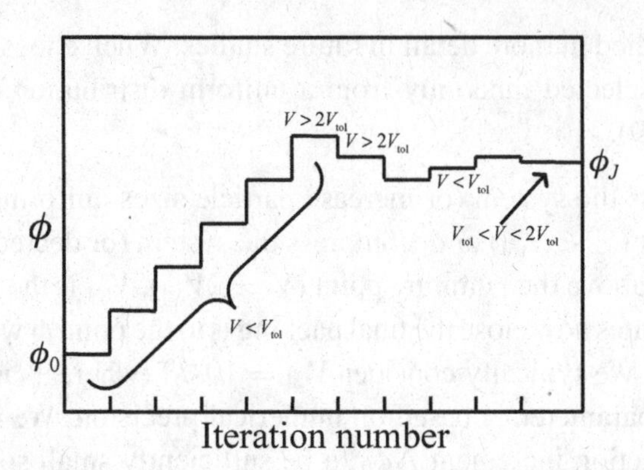

Figure 2.3 Schematic of the isotropic compression method for creating jammed particle packings. The system is initialized at a packing fraction ϕ_0 below the jamming onset ϕ_J. After a series of compressions and decompressions of decreasing amplitude, the jamming onset ϕ_J is identified with total potential energy per particle at a local energy minimum that satisfies $V_{\rm tol} < V < 2V_{\rm tol}$.

results. We note, however, that even in this limit the two methods are not equivalent because there may be more than one energy minimum accessible from a given point in configuration space without traversing an energy barrier.

For the conjugate-gradient method, we terminate the minimization process when either of the following two conditions on the potential energy per particle V is satisfied: (a) two successive conjugate-gradient steps t and $t+1$ yield nearly the same energy value, $(V_{t+1} - V_t)/V_t < \delta = 10^{-16}$; or (b) the potential energy per particle at the current step is extremely small, $V_t < V_{\rm min} = 10^{-16}$. Since the potential energy oscillates in time in the molecular-dynamics method, condition (a) is replaced by the requirement that the relative potential-energy fluctuations satisfy the inequality $\langle (V - \langle V \rangle)^2 \rangle^{1/2} / \langle V \rangle < \delta$. Stopping criteria based on the rms or maximum total force on the particles can also be implemented.

Step (4) The packing-generation procedure terminates when the potential energy at a local energy minimum satisfies $V_{\rm tol} < V < 2V_{\rm tol}$. Using this method, we are able to locate the jamming threshold in packing fraction ϕ_J to within 10^{-8} for each static packing. A schematic of the dynamics of the packing fraction during the compression method is shown in Figure 2.3. This method can be performed for many different initial configurations to generate an ensemble of jammed packings.

2.2.2 Applied shear strain

Athermal jammed systems can also be generated in the presence of applied shear stress or strain. In fact, recent experimental [21], simulation [22], and theoretical [23, 24] studies have emphasized that the form of stress correlations in jammed

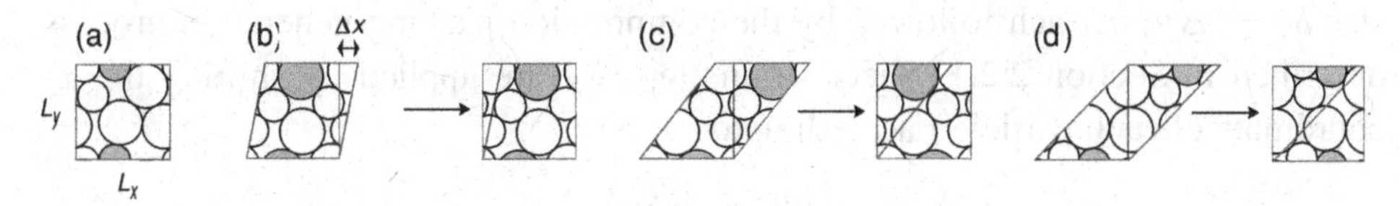

Figure 2.4 Schematic diagram of shear-periodic boundary conditions. (a) Static packing with $N = 6$ particles confined to a $L_x \times L_y$ box with shear strain $\gamma = \Delta x / L_y = 0$. The shading indicates a given particle in the primary cell and its image. (b) Static packing in a unit cell with $\gamma = 0.2$. Note that at arbitrary γ, a given particle in the primary cell is not directly above (or below) its image. (c) Static packing at $\gamma = 0.8$, which shows that configurations at γ and $1 - \gamma$ are related by an inversion about the vertical (shear-gradient) direction. (d) Static packing in (a) at $\gamma = 1$, which is identical to standard periodic boundary conditions. Thus shear-periodic boundary conditions have unit period. Adapted from reference [28].

packings created by isotropic compression and shear are fundamentally different, with longer ranged spatial correlations along the compressive direction in sheared systems. In experiments, simple shear can be implemented using planar [25] or Couette geometries [26]. In simulations, bulk planar shear flow can be realized using Lees–Edwards or shear-periodic boundary conditions [27]. In Figure 2.4, we show a schematic diagram of shear-periodic boundary conditions implemented in 2D. In each panel, the top (bottom) image cells are shifted by $\Delta x = \gamma L_y$ to the right (left), where γ is the shear strain and L_y is the dimension of the simulation cell in the shear gradient direction.

Note that shear-periodic boundary conditions are identical at $\gamma = x$ and $1 - x$ and have unit period as shown in Figure 2.4; thus, we only need to study the range $\gamma = 0$ to 0.5 to generate static packings over the full range of shear strain. Prior studies of jammed systems have focused on isotropically compressed packings at $\gamma = 0$ [13], whereas there are relatively few studies of sheared packings at nonzero γ [28]. To generate anisotropic static packings, the region $\gamma = [0, 0.5]$ can be divided into small shear strain intervals, e.g. $\Delta\gamma = 10^{-2}$. At each sampled shear strain γ_s, the four-step compression/decompression method discussed in Section 2.2.1 can be implemented to generate static, just-touching particle packings. That is, the particles' positions and orientations are chosen randomly, and the system is subjected to a sequence of compressions and decompressions with decreasing amplitude, each followed by energy relaxation at a *fixed shear strain*, until the energy of the system falls within a prescribed window. When this procedure is repeated many times for different γ_s and independent initial conditions, one can generate ensembles of packings over a series of discrete strains γ_s.

To create continuous maps of static packings between sampled shear strains γ_s and γ_{s+1}, one can apply n successive shear strains to each static packing at γ_s of size $\delta\gamma = \Delta\gamma/n$, each followed by the compression packing-generation process described in Section 2.2.1. Shear strain steps $\delta\gamma$ are applied by shifting the x-coordinate of each particle i according to:

$$x_i \rightarrow x_i + \delta\gamma y_i \tag{2.4}$$

in conjunction with shear-periodic boundary conditions. A similar procedure can also be performed to study the continuous set of static packings for shear strains in the opposite direction between γ_s and γ_{s-1}. Note that this procedure generates a series of static packings over a range of shear strain at fixed *zero pressure*, not at fixed volume as in previous studies of quasistatically sheared Lennard–Jones [29] and other model glasses [30].

A schematic of the process for generating the continuous set of static packings between shear strains γ_s and γ_{s+1} (or γ_s and γ_{s-1}) is shown in Figure 2.5. Panels (a) and (b) show two possible evolutions of the potential energy landscape following a shear strain step $\delta\gamma$. In (a), no particle rearrangement event occurs and the local minimum (1) can be continuously deformed into local minimum (4) by applying shear strain $\delta\gamma$. The dynamics of the system can be summarized as follows: we apply a shear strain $\delta\gamma$ to the initial static packing with strain γ (1), which shifts the point in configuration space and the potential energy landscape (2). The energy is then minimized at fixed shear strain $\gamma + \delta\gamma$ (3), and the system is decompressed (or compressed, followed by energy minimization) to bring it to a static packing with infinitesimal overlap (4). The static packings at shear strains γ (1) and $\gamma + \delta\gamma$ (4) are overlayed in panel (c). Note that the particle contact networks are identical at γ (gray) and $\gamma + \delta\gamma$ (black).

In contrast, Figure 2.5(b) shows the evolution of the system when at least one particle rearrangement occurs following a shear strain step. In this case, a shear strain $\delta\gamma$ is applied to a static packing at shear strain γ (1), and the system moves in configuration space so that it exists in the basin of a new local minimum (2). Upon energy minimization at fixed $\gamma + \delta\gamma$, the system moves to an unjammed packing (3). Following compression and energy minimization, the system resides in a new static packing at $\gamma + \delta\gamma$ (4) that is not continuously related to the static packing at γ. In Figure 2.5(d), we overlay the static packings at γ (gray) and $\gamma + \delta\gamma$ (black) and show that the particle contact networks are not the same.

For each distinct static packing at γ_s, one can monitor the particle contact network as the system evolves toward γ_{s+1} (and γ_{s-1}) and identify any changes that occur. If there are changes in the particle contact network, physical quantities such as the jamming packing fraction ϕ_J, pressure, and shear stress (and their derivatives) are discontinuous. For example, in Figure 2.6(b), we show a discontinuity in ϕ_J at $\gamma_s < \gamma^* < \gamma_{s+1}$, where a particle rearrangement event occurs. $\delta\gamma$ can be tuned to eliminate as many of the particle rearrangement events as possible.

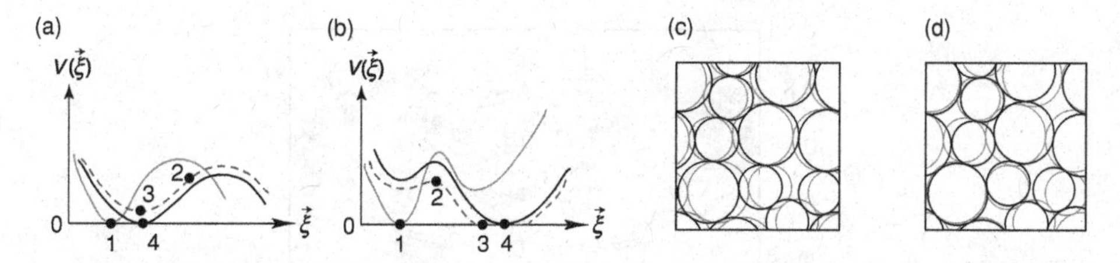

Figure 2.5 Schematic of the evolution of the potential energy landscape during quasistatic shear at fixed zero pressure from shear strain γ to $\gamma + \delta\gamma$. In (a), the system evolves continuously from the local minimum at shear strain γ (1) to the one at $\gamma + \delta\gamma$ (4) because there are no particle rearrangement events during the shear strain interval. In contrast, in (b) the system undergoes particle rearrangement events during the strain interval $\delta\gamma$, and therefore it resides in a fundamentally different potential energy minimum at $\gamma + \delta\gamma$ (4) compared to the one at γ (1). Snapshots of the static packings at shear strain γ (gray) and $\gamma + \delta\gamma$ (black) are superimposed in (c) and (d), which correspond to the potential energy landscape dynamics in (a) and (b), respectively. In (d), three of the original contacts are removed and four new contacts are generated as a result of the particle rearrangements that occurred during the strain interval $\delta\gamma$. Adapted from reference [28].

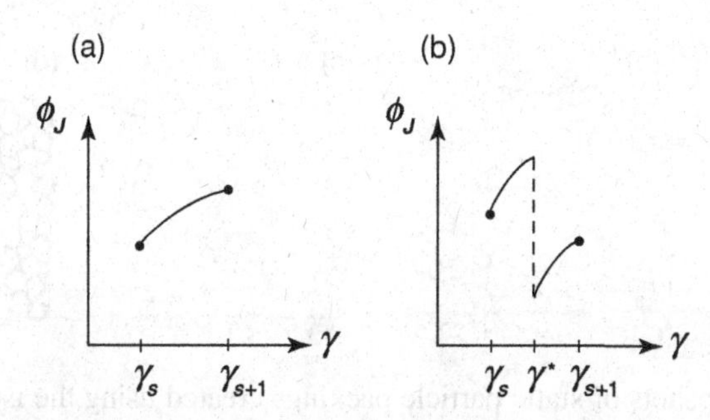

Figure 2.6 Schematic of the evolution of the jamming packing fraction ϕ_J during the shear strain interval γ_s to γ_{s+1}. In (a) the particle contact network does not change from γ_s to γ_{s+1}, while in (b), it does at γ^*. Adapted from reference [28].

In Figure 2.6(a), we show the continuous evolution of ϕ_J between γ_s and γ_{s+1} when there are no particle rearrangement events. This continuous evolution of ϕ_J represents a portion of a 'geometric family' of static packings all with the same particle contact network that exist over a continuous range of shear strain from γ_s to γ_{s+1}. In Figure 2.7, we show results where we have pieced together these continuous segments to construct nearly all of the geometric families over the full range of shear strain for a small $N = 10$ frictionless granular system [28]. Note that even though there are an infinite number of static packings over the continuous range of shear strain, there are a *finite* number of geometric families that can be counted.

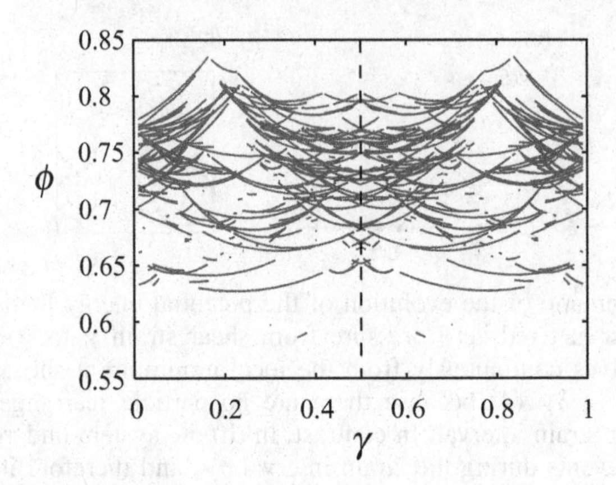

Figure 2.7 The solid lines show the jammed packing fraction ϕ_J versus shear strain γ for different geometric families. The continuous geometric families of static packings possess the same contact network over a given range of shear strain. When the solid line breaks, the continuous family becomes unstable and the particle contact network changes. The families are symmetric with respect to reflection about $\gamma = 0.5$ (dashed vertical line). This figure is adapted from reference [28].

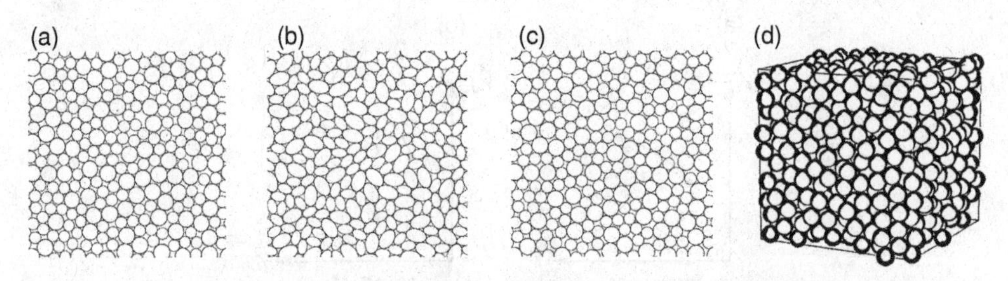

Figure 2.8 Snapshots of static particle packings created using the isotropic compression method in Section 2.2.1 (a) 2 : 1 (fraction of small particles to fraction of large particles) mixture of bidisperse disks with diameter ratio $\sigma_l/\sigma_s = 1.4$, (b) 2 : 1 mixture of bidisperse ellipses with ratio of major axes $a_l/a_s = 1.4$, (c) polydisperse mixture of disks with a flat probability distribution for the diameters between σ_s and $1.4\sigma_s$, and (d) monodisperse spheres.

2.2.3 Other important variables

The packing-generation methods described in Sections 2.2.1 and 2.2.2 can be generalized to study packings as a function of a number of important variables that affect their structural and mechanical properties. These parameters include the dimension (2D, 3D, and higher dimensions) [31], particle size distribution [32], boundary conditions [28], and particle shape [16]. In Figure 2.8, we show several different types of static granular packings that we generated in prior studies, including (a) bidisperse ($2N/3$ small and $N/3$ large) disks with diameter ratio

$\sigma_l/\sigma_s = 1.4$ [13], (b) bidisperse (same composition as in (a)) ellipses with ratio of the major axes $a_l/a_s = 1.4$ [33], (c) polydisperse disks with a uniform distribution of particle sizes with width $0.4\sigma_s$ and mean $1.2\sigma_s$ [34] (labeled polydispersity $p = 1.4$), and (d) monodisperse spheres [13].

Important control variables for the particle size include the diameter ratio and relative number of large and small particles for bidisperse, tridisperse, or other systems with discrete particle species, and properties of the particle size distribution for continuously polydisperse systems, i.e. the mean and rms width for uniform, normal, and log-normal size distributions.

The packing-generation procedures we discussed in Sections 2.2.1 and 2.2.2 employ periodic boundary conditions at either fixed zero pressure (Section 2.2.1) or fixed shear strain (Section 2.2.2). However, it is straightforward to generalize these methods to systems with smooth or rough fixed walls [2], constraints imposed at the boundaries, such as constant stress [35], and gravity [36].

In addition, particle shape is a key variable that significantly affects the jamming transition in particle packings. Introducing nonspherical shapes does not significantly alter the steps in the packing-generation procedures in Sections 2.2.1 and 2.2.2. However, even for simple anisotropic shapes, such as ellipses, the calculation of the contact distance between grains i and j is nontrivial since it depends on $\hat{r}_{ij}$ and grain orientations.

2.2.3.1 *Particle shape*

Recent studies have suggested that particle shape strongly influences the nature of the jamming transition [33,37]. Specifically, the scaling of the shear modulus with packing fraction and the shape of the vibrational spectrum are fundamentally different from that for spherical particles. Thus, the ability to create jammed packings of nonspherical particles and study the effects of particle shape on the jamming transition are clearly timely and important research efforts.

Particle shape is an enormous parameter space. In this discussion, we will focus on three relevant shape variables, all of which are captured in Figure 2.9: (1) convex (a) versus concave (b) – (d) shapes, (2) rigid anisotropic grains with different degrees of symmetry constructed by fusing spherical particles together (b) – (d), and (3) ellipsoidal particles (a). The jamming behavior of convex vs. concave particle shapes can be contrasted by studying ellipsoidal vs. rigid linear n-mers, where n is the number of spherical particles that have been fused together. Fused colloidal silica spheres can now be made reliably in a variety of shapes, including dimers and trimers. The benefit of studying rigid anisotropic particles formed from fusing spherical particles together, such as linear n-mers, asymmetric n-mers, and nonlinear n-mers, is that the contact distance, forces, and interaction energy between grains i and j can be calculated by assuming that the grains are

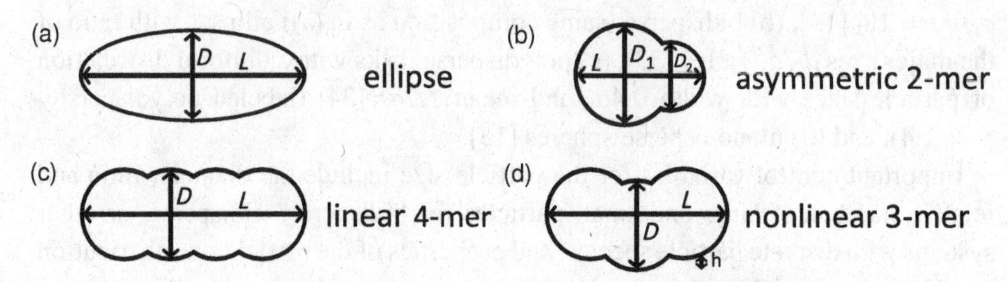

Figure 2.9 The packing-generation methods described in Section 2.2 can be employed in systems with anisotropic particle shapes, such as ellipsoids, linear n-mers, asymmetric n-mers, and nonlinear n-mers. Examples of geometric parameters that can be tuned are the aspect ratio $\alpha = L/D$, asymmetry $s = D_2/D_1$ for asymmetric dimers, and bending angle $c = h/L$ for nonlinear trimers.

composed of a series of spherical particles. The calculation of the contact distance and all other quantities is trivial for spherical particles. In contrast, ellipsoidal particles have simple nonspherical shapes, yet the calculation of the contact distance, interaction forces, and other quantities is quite complicated.

The packing-generation procedures discussed in Sections 2.2.1 and 2.2.2 require the specification of a potential energy function $V(\vec{\xi})$ and its first derivatives $\partial V(\vec{\xi})/\partial\vec{\xi}$, where $\vec{\xi}$ represents the configurational degrees of freedom, e.g. $\vec{\xi} = \{x_i, y_i, \theta_i\}$ with $i = 1, \ldots, N$ for ellipses, where θ_i is the angle between $\hat{x}$ and the long axis of grain i. We also assume that the potential energy only depends on r_{ij}/σ_{ij}, where σ_{ij} is the contact distance between grains i and j. For spherical particles, the contact distance $\sigma_{ij} = (\sigma_i + \sigma_j)/2$ is a constant (average diameter) that depends on the particle species in polydisperse systems and the only relevant derivatives of the potential energy are $\partial r_{ij}/\partial r_i$, where $r = x, y, z$. For nonspherical particles, the contact distance is not constant and depends on $\hat{r}_{ij}$ and the orientations of particles i and j. For ellipsoidal particles, we define σ_{ij} as the true contact distance: the center-to-center separation r_{ij} at which two particles come into contact at fixed orientation. The definition of σ_{ij} for two ellipses i and j with orientations $\hat{\mu}_i$ and $\hat{\mu}_j$ and center-to-center separation r_{ij} is shown in Figure 2.10. The method for calculating σ_{ij} for ellipsoidal particles is described in detail below.

2.2.3.2 *Contact distance for ellipsoidal particles*

To generate static packings of ellipsoidal particles using soft-particle methods, one must be able to calculate the potential energy $V(\vec{\xi})$ and its derivatives for general configurations $\vec{\xi}$, which involves determining the contact distance σ_{ij} and its derivatives. This section will provide a survey of the techniques for calculating σ_{ij} in both 2D and 3D. In 3D, we will limit the discussion to spheroids – ellipsoids with one axis of symmetry. A spheroid in which the long (short) axis is the axis of

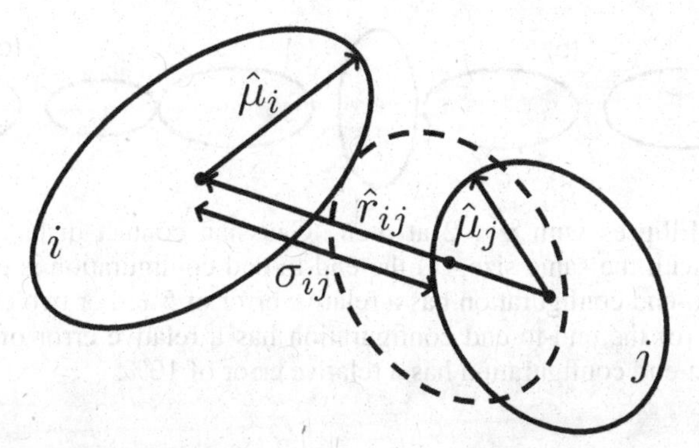

Figure 2.10 Definition of the contact distance σ_{ij} for ellipsoidal particles i and j with unit vectors $\hat{\mu}_i$ and $\hat{\mu}_j$ that characterize the orientations of their major axes. σ_{ij} is the center-to-center separation r_{ij} at which ellipsoidal particles first touch when they are brought together along $\vec{r}_{ij}$ at fixed orientation.

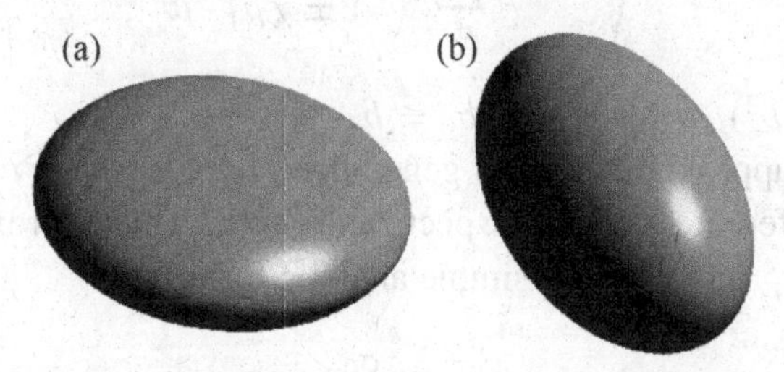

Figure 2.11 (a) Oblate and (b) prolate spheroids with aspect ratios $\alpha = 0.375 < 1$ and $\alpha = 1.7 > 1$.

symmetry is termed an oblate (prolate) spheroid, as shown in Figure 2.11. We will refer to both ellipses and spheroids as ellipsoidal particles and characterize the ratio of the major to minor axes by $\alpha = a/b$ in 2D or the ratio of the polar to equatorial lengths in 3D. As shown in Figure 2.10, configurations of ellipsoidal particles are specified by the centers of mass $\vec{r}_i$ and the orientation of the major axis $\hat{\mu}_i$ of each particle i.

One of the simplest methods for obtaining an approximate expression for the contact distance between ellipsoidal particles is the 'Gaussian approximation' introduced by Berne and Pechukas [38]. In this method, the contact distance between two ellipsoidal particles of the *same* size is approximated by finding the overlap between two Gaussian functions, $G_i(\vec{r})$ and $G_j(\vec{r})$, whose contour surfaces at $1/e$ coincide with the surfaces of particles i and j. $G_i(\vec{r}) = e^{-((x-x_i)^2+(y-y_i)^2)/a_i^2-(z-z_i)^2/b_i^2}$ for an ellipsoidal particle centered at (x_i, y_i, z_i) with orientation $\hat{\mu}_i = \hat{z}$ and major (minor) axis a (b). More generally, $G_i(\vec{r}) = e^{-(r-r_i)_k(\gamma^{-1})_{kl}(r-r_i)_l}$, where $\gamma_{kl} = (a^2 - b^2)\mu_k\mu_l + b^2\delta_{kl}$, δ_{kl} is the Kronecker delta, and $k, l \in \{x, y, z\}$. Integrating $\int G_i(\vec{r})G_j(\vec{r})d^3r$, we obtain another Gaussian function, $G_0 e^{-2(r_{ij}/\sigma_{ij}^g)^2}$ [39].

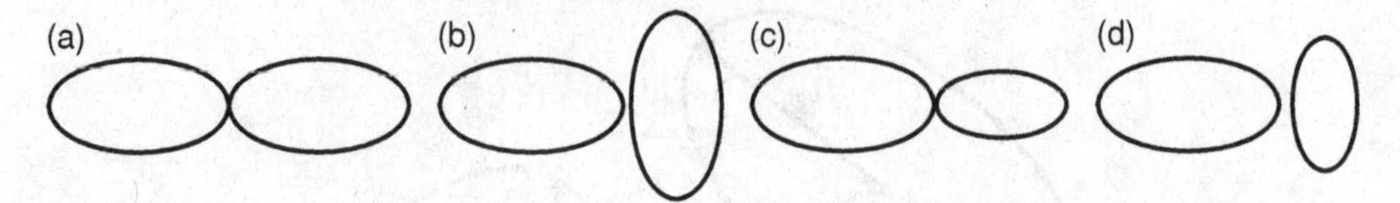

Figure 2.12 Ellipses with $\alpha = 2$ at their 'Gaussian contact distance' σ_{ij}^c. For two ellipses with the same size, (a) the end-to-end configuration is exact, while (b) the side-to-end configuration has a relative error of 5%. For two ellipses with $a_j/a_i = 1.4$, (c) the end-to-end configuration has a relative error of 1%, while (d) the side-to-end configuration has a relative error of 10%.

$\sigma_{ij}^g > \sigma_{ij}$ is an approximation to the true contact distance, and is given by:

$$\sigma_{ij}^g = \frac{\sigma_0}{\sqrt{1 - \frac{\chi}{2}\sum_{\pm} \frac{(\hat{r}_{ij}\cdot\hat{\mu}_i \pm \hat{r}_{ij}\cdot\hat{\mu}_j)^2}{1 \pm \chi\hat{\mu}_i\cdot\hat{\mu}_j}}}, \tag{2.5}$$

where $\chi = (a^2 - b^2)/(b^2 + a^2)$ and $\sigma_0 = b$.

The Gaussian approximation was generalized by Cleaver *et al.* for ellipsoidal particles with different sizes and aspect ratios [40]. This approximation for the contact distance σ_{ij}^c has a similar simple analytic form:

$$\sigma_{ij}^c = \frac{\sigma_0}{\sqrt{1 - \frac{\chi}{2}\sum_{\pm} \frac{(\beta\hat{r}_{ij}\cdot\hat{\mu}_i \pm \beta^{-1}\hat{r}_{ij}\cdot\hat{\mu}_j)^2}{1 \pm \chi\hat{\mu}_i\cdot\hat{\mu}_j}}}, \tag{2.6}$$

where $\sigma_0 = \sqrt{(b_i^2 + b_j^2)/2}$, $\chi = \left(\frac{(a_i^2-b_i^2)(a_j^2-b_j^2)}{(a_j^2+b_i^2)(a_i^2+b_j^2)}\right)^{1/2}$, and $\beta = \left(\frac{(a_i^2-b_i^2)(a_j^2+b_i^2)}{(a_j^2-b_j^2)(a_i^2+b_j^2)}\right)^{1/4}$ in the Cleaver form. The Gaussian approximation to the contact distance behaves poorly for ellipsoidal particles with different sizes. In Figure 2.12, we show σ_{ij}^c for different relative orientations of bidisperse ellipses; the relative deviation from the true contact distance can be as large as $e \sim 10\%$ for $a_j/a_i = 1.4$ and $\alpha = 2.0$. *The Gaussian approximation should therefore not be used to study 2D amorphous ellipse packings since polydispersity is required to suppress bond orientational order.* For monodisperse ellipses with $\alpha = 2.0$, $0\% < e < 5\%$, and similar results are expected for 3D systems.

Perram and Wertheim pursued a related approach, yet their formulation yields the exact contact distance for ellipsoidal particles with different sizes in 2D and 3D [41]. They define $F_i^{\text{single}}(\vec{r}) = (r - r_i)_k(\gamma^{-1})_{kl}(r - r_i)_l$, where $F_i^{\text{single}}(\vec{r}) > 1 (< 1)$ for a point $\vec{r}$ outside (inside) ellipsoidal particle i. They were able to show that for a given λ, the function $F_{ij}^{\text{pair}}(\vec{r}, \lambda) = \lambda F_i^{\text{single}}(\vec{r}) + (1 - \lambda)F_j^{\text{single}}(\vec{r})$ has a unique minimum at $\vec{r} = \vec{r}_{\min}(\lambda)$. Note that $F_{ij}^{\text{pair}}(\vec{r}_{\min}(\lambda), \lambda) = 0$ when $\vec{r}_{\min}(\lambda = 0) = \vec{r}_j$

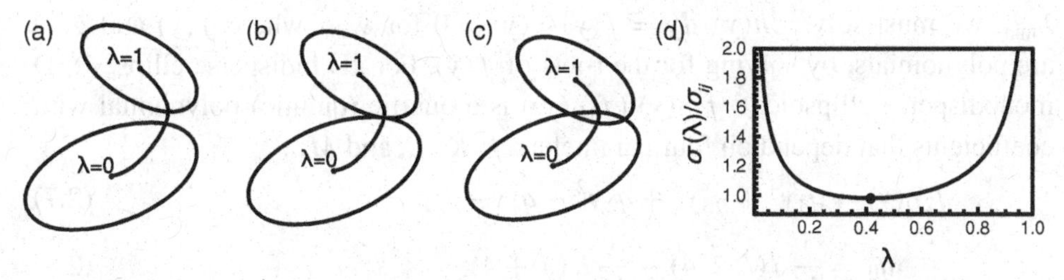

Figure 2.13 Three configurations with the same $\hat{r}_{ij}$, $\hat{\mu}_i$, and $\hat{\mu}_j$, but different r_{ij} are shown in panels (a)–(c). The solid curve $\vec{r}_{\min}(\lambda)$ starts from the center of ellipse j at $\lambda = 0$ and ends at the center of ellipse i at $\lambda = 1$. In (a), $F_{ij} > 1$ for nontouching ellipses; in (b), $F_{ij} = 1$ for 'just-touching' ellipses; and in (c), $F_{ij} < 1$ for overlapping ellipses. σ_{ij} has the same value for (a)–(c). In (d), we show that $\sigma(\lambda)/\sigma_{ij} = [F_{ij}/F_{ij}^{\text{pair}}(\vec{r}_{\min}(\lambda), \lambda)]^{1/2}$ has a minimum at $\lambda^* = 0.415$ (filled circle) for this particular choice of $\hat{r}_{ij}$, $\hat{\mu}_i$, and $\hat{\mu}_j$.

and $\vec{r}_{\min}(\lambda = 1) = \vec{r}_i$. Thus, λ parameterizes a curve $\vec{r}_{\min}(\lambda)$ for which the function $F_{ij}^{\text{pair}}(\vec{r}_{\min}(\lambda), \lambda))$ goes from 0 at $\lambda = 0$ back to 0 at $\lambda = 1$ and is concave down for λ in the range [0, 1] [41].

For the nonoverlapping ellipsoidal particles i and j in Figure 2.13(a), $F^{\text{single}}(\vec{r}_{\min}(\lambda), \lambda) > 1$ for both i and j over some range of λ, which implies that $F_{ij}^{\text{pair}}(\vec{r}_{\min}(\lambda), \lambda) > 1$ for the same range of λ. For the overlapping ellipsoidal particles in Figure 2.13(c), $F^{\text{single}}(\vec{r}_{\min}(\lambda), \lambda) < 1$ for either ellipses i or j for all λ, and thus $F_{ij}^{\text{pair}}(\vec{r}_{\min}(\lambda), \lambda) < 1$ for all λ [41]. Since the maximum over λ of $F_{ij}^{\text{pair}}(\vec{r}_{\min}(\lambda), \lambda) < 1$ for overlapping ellipsoidal particles and is greater than unity for nonoverlapping ellipsoidal particles, the two particles will come into contact when $F_{ij} \equiv \max_\lambda F_{ij}^{\text{pair}}(\vec{r}_{\min}(\lambda), \lambda) = 1$. Perram and Wertheim further showed that $F_{ij} = (r_{ij}/\sigma_{ij})^2$, where the contact distance $\sigma_{ij} = \min_\lambda \sigma(\lambda)$ and $\sigma(\lambda) = \sigma_{ij}^c$ with λ-dependent parameters $\sigma_0(\lambda) = \frac{1}{2}\sqrt{\frac{b_i^2}{\lambda} + \frac{b_j^2}{1-\lambda}}$, $\chi(\lambda) = \left(\frac{(a_i^2-b_i^2)(a_j^2-b_j^2)}{\left(a_j^2+\frac{1-\lambda}{\lambda}b_i^2\right)\left(a_i^2+\frac{\lambda}{1-\lambda}b_j^2\right)}\right)^{1/2}$, and $\beta(\lambda) = \left(\frac{(a_i^2-b_i^2)\left(a_j^2+\frac{1-\lambda}{\lambda}b_i^2\right)}{(a_j^2-b_j^2)\left(a_i^2+\frac{\lambda}{1-\lambda}b_j^2\right)}\right)^{1/4}$. We obtain the Cleaver–Gaussian approximation for σ_{ij}, if instead of minimizing $\sigma(\lambda)$ over λ, we set $\lambda = 1/2$ [40,42].

The final step in determining the contact distance is to calculate $\lambda_{\min}$ at which $\sigma(\lambda)$ is a minimum. This task can be approached in two ways: either minimize $\sigma(\lambda)$ numerically or derive an analytical expression for $\lambda_{\min}$. We prefer the latter because it improves the efficiency and accuracy of the calculation. Determining $\lambda_{\min}$ involves solving a quartic (for 2D bidisperse ellipses) or quintic (for 3D monodisperse ellipsoids) polynomial equation.

We now sketch an outline for deriving these polynomial equations since this has not yet appeared in the literature. First, minimizing $\sigma(\lambda)$ with respect to λ is the same as maximizing $\sigma^{-2}(\lambda)$ over λ. We then make the substitution $y = \lambda - 1/2$ (because it simplifies the algebra) and define $h(y) = [\sigma(y + 1/2)]^{-2}$. Thus, to find

$\lambda_{\min}$, we must solve $dh(y)/dy = f(y)/g(y) = 0$ for $y_{\min}$, where $f(y)$ and $g(y)$ are polynomials, by solving for the roots of $f(y)$. For 2D bidisperse ellipses (3D monodisperse ellipsoids), $f_{2D}(y)$ ($f_{3D}(y)$) is a quartic (quintic) polynomial with coefficients that depend on four parameters J, K, L, and M:

$$f_{2D}(y) = q_4 y^4 + q_3 y^3 + q_2 y^2 + q_1 y + q_0 \tag{2.7}$$

$$q_0 = \frac{1}{16} J(L+4) - \frac{1}{16} K(J+4) \tag{2.8}$$

$$q_1 = -(1 + K/2)(J+2) \tag{2.9}$$

$$q_2 = -\frac{1}{2} J(L+6) - \frac{1}{2} K(3J+2) \tag{2.10}$$

$$q_3 = -2JK \tag{2.11}$$

$$q_4 = J(L-K) \tag{2.12}$$

$$J = \frac{a_j^2 - (a_j^2 - b_j^2)(\hat{r}_{ij} \cdot \hat{\mu}_j)^2}{a_i^2 - (a_i^2 - b_i^2)(\hat{r}_{ij} \cdot \hat{\mu}_i)^2} - 1 \tag{2.13}$$

$$K = \left(\frac{a_j b_j}{a_i b_i}\right)^2 - 1 \tag{2.14}$$

$$L = \frac{1}{a_i^2 b_i^2}(a_i^2 a_j^2 + b_i^2 b_j^2 + (a_i^2 - b_i^2)(a_j^2 - b_j^2)(\hat{\mu}_i \cdot \hat{\mu}_j)^2) - 2 \tag{2.15}$$

$$f_{3D}(y) = q_5 y^5 + q_4 y^4 + q_3 y^3 + q_2 y^2 + q_1 y + q_0 \tag{2.16}$$

$$q_0 = \frac{1}{16} J(L+4) \tag{2.17}$$

$$q_1 = -(J+2) + \frac{1}{8} M(L+8) \tag{2.18}$$

$$q_2 = -\frac{1}{2} J(L+6) \tag{2.19}$$

$$q_3 = -M(L+4) \tag{2.20}$$

$$q_4 = JL \tag{2.21}$$

$$q_5 = 2ML \tag{2.22}$$

$$J = \frac{\alpha^2 - 1}{\alpha^2} \frac{((\hat{r}_{ij} \cdot \hat{\mu}_i)^2 - (\hat{r}_{ij} \cdot \hat{\mu}_j)^2)}{1 - \frac{\alpha^2 - 1}{\alpha^2}(\hat{r}_{ij} \cdot \hat{\mu}_i)^2} \tag{2.23}$$

$$L = \frac{(\alpha^2 - 1)^2}{\alpha^2}(1 - (\hat{\mu}_i \cdot \hat{\mu}_j)^2) \tag{2.24}$$

$$M = \frac{(\alpha^2 - 1)^2}{\alpha^2} \frac{1}{1 - \frac{\alpha^2 - 1}{\alpha^2}(\hat{r}_{ij} \cdot \hat{\mu}_i)^2}((\hat{r}_{ij} \cdot \hat{\mu}_i)^2 + (\hat{r}_{ij} \cdot \hat{\mu}_j)^2 - \hat{\mu}_i \cdot \hat{\mu}_j(2(\hat{r}_{ij} \cdot \hat{\mu}_i)(\hat{r}_{ij} \cdot \hat{\mu}_j) - \hat{\mu}_i \cdot \hat{\mu}_j) - 1). \tag{2.25}$$

Table 2.1 *The base 10 logarithm of the relative error of the Gaussian, linear, and quadratic approximations for the contact distance from the exact value,* $\log_{10}(|\sigma_{\text{approx}} - \sigma_{\text{exact}}|/\sigma_{\text{exact}})$, *for bidisperse ellipses with size ratio 1.4 over a range of aspect ratios* α. *These estimates were averaged over* $\hat{r}_{ij}$ *and orientations.*

Approximation	$\alpha = 1.0$	1.2	1.5	2.0	3.0	10.0
Gaussian	−2.0	−2.0	−1.9	−1.6	−1.3	−0.4
Linear	−3.9	−3.8	−3.7	−3.6	−1.5	−0.1
Quadratic	−6.3	−6.2	−6.3	−5.0	−3.8	−1.4

The coefficients J, K, L, and M reveal the symmetry of the polynomial $f(y)$. $M = 0$ in two dimensions, $L = 0$ when ellipsoidal particles i and j of the same aspect ratio have parallel orientations, $K = 0$ when i and j have the same size, and $J = 0$ when the orientations of ellipsoids i and j of the same size and aspect ratio have the same angle with respect to their center-to-center vector $\vec{r}_{ij}$. The symmetries found for M, L, and K are obvious (2D, parallel, and same size), but the symmetry found in J is not. When $J = 0$, $f_{3D}(y)$ only contains odd terms in y, which implies that $\lambda_{\min} = 1/2$ is a solution, and one has a simple expression for the contact distance in a nontrivial case!

We are now in a position to quantitatively compare the exact results for the contact distance to various levels of approximation. For example, we can truncate $f(y)$ at either linear or quadratic order, which yield:

$$\lambda_{\min}^{\text{linear}} = -\frac{q_0}{q_1}; \qquad \lambda_{\min}^{\text{quadratic}} = -\frac{\left(q_1 - \sqrt{q_1^2 - 4q_0q_2}\right)}{2q_2}. \tag{2.26}$$

Even at this level of approximation, the solution is significantly more precise than the commonly used Gaussian approximation, and is more efficient than numerically solving a quartic or quintic equation at each step in the packing-generation process. To demonstrate the improved precision, we include Table 2.1 that lists the relative error in finding the true contact distance for the Gaussian, linear, and quadratic approximation methods for two ellipses i and j with $a_j/a_i = 1.4$ over a range in aspect ratios from $\alpha = 1$ to 10. These estimates were averaged over $\hat{r}_{ij}$ and orientations.

2.2.3.3 *Particle shape annealing*

Static packings of anisotropic particles can also be generated using a 'particle shape annealing' method. This method involves starting with a static packing of spherical particles, changing a shape parameter at a given rate to cause the particles to

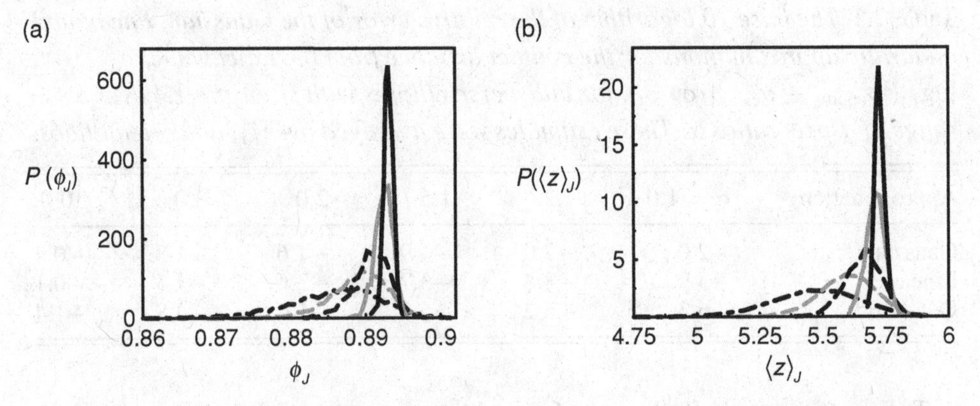

Figure 2.14 Distribution of (a) packing fractions ϕ_J and (b) mean contact numbers $\langle z \rangle_J$ at jamming onset for ellipses at $\alpha = 1.6$ using the isotropic compression packing-generation method as a function of system size $N = 32$ (black dot dashed), 64 (gray dashed), 256 (black dashed), 512 (gray solid), and 1024 (black solid).

become nonspherical, and using the compression method described in Section 2.2.1 to obtain a packing of just-touching particles with the new shape parameter.

We provide specific details of the particle shape annealing method for creating packings of ellipsoidal particles at a given aspect ratio α. The process starts with a static packing of spherical particles. The aspect ratio of each disk/sphere is then increased from $\alpha = 1$ to $1 + \Delta\alpha$ with the direction of the major axis chosen randomly. A static packing of ellipsoidal particles at $\alpha = 1 + \Delta\alpha$ is formed from this initial state using the compression method from Section 2.2.1. The ellipsoidal particles in this new packing are further elongated, and the protocol is repeated until a packing with the desired aspect ratio is reached. From our previous studies of ellipse packings [33], we find that the average jammed packing fraction $\langle \phi_J \rangle$ is larger than that obtained using the isotropic compression method, even though the packings do not possess increased spatial or orientational order as shown in Figure 2.18(b). In addition, the particle shape annealing procedure does not depend sensitively on the step size $\Delta\alpha$, at least for sufficiently small $\Delta\alpha$.

2.2.4 *Distributions of jamming onsets*

In previous sections, we described several methods for generating jammed particle packings. By creating packings for large numbers of independent, random initial conditions, one can create an ensemble of static packings and measure the distribution of jammed packing fractions $P(\phi_J)$ as shown in Figure 2.14(a) for ellipses at $\alpha = 1.6$ and several system sizes from $N = 32$ to 1024. Note that the distribution is broad for small systems, but approaches a δ-function in the large-N limit.

Packing-generation procedures for frictionless spherical and nonspherical grains give rise to jammed packings with well-defined mean packing fractions $\langle \phi_J \rangle$ in the large-system limit. In previous studies, we found that the width $\mathcal{W}$ of the packing fraction distribution for spherical particles scaled as $\mathcal{W} \sim N^{-\nu}$, with $\nu \approx 0.5$ [13]. Preliminary studies indicate that the scaling exponent depends on the aspect ratio α for ellipsoidal particles [43].

2.3 Mechanical stability

After static particle packings are generated, one can determine whether they are *mechanically stable*, i.e. in a state of total force and torque balance for each particle (except floater particles that are not locally stable) and stable with respect to infinitesimal deformations. Mechanical stability can be assessed by calculating the dynamical matrix of second derivatives of the total potential energy V [11]:

$$M_{kl} = \frac{\partial^2 V}{\partial \vec{\xi}_k \partial \vec{\xi}_l}, \tag{2.27}$$

where $\vec{\xi}_k$ are the relevant configurational degrees of freedom for particle k. As examples, $\vec{\xi}_k = \{x_k, y_k\}$ for disks and $\vec{\xi}_k = \{x_k, y_k, d_k\theta_k\}$ for ellipses, where x_k and y_k are the center of mass coordinates, θ_k is the angle between the major axis and $\hat{x}$, and d_k is a lengthscale for dimensional consistency that is typically obtained from the second moment of the mass distribution of the grain about its major axis. For ellipses, $d_k = \frac{1}{4}\sqrt{a_k^2 + b_k^2}$. In this discussion, we will assume that all grains have mass $m = 1$ with a uniform mass distribution. The dimension of the dynamical matrix is determined by the number of degrees of freedom (DOF) $d_f \mathcal{N}$ for a given system, where $\mathcal{N} = N - N_f$ and N_f is the number of floaters in the system. For example, $d_f = 2$ (3) for disks (ellipses). The dynamical matrix is real and symmetric with dimension $d_f \mathcal{N} \times d_f \mathcal{N}$.

To determine whether or not a static packing is mechanically stable, we diagonalize M_{kl} to find its $d_f \mathcal{N}$ eigenvalues e_i and associated eigenvectors $\hat{e}_i$, with $\hat{e}_i^2 = 1$. For systems with periodic boundary conditions, d of these eigenvalues are zero due to translational invariance. For mechanically stable states, the dynamical matrix possesses $d_f \mathcal{N} - d$ nontrivial eigenvalues with $e_i > 0$. This implies that all nontrivial deformations give rise to particle overlap and an increase in the potential energy to second order. In practice, one typically uses a threshold (e.g. $e_{\min} = 10^{-6}$) above which eigenvalues are deemed nonzero [28].

To illustrate the importance of testing mechanical stability, we show two $N = 7$ monodisperse static disk packings (solid circles) in Figures 2.15(a) and (b) at nearly the same packing fraction $\phi \approx 0.73$. The packing in (a) is mechanically stable,

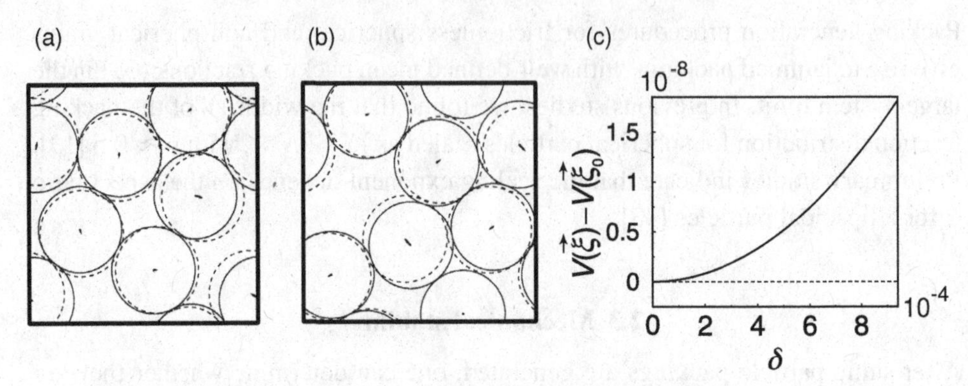

Figure 2.15 (a) Mechanically and (b) locally stable packings at nearly identical packing fractions. The original (perturbed) packings are shown using solid (dashed) circles. The packings are deformed according to: $\vec{\xi} = \vec{\xi}_0 + \delta\hat{e}_1$, where $\vec{\xi}_0$ is the unperturbed configuration, δ is the amplitude of the perturbation, and $\hat{e}_1$ is the eigenvector of the dynamical matrix (evaluated at $\vec{\xi}_0$) corresponding to the lowest nontrivial eigenvalue. Solid lines connecting the centers of corresponding particles in configurations $\vec{\xi}$ and $\vec{\xi}_0$ at $\delta = 0.2$ are also shown. (c) Change in potential energy per particle $V(\vec{\xi}) - V(\vec{\xi}_0)$ versus δ for the (a) mechanically stable (solid line) and (b) locally stable (dashed line) configurations. Note that the locally stable configuration can be deformed along $\hat{e}_1$ without energy cost.

while that in (b) is only locally stable [44]. To show this, we diagonalized the dynamical matrix for these configurations and then deformed them by δ along the eigenvector $\hat{e}_1$ corresponding to the smallest nontrivial eigenvalue: $\vec{\xi} = \vec{\xi}_0 + \delta\hat{e}_1$. In Figure 2.15(c), we plot the change in energy $\Delta V \equiv V(\vec{\xi}) - V(\vec{\xi}_0)$ versus δ. $\Delta V \sim \delta^2$ for the mechanically stable packing, while $\Delta V = 0$ for the locally stable packing. In Figure 2.15(b), one can see that for locally stable packings there are collective modes that do not give rise to particle overlap and thus do not change the potential energy.

The spectrum of normal modes (or vibrational frequencies) $\omega_i = \sqrt{e_i/M}$, where $M = Nm$, yields significant insight into the structural and mechanical properties of MS packings. The normal mode spectra are typically visualized in two ways: (1) the ordered list of $d_f\mathcal{N} - d$ nonzero frequencies and (2) the density of states (DOS) $D(\omega) = (N(\omega + \delta\omega) - N(\omega))/\delta\omega$, where $N(\omega)$ is the number of modes with frequency ω. The spectrum of normal modes for jammed bidisperse disk packings is shown in Figure 2.16, where we plot the sorted list of frequencies and DOS in (a) and (b), respectively. For jammed disk packings, we see the characteristic continuous increase in the sorted frequencies in (a) and plateau in the DOS at low frequencies in (b) [13]. In contrast to jammed sphere packings, we find three distinct regimes, separated by two gaps, in the normal mode spectrum for jammed ellipse packings over a range of aspect ratios [33, 37] as shown in

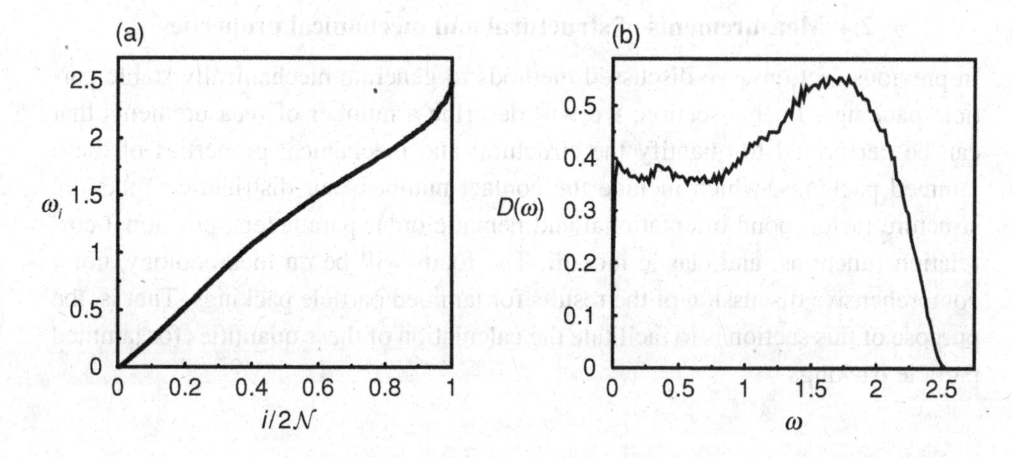

Figure 2.16 (a) Vibrational modes ω_i sorted in increasing order versus index $i/2\mathcal{N}$ normalized by the number of DOF and (b) density of vibrational modes $D(\omega)$ (averaged over 100 configurations) for jammed 50–50 bidisperse disk packings with diameter ratio $d = 1.4$ at $\phi - \phi_J = 10^{-4}$.

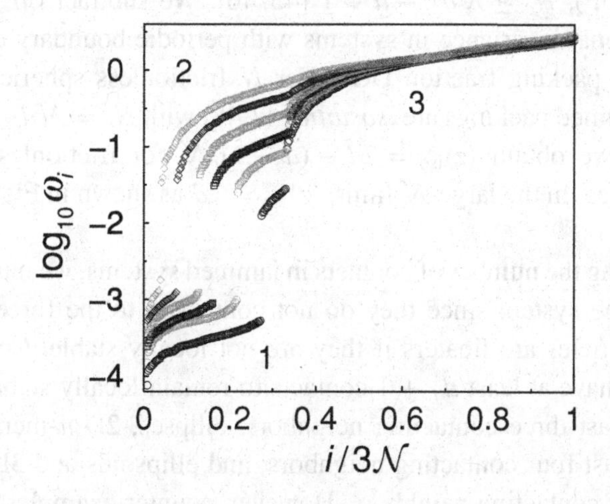

Figure 2.17 Normal mode frequencies ω_i vs. index $i/3\mathcal{N}$ normalized by the number of DOF, sorted by increasing frequency for $N = 120$ ellipse packings at six aspect ratios, $\alpha = 1.02$ (black circles), 1.06 (gray circles), 1.1 (black squares), 1.2 (gray squares), 1.4 (black diamonds), and 1.8 (gray diamonds). The sorted frequency spectrum possesses three distinct branches (numbered 1, 2, and 3). Adapted from reference [33].

Figure 2.17. A novel feature of ellipse packings is that in the just-touching limit perturbations along modes in region 1 give rise to *quartic*, not quadratic, increases in potential energy. Moreover, modes in regions 1 and 2 are primarily rotational in character.

2.4 Measurements of structural and mechanical properties

In previous sections, we discussed methods to generate mechanically stable particle packings. In this section, we will describe a number of measurements that can be performed to quantify the structural and mechanical properties of these jammed packings, which include the contact number, pair distribution function, structure factor, bond orientational and nematic order parameters, positional correlation functions, and elastic moduli. The focus will be on methodology, not a comprehensive discussion of the results for jammed particle packings. That is, the purpose of this section is to facilitate the calculation of these quantities for jammed particle packings.

2.4.1 Contact number

In a static granular packing, mechanical stability (to second order) can be achieved only if the number of contacts is greater than or equal to the number of degrees of freedom (DOF), $N_c \geq \mathcal{N} d_f - d + 1$ [45, 46]. We subtract off d trivial DOF due to translational invariance in systems with periodic boundary conditions and add 1 from the packing fraction DOF. For N frictionless spherical grains in d dimensions, jammed packings are *isostatic* [47,48] with $N_c = \mathcal{N} d - d + 1$. Using $N_c = \mathcal{N} \langle z \rangle / 2$, we obtain $\langle z \rangle_{\text{iso}} = 2d - (d-1)/\mathcal{N}$ for frictionless packings of spherical particles. In the large-$\mathcal{N}$ limit, $\langle z \rangle_{\text{iso}} = 2d$ as shown in Figure 2.18(a) for $\alpha = 1$.

When counting the number of contacts in jammed systems, we must first remove all floaters in the system since they do not contribute to the force network. We assume that particles are floaters if they are not locally stable. Convex particles must generally have at least $d_f + 1$ contacts to remain locally stable. Thus, disks must have at least three contacting neighbors; ellipses, 2D n-mers, and spheres must have at least four contacting neighbors; and ellipsoids and 3D n-mers must have at least six contacting neighbors. However, counter-examples include a disk with three contacting neighbors on one side of its equator and an ellipse with three contacts, whose normals at the points of contact intersect at a single point [16] as shown in Figure 2.19.

Once all floaters have been removed, N_c is obtained by counting the number of just-touching (slightly overlapped) pairs of particles. Since there are two contacting particles for each contact, the average number of contacts per particle $\langle z \rangle = 2N_c/\mathcal{N}$. Note that the constraint of mechanical stability is global, i.e. on N_c or $\mathcal{N} \langle z \rangle / 2$, not the contact numbers of individual grains z_i. We show the distribution of z_i for bidisperse disk packings in Figure 2.20. Note that a large fraction of particles have three (and five) contacts, not four.

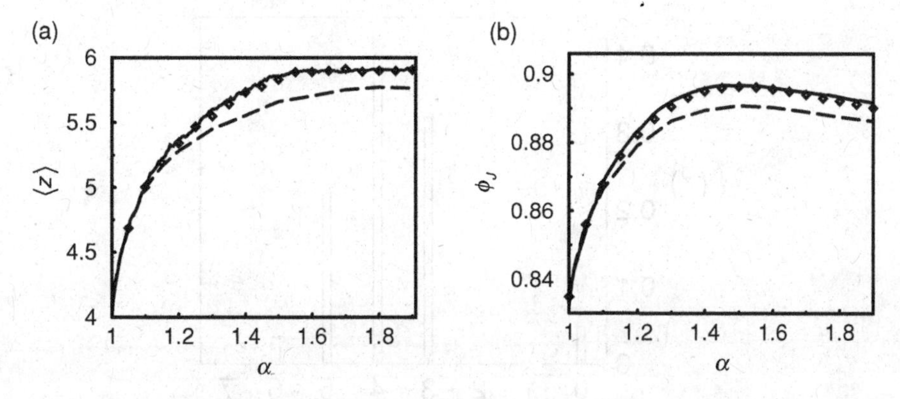

Figure 2.18 Average (a) contact number $\langle z \rangle$ and (b) packing fraction at jamming ϕ_J versus aspect ratio α for bidisperse ellipse packings using the isotropic compression (dashed line) and particle shape annealing methods. Two annealing rates are shown, $\Delta\alpha = 0.05$ (solid line) and 0.005 (diamonds). Adapted from reference [33].

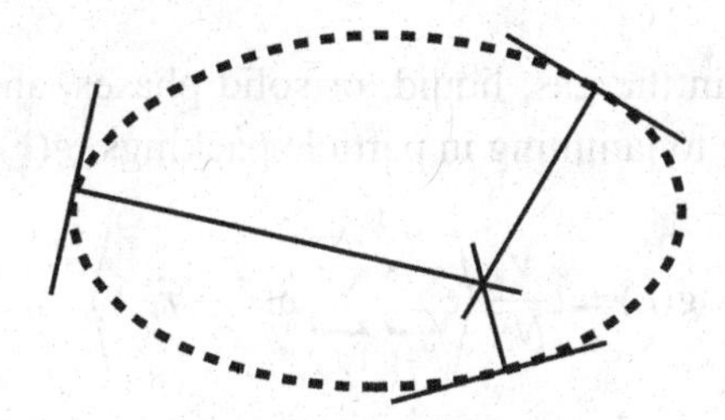

Figure 2.19 Ellipses are generically locally jammed, i.e. the ellipse can neither translate nor rotate without causing particle overlap with other particles held fixed, when they possess four or more contacts. However, ellipses can be locally jammed with three contacts if the normals to the tangent lines at contact happen to intersect at a single point [16].

Unlike spherical particles, ellipsoidal particles are *hypostatic* with $N_c < \mathcal{N} d_f - d + 1$. Packings of ellipsoidal particles can have nearly any value of contact number from the isostatic value for spherical particles $\langle z \rangle_{\text{iso}} = 2d$ to the isostatic value for ellipsoidal particles $\langle z \rangle_{\text{iso}} = 2d_f$. In Figure 2.18(a), we show that $\langle z \rangle$ ranges from $\langle z \rangle = 4$ when $\alpha = 1$ to $\langle z \rangle \lesssim 6$ (for particle shape annealing) near $\alpha = 2$ for ellipse packings [16, 33, 37].

2.4.2 Pair distribution function and structure factor

The pair distribution function $g(\vec{r})$ gives the probability for finding two particles $\vec{r}$ apart in a given system normalized by the probability for finding two particles separated by $\vec{r}$ in an ideal gas at the same density. In isotropic systems, the pair distribution function only depends on separation r. Here, $g(r)$ can be used to detect

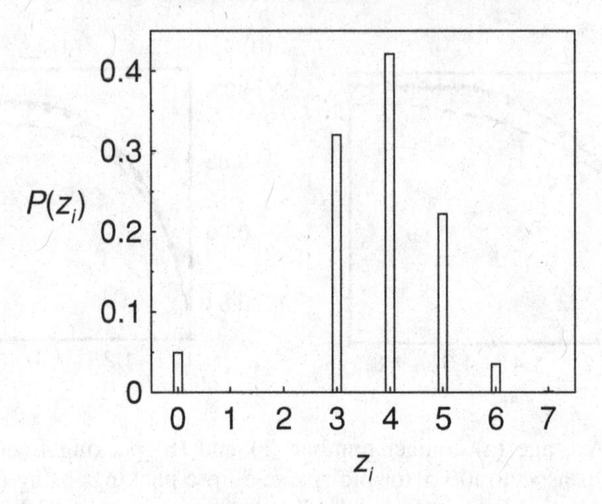

Figure 2.20 Histogram of the number contacts for individual particles z_i for jammed packings of bidisperse disks. Note that there are significant fluctuations away from four contacts, but all particles have $z_i \geq 3$ except for floaters.

whether systems exist in the gas, liquid, or solid phases, and provide a sensitive measure of the distance to jamming in particle packings. $g(r)$ can be expressed as:

$$g(\vec{r}) = \frac{V}{N^2}\left\langle \sum_{i=1}^{N}\sum_{i \neq j}^{N} \delta(\vec{r} - \vec{r}_{ij}) \right\rangle, \tag{2.28}$$

where V is the volume of the system [27]. In simulations, the δ-function is replaced by a function that is nonzero over a small range δr. Also, in periodic systems with simulation cell size $\mathcal{L}$, the maximum separation is $\mathcal{L}/\sqrt{2}$; however, the statistics are greatly reduced for $r > \mathcal{L}/2$. In liquids and solids, $g(r)$ is a function ϕ and T, but in model hard sphere systems, $g(r)$ is only a function ϕ.

The pair distribution function has strikingly different features for fluids, jammed packings, and crystalline systems as shown in Figure 2.21. Note that we normalize r by the rms contact distance $\langle \sigma^2 \rangle^{1/2}$ averaged over all particle species to account for polydispersity and anisotropic particles. For monodisperse systems, $\langle \sigma^2 \rangle^{1/2}$ is the particle diameter. For fluids (Figure 2.21(a)) with only short-range correlations, $g(r)$ has noticeable first and second neighbor peaks and then decays to 1 beyond 3–4 particle diameters.

For crystalline solids, $g(r)$ possesses sharp peaks that correspond to the different interparticle distances allowed by the symmetry of the lattice. In Figure 2.21(c), we show $g(r)$ for a hexagonal crystal in 2D. The hexagonal lattice is composed of linear combinations of the vectors $\hat{n}_1 = (1, 0)$ and $\hat{n}_2 = (1/2, \sqrt{3}/2)$. In units of the particle diameter, the lattice vectors are $\vec{n}_{kl} = k\hat{n}_1 + l\hat{n}_2$, where k, l are nonnegative integers, and the corresponding distances

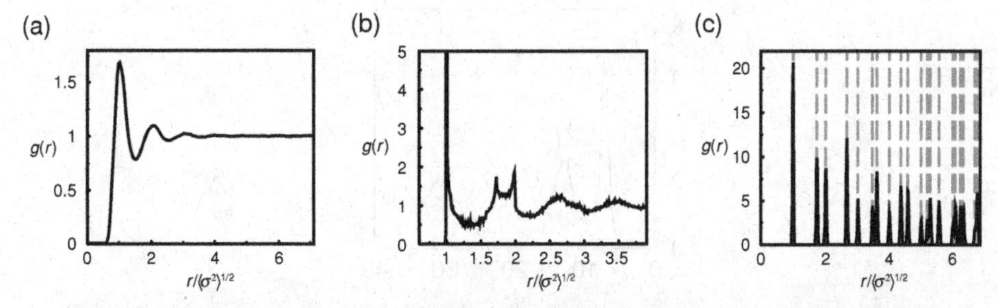

Figure 2.21 Pair distribution function, $g(r)$, for (a) liquid (b) jammed, and (c) crystalline systems. In (a), we show $g(r)$ for a 2D bidisperse system at $\phi = 0.7$ and temperature $T/\epsilon = 0.004 \gg T_g$ (where $k_b = 1$ and T_g is the glass transition) with repulsive linear spring interactions. In (b) (adapted from reference [13]), we show $g(r)$ for jammed 3D monodisperse packings at $\phi - \phi_J = 10^{-2}$ using the repulsive linear spring potential. Note that the first peak in (b) extends to $g(r \to \sigma) \approx 40$. $g(r)$ for a hexagonal crystal in 2D near $\phi_{\text{hex}} = \pi/2\sqrt{3}$ and $T = 0$ is plotted in (c). The dashed vertical lines indicate the possible separations for the hexagonal lattice.

are $n_{kl} = \sqrt{k^2 + lk + l^2} = 1, \sqrt{3}, 2, \sqrt{7}, 3, 2\sqrt{3}, \sqrt{13}, \ldots$. A similar procedure can be followed for other crystals, e.g. the face centered cubic (FCC) lattice, with $\hat{n}_1 = (1, 0, 0)$, $\hat{n}_2 = (1/2, \sqrt{3}/2, 0)$, and $\hat{n}_3 = (1/2, 1/2\sqrt{3}, \sqrt{2/3})$, and $n_{klm} = \sqrt{k^2 + l^2 + m^2 + kl + lm + km} = 1, \sqrt{3}, 2, \sqrt{6}, \sqrt{7}, \ldots$.

Even though $g(r)$ for glassy systems does not show strong signatures that signal the glass transition [49], there are several key features of $g(r)$ that signal the onset of jamming. For example, as systems approach the jamming transition, the height of the first peak diverges as its width tends to zero and the broad second peak found in liquids splits into two peaks, both of which become singular near jamming [13,50]. In Figure 2.21(b), we show these features for a 3D monodisperse system with $\phi - \phi_J = 10^{-2}$.

The pair distribution function can be measured easily in granular and colloidal systems via direct visualization of particles. On smaller lengthscales or in cases where direct visualization is not possible, one can obtain structural information, such as the structure factor $S(\vec{k})$ from light or x-ray scattering. $S(\vec{k})$ is the autocorrelation function of Fourier transformed density $\rho(\vec{k}) = \sum_{i=1}^{N} e^{\imath \vec{k}\cdot\vec{r}_i}$, $S(\vec{k}) = N^{-1}\langle \rho(\vec{k})\rho(-\vec{k})\rangle$, which can be written as:

$$S(\vec{k}) = \frac{1}{N}\left\langle \sum_{i=1}^{N}\sum_{i\neq j}^{N} e^{\imath \vec{k}\cdot(\vec{r}_i - \vec{r}_j)} \right\rangle, \tag{2.29}$$

where $\vec{k} = (2\pi/\mathcal{L})(n_x\hat{x} + n_y\hat{y} + n_z\hat{z})$, with $n_x, n_y, n_z = 0, 1, \ldots$, are the allowed wavevectors and angle brackets denote an ensemble average. The isotropic $S(k)$

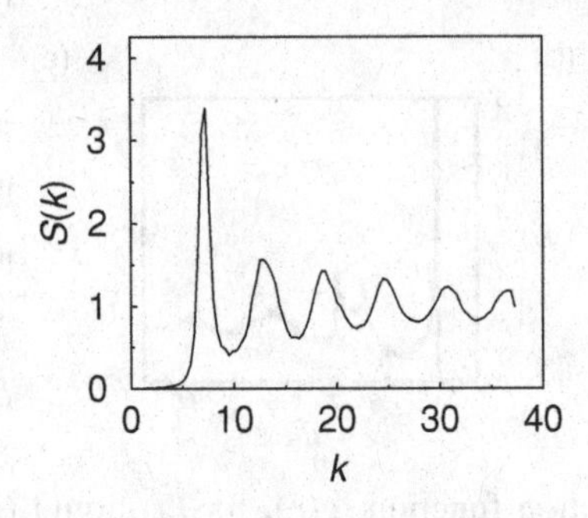

Figure 2.22 Angular averaged structure factor $S(k)$ for jammed monodisperse sphere packings at $\phi - \phi_J = 10^{-4}$. Adapted from reference [13].

can be obtained by angular averaging over $\vec{k}/k$ at fixed k. $S(k)$ is also related to the spatial Fourier transform of $g(r)$, but since simulations have access to particle positions, both are typically calculated directly. In Figure 2.22, we show $S(k)$ for jammed packings of monodisperse spheres at $\phi - \phi_J = 10^{-4}$. A key feature of $S(k)$ for jammed packings is the long-lived fluctuations that occur at $k_n^* = 2\pi n/a$, where $n = 1, 2, 3, \ldots$ and a is the particle diameter, which are a direct result of the divergent first peak in $g(r)$. Recent studies have also investigated the novel power-law scaling of $S(k)$ at low k in jammed sphere packings [51].

2.4.3 Order parameters

It is important to characterize the translational and orientational order as systems approach the jamming transition. For example, in granular shear flows one must distinguish between crystallization kinetics and jamming behavior [52,53]. Order parameters that identify various symmetries can be measured and used to quantify order or disorder. We will discuss three commonly used order parameters: the nematic order parameter P_2, which evaluates to one in the case that all orientations of uniaxial particles are parallel; the bond orientational parameter Q_6, which evaluates to one for systems in which the positional degrees of freedom have perfect six-fold symmetry; and a translation order parameter $\mathcal{G}$, which is the ratio of the first minimum to the first maximum in $g(r)$ and tends to zero as systems crystallize.

The nematic order parameter is defined as $P_2 = \langle 2\cos^2\theta - \frac{1}{2}\rangle$ in 2D and $P_2 = \langle \frac{3}{2}\cos^2\theta - \frac{1}{2}\rangle$ in 3D, where angle brackets denote an average over all particles in the system, θ is the angle of the particle director (i.e. the unit vector along the long axis, $\hat{n}$) relative to the average nematic director $\hat{N}$. $\hat{N}$ can be obtained by calculating $\langle \hat{n} \rangle$ or by maximizing P_2 with respect to $\hat{n}$. $P_2 = 1$ for completely aligned systems, whereas $P_2 \sim 1/\sqrt{N}$ $(1/\sqrt{N_d})$ in the absence (presence) of nematic domains, where N_d is the number of domains. In Figure 2.23(c) and (d) we show

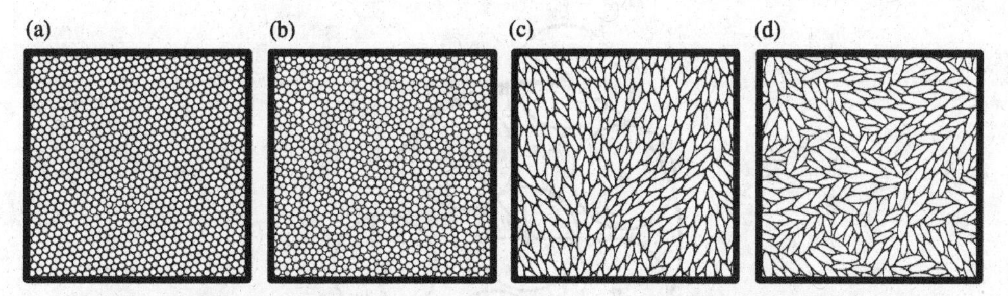

Figure 2.23 Packings that display varying degrees of bond orientational Q_6 and nematic P_2 order. In (a) and (b), we show $N = 1024$ disk packings with polydispersity $p = 1.1$ (1.4), $Q_6^g = 0.93$ (0.03), and $Q_6^l = 0.95$ (0.63). In (c) and (d), we show $N = 256$ monodisperse ellipse packings with $\alpha = 3$ and $P_2 = 0.64$ (0.11) using two different packing protocols.

monodisperse ellipse packings at aspect ratio $\alpha = 3$ and roughly the same packing fraction that were created using two different packing-generation methods. The method used in (c) produces roughly aligned ellipses with $P_2 = 0.64$, while the method in (d) produces randomly oriented ellipses with $P_2 = 0.11$.

The bond orientational order parameter Q_6, which measures hexagonal registry of nearest neighbors, can be calculated "locally," which does not consider phase information, or "globally," which allows phase cancellations. A polycrystal will yield a large value for the local bond orientational order parameter Q_6^l, even though the global order parameter $Q_6^l \sim 1/\sqrt{N_d}$, where N_d is the number of polycrystalline domains. The expressions for the global and local definitions of Q_6 are given below. Equations (2.30) (global), (2.31) (local), (2.32) (global), and (2.33) (local) provide expressions for the bond orientational order parameters in 2D and 3D, respectively:

$$Q_6^g = \frac{1}{N}\left|\sum_{i=1}^{N}\frac{1}{n_i}\sum_{j=1}^{n_i} e^{6\imath\theta_{ij}}\right| \tag{2.30}$$

$$Q_6^l = \frac{1}{N}\sum_{i=1}^{N}\frac{1}{n_i}\left|\sum_{j=1}^{n_i} e^{6\imath\theta_{ij}}\right| \tag{2.31}$$

$$Q_6^g = \left(\frac{4\pi}{13}\sum_{m=-6}^{6}\left|\frac{1}{N}\sum_{i=1}^{N}\frac{1}{n_i}\sum_{j=1}^{n_i} Y_6^m(\theta_{ij}, \phi_{ij})\right|^2\right)^{1/2} \tag{2.32}$$

$$Q_6^l = \left(\frac{4\pi}{13}\sum_{m=-6}^{6}\frac{1}{N}\sum_{i=1}^{N}\frac{1}{n_i}\left|\sum_{j=1}^{n_i} Y_6^m(\theta_{ij}, \phi_{ij})\right|^2\right)^{1/2}, \tag{2.33}$$

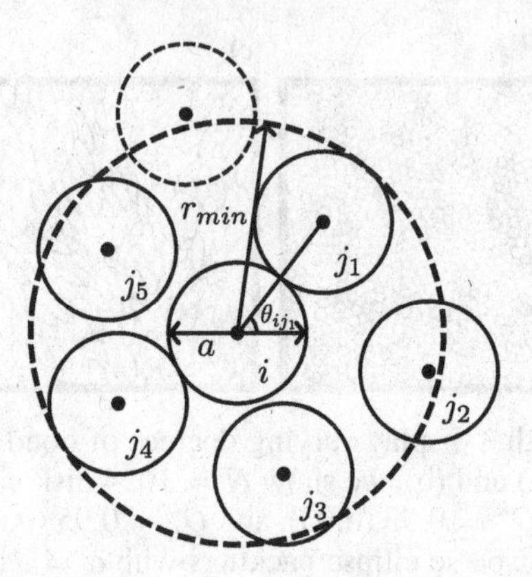

Figure 2.24 The nearest-neighbor range employed for calculations of Q_6. The dashed particle is outside the nearest-neighbor radius $r_{\rm min}$, which is defined as the first minimum of $g(r)$.

where θ_{ij} is the angle (axial angle in 3D) between a central particle i and neighbors j, ϕ_{ij} is the polar angle between i and j, Y_m^l are the spherical harmonics, and n_i denotes the number of nearest neighbors of i. Two particles are deemed nearest neighbors if their center-to-center separation $r_{ij} < r_{\rm min}$, which we set as the first minimum of $g(r)$ as shown in Figure 2.24. The spherical harmonics $Y_l^m(\theta_{ij}, \phi_{ij}) = \sqrt{\frac{2l+1}{4\pi}\frac{(l-6)!}{(l+6)!}} e^{\imath m\phi_{ij}} P_l^m(\cos\theta_{ij})$, where $P_l^m(\cos\theta_{ij})$ are Legendre polynomials [54].

In Figure 2.23 we show two jammed disk packings with different polydispersities at roughly the same packing fraction. The lower polydispersity $p = 1.1$ is polycrystalline with a large Q_6^g, while the packing at $p = 1.4$ is amorphous with a negligible Q_6^g. This shows that weakly polydisperse disk packings are prone to crystallization. In Figure 2.25, we show that both finite aspect ratio α and polydispersity p give rise to disorder. $\alpha > 1.2$ and $p > 1.2$ both lead to amorphous packings with small Q_6^g. In Figure 2.26, we show snapshots of dimer packings as a function of increasing α and disorder. Panels (a)–(f) correspond to the dashed line in Figure 2.25.

We also mention briefly the order parameter $\mathcal{G} = g(r_{\rm min})/g(r_{\rm max})$ that is sensitive to translational order. It is zero for crystalline systems, but is a finite constant for jammed systems. It has been used for example in studies of metallic and structural glasses to determine the onset of crystallization as a function of cooling rate [55].

2.4.4 Correlation functions and lengths

Although jammed systems are amorphous on macroscopic scales, they can possess order over short lengthscales that is averaged out when calculating global order parameters such as Q_6 and P_2. For instance, jammed systems of monodisperse disks

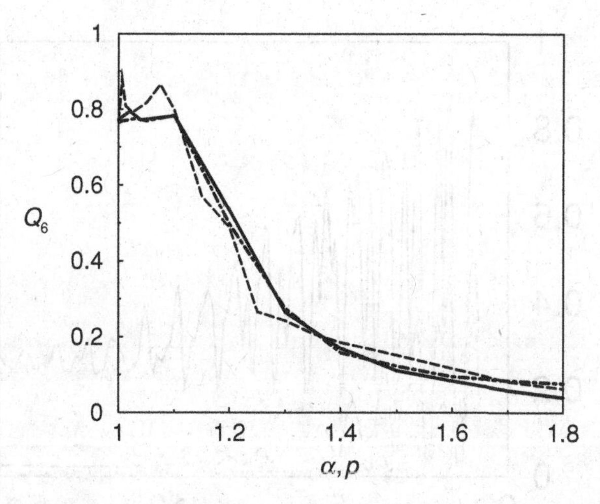

Figure 2.25 Q_6 (global) versus aspect ratio α for monodisperse ellipse packings (solid line), α for 2D dimer packings (dashed line), and polydispersity p for disk packings (dot-dashed line). All data are for $N = 256$ particles.

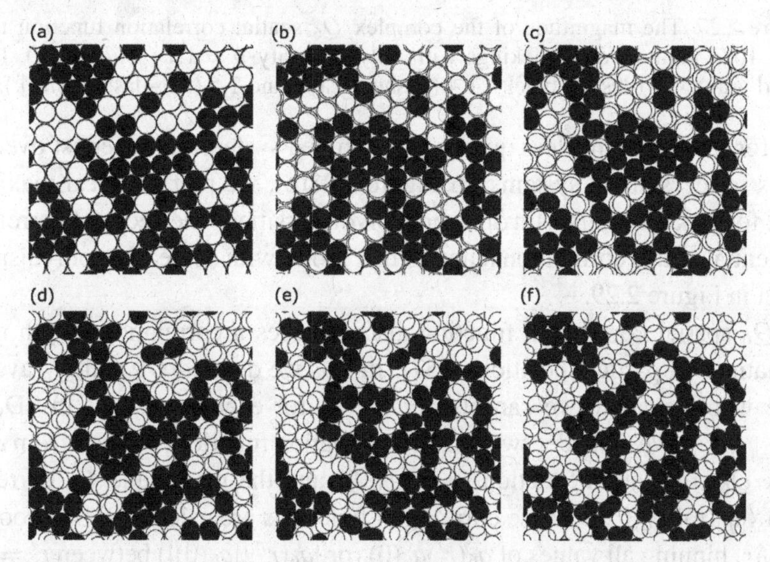

Figure 2.26 Jammed packings of $N = 256$ dimers with aspect ratios (a) $\alpha = 1.04$, (b) 1.1, (c) 1.2, (d) 1.3, (e) 1.4, and (f) 1.6. Corresponding global Q_6 values are plotted in Figure 2.25 (dashed line).

form polycrystals, which yield a low value of Q_6^g because the ordered domains are out of phase with each other. Similarly, uniaxial objects form nematic domains when quenched rapidly [56]. We will now review several spatial correlation functions that provide correlation lengths related to the size of ordered subregions.

We will first measure a spatial correlation length from the decay of correlations in $g(r)$. As we showed earlier in Figure 2.21, the fluctuations in $g(r) - 1$ die out

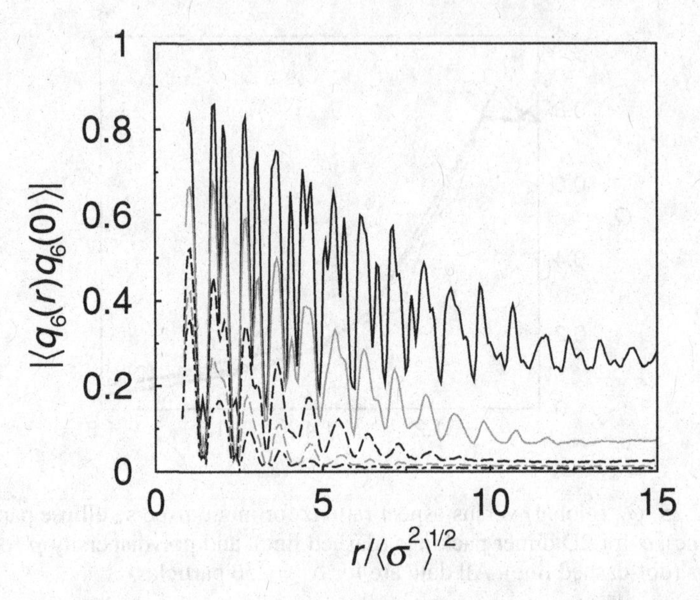

Figure 2.27 The magnitude of the complex Q_6 spatial correlation function for $N = 1024$ jammed disk packings with polydispersity $p = 1.1$ (solid black), 1.2 (solid gray), 1.3 (dashed black), 1.4 (dashed gray), and 1.5 (dot–dashed black).

quickly for fluid systems, less quickly for jammed systems, and persist over large distances for crystalline systems. In Figure 2.28(b), we plot the local maxima of $g(r) - 1$ for disk packings with different polydispersities. The decay of correlations is exponential with a correlation length ξ that grows with decreasing polydispersity as shown in Figure 2.29.

The Q_6 spatial correlation function also provides a correlation length related to fluctuations in particle positions [57]. Just as we defined Q_6^g and Q_6^l averaged over the whole system, we can also define q_6^i for each particle i. In 2D, $q_6^i = n_i^{-1} \sum_{j=1}^{n_i} e^{6\imath\theta_{ij}}$. We consider two spatial correlation functions formed from q_6^i: the complex Q_6 correlation function $\langle q_6(r) q_6(0) \rangle$ and the magnitude Q_6 correlation function $\langle |q_6(r)||q_6(0)| \rangle$. These correlation functions are calculated by choosing a bin size δr, binning all values of $q_6(r')q_6(0)$ (or $|q_6(r')||q_6(0)|$) between $r' = r$ and $r' = r + \delta r$, and then dividing by the number of pairs between r and $r + \delta r$. Since $q_6(r)q_6(0)$ is a complex number and contains phase information, the difference between these two correlation functions is analogous to the difference between the order parameters Q_6^g and Q_6^l. The complex Q_6 correlation function is sensitive to fluctuations in phase. Therefore, it should decay to $1/\sqrt{N_b}$ in an amorphous system, where N_b is the number of bonds, or $1/\sqrt{N_{\rm d}}$ in a polycrystal. In contrast, the magnitude Q_6 correlation function is not sensitive to fluctuations in phase.

In Figure 2.27, we show the magnitude of the complex Q_6 correlation function $|\langle q_6(r) q_6(0) \rangle|$ as a function of polydispersity for jammed disk packings.

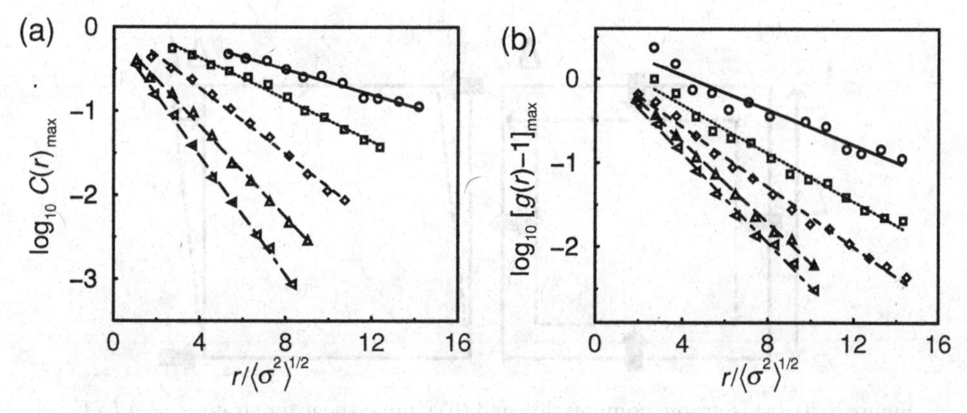

Figure 2.28 Local maxima of (a) $C(r)$ (defined in the text) and (b) $g(r) - 1$ for disk packings with polydispersities $p = 1.1$ (circles), 1.2 (squares), 1.3 (diamonds), 1.4 (upward triangles), and 1.5 (leftward triangles). The lines show least-square fits to $e^{-r/\xi}$.

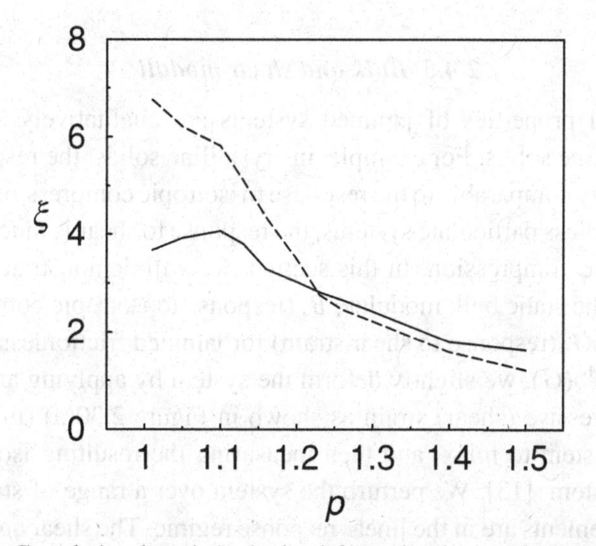

Figure 2.29 Correlation length ξ obtained from the decay of the local maxima of $C(r) = |\langle q_6(r)q_6(0)\rangle - \langle q_6(\infty)q_6(0)\rangle|$ (dashed line) and $g(r) - 1$ (solid line) shown in Figure 2.28 for disk packings as a function of polydispersity p.

$|\langle q_6(r)q_6(0)\rangle|$ decays more slowly as the polydispersity decreases due to the increase in polycrystalline domain size as shown in the snapshots in Figure 2.23(a) and (b). To extract a correlation length from the decay of correlations in Figure 2.27, we plot the local maxima of $C(r) = |\langle q_6(r)q_6(0)\rangle - \langle q_6(r)q_6(\infty)\rangle|$ in Figure 2.28(a). The correlations decay exponentially with distance over a length-scale that increases with decreasing polydispersity p. The dependence of the correlation length from $|\langle q_6(r)q_6(0)\rangle|$ on p appears in Figure 2.29. Note that the

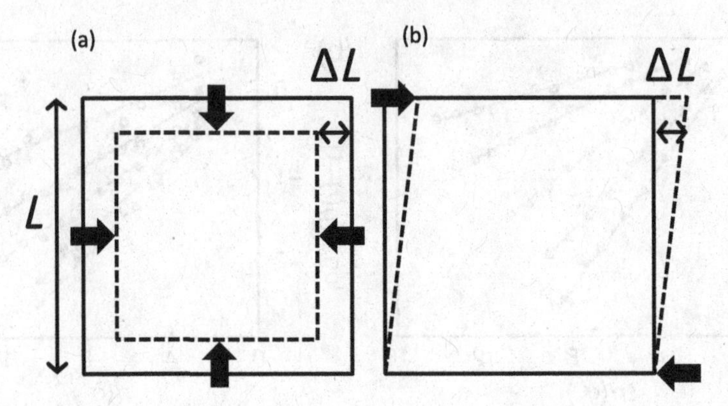

Figure 2.30 (a) Isotropic compression and (b) simple shear for strain $\gamma = \Delta L/L$.

lengthscales from the complex Q_6 and $g(r)$ correlation functions show a marked departure at low polydispersities.

2.4.5 *Bulk and shear moduli*

The mechanical properties of jammed systems are qualitatively different from those of crystalline solids. For example, in crystalline solids, the response to shear strain is generally comparable to the response to isotropic compression. However, in jammed, frictionless particulate systems, the response to shear is much weaker than that for isotropic compression. In this section, we will demonstrate this property by calculating the static bulk modulus, B, (response to isotropic compression) and shear modulus, G, (response to shear strain) for jammed frictionless packings.

To measure B (G), we slightly deform the system by applying an infinitesimal isotropic compressive (shear) strain as shown in Figure 2.30(a) (Figure 2.30(b)), allowing the system to relax, and then measuring the resulting isotropic (shear) stress in the system [13]. We perturb the system over a range of strains to verify that the measurements are in the linear response regime. The shear and bulk moduli are obtained by measuring the response of the pressure tensor, $P_{\alpha\beta}$ to the applied strain, where:

$$P_{\alpha\beta} = -L^{-d} \sum_{i>j} r_{ij\alpha} \frac{r_{ij\beta}}{r_{ij}} \frac{dV}{dr_{ij}} \tag{2.34}$$

and $\alpha, \beta \in \{x, y, z\}$. The bulk and shear moduli are defined by $B = \phi dP/d\phi$ and $G = d\Sigma/d\gamma$, where P is the pressure and $\Sigma = -P_{xy}$ is the shear stress (when x (y) is the shear (gradient) direction). In Figure 2.31, we show that Σ is linear in γ and P is linear in $\phi - \phi_J$ over several orders of magnitude in jammed disk packings.

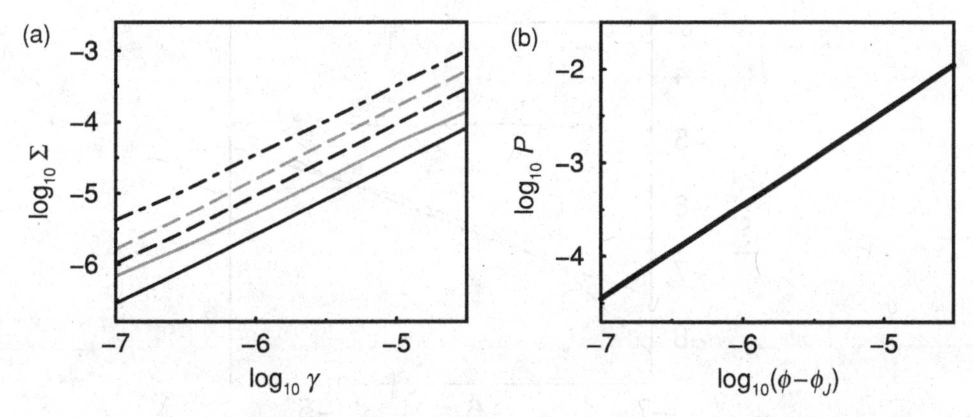

Figure 2.31 (a) Shear stress Σ versus shear strain γ in jammed disk packings at $\phi - \phi_J = 6 \times 10^{-5}$ (solid black), 3×10^{-4} (solid gray), 1×10^{-3} (dashed black), 4×10^{-3} (dashed gray), 2×10^{-2} (dot–dashed black). (b) Pressure P versus $\phi - \phi_J$ for jammed disk packings. P is linear in $\phi - \phi_J$, and Σ is linear in γ over several orders of magnitude.

The magnitude of the shear stress versus strain curve has strong $\phi - \phi_J$ dependence, showing a power law dependence $G \propto (\phi - \phi_J)^{1/2}$ [13]. In contrast, the bulk modulus does not depend strongly on $\phi - \phi_J$. Thus, in the limit $\phi \to \phi_J$, the bulk modulus remains finite, while the shear modulus goes to zero. This behavior has been related to the depletion of low frequency modes in the jamming density of states [58,59], revealing fundamental physics not found in crystalline solids.

A subtle aspect of the shear modulus calculations especially for packings of spherical particles is the wide distribution of yield strains. In Figure 2.31, the yield strain $\gamma_y > 10^{-4}$ for all $\phi - \phi_J$ shown. However, γ_y even at fixed $\phi - \phi_J$ has large fluctuations, and thus for measurements on some configurations it is difficult to be simultaneously below the yield strain, above numerical noise, and within the linear regime. Several examples of nonlinear stress versus strain curves are shown in Figure 2.32. We have noticed that this behavior is diminished in large systems and systems composed of ellipsoidal particles.

Though B is nearly the same for jammed packings of ellipses and disks (Figure 2.33, main panel), G is much smaller for ellipses than for disks (Figure 2.33, inset). In fact, $G \propto (\phi - \phi_J)$ for ellipses at sufficiently low $\phi - \phi_J$, whereas $G \propto (\phi - \phi_J)^{1/2}$ for disks. For low aspect ratio ($\alpha < 1.01$), the $\phi - \phi_J$ scaling intersects the $(\phi - \phi_J)^{1/2}$ scaling for disks at $\phi^*(\alpha)$, below which the system has sphere-like $(\phi - \phi_J)^{1/2}$ scaling, and above which the system has $\phi - \phi_J$ scaling [33]. This new scaling behavior has been linked to quartic vibrational modes found in "just-touching" ellipse packings. Thus, near-jamming ellipse packings are much more susceptible to shear than packings of spherical particles.

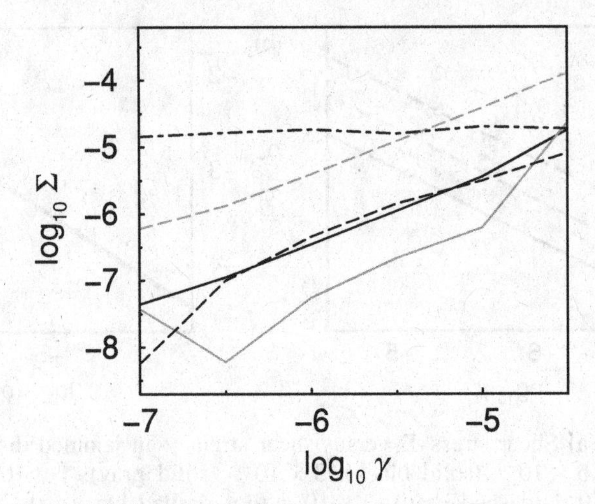

Figure 2.32 (a) Shear stress Σ versus shear strain γ for the same $\phi - \phi_J$ values in Figure 2.31. For some of these configurations, the yield strain is extremely small, which causes nonlinearity in $\Sigma(\gamma)$.

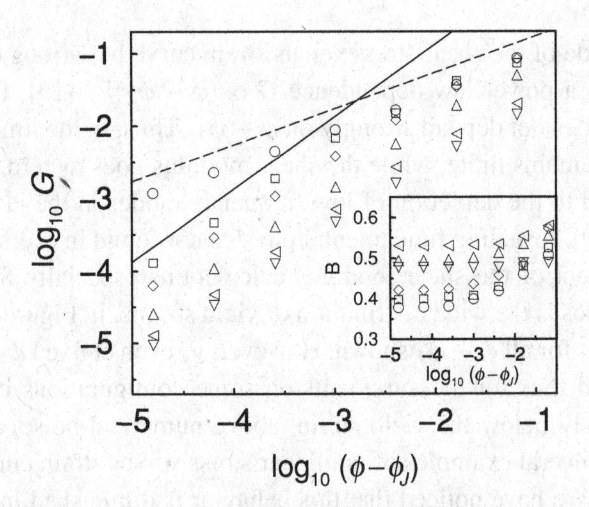

Figure 2.33 Shear (main plot) and bulk (inset) moduli versus $\phi - \phi_J$ for ellipse packings at $\alpha = 1.0$ (circles), 1.002 (squares), 1.01 (diamonds), 1.1 (upward triangles), 1.5 (leftward triangles), 2.0 (downward triangles). Notice that over the range $\phi - \phi_J = 10^{-5}$ to $\phi - \phi_J = 10^{-1}$, B changes by a factor of 1.5, while G changes by a factor of 100. The solid (dashed) line has slope 1 (0.5). Adapted from reference [33].

Acknowledgments

Support from NSF grant numbers DMR-0448838 (CSO) and DMS-0835742 (CS) is acknowledged. We gratefully acknowledge helpful discussions with Bulbul Chakraborty, Mitch Mailman, and Gregg Lois.

References

[1] S. Torquato, T. M. Truskett, and P. G. Debenedetti, "Is random close packing of spheres well defined?" *Phys. Rev. Lett.* **84**, 2064 (2000).
[2] N. Xu and C. S. O'Hern, "Measurements of the yield stress in frictionless granular systems," *Phys. Rev. E* **73**, 061303 (2006).
[3] W. Kob and H. C. Andersen, "Testing mode-coupling theory for a supercooled binary Lennard–Jones mixture I: the van Hove correlation function," *Phys. Rev. E* **51**, 4626 (1995).
[4] P. N. Pusey and W. van Megan, "Observation of a glass transition in suspensions of spherical colloidal particles," *Phys. Rev. Lett.* **59**, 2083 (1987).
[5] K. N. Pham, A. M. Puertas, J. Bergenholtz, S. U. Egelhaaf, A. Mousaïd, P. N. Pusey, A. B. Schofield, M. E. Cates, M. Fuchs, and W. C. K. Poon, "Multiple glassy states in a simple model system," *Science* **296**, 104 (2002).
[6] L. E. Silbert, D. Ertas, G. S. Grest, T. C. Halsey, and D. Levine, "Geometry of frictionless and frictional sphere packings," *Phys. Rev. E* **65**, 031304 (2002).
[7] D. J. Durian and D. A. Weitz, "Foams," in J. I. Kroschwitz (ed.), *Kirk-Othmer Encyclopedia of Chemical Technology*, Vol. 11, p. 783, 4th edition (Wiley, New York, 1994).
[8] T. G. Mason, J. Bibette, and D. A. Weitz, "Elasticity of compressed emulsions," *Phys. Rev. Lett.* **75**, 2051 (1995).
[9] Z. Zhang, N. Xu, D. D. N. Chen, P. Yunker, A. M. Alsayed, K. B. Aptowicz, P. Habdas, A. J. Liu, S. R. Nagel, and A. G. Yodh, "Thermal vestige of the zero-temperature jamming transition," *Nature* **459**, 230 (2009).
[10] P. G. Debenedetti and F. H. Stillinger, "Supercooled liquids and the glass transition," *Nature* **410**, 259 (2001).
[11] A. Tanguy, J. P. Wittmer, F. Leonforte, and J.-L. Barrat, "Continuum limit of amorphous elastic bodies: a finite-size study of low frequency harmonic vibrations," *Phys. Rev. B* **66**, 174205 (2002).
[12] B. D. Lubachevsky, F. H. Stillinger, and E. N. Pinson, "Disks vs. spheres: contrasting properties of random packings," *J. Stat. Phys.* **64**, 501 (1991).
[13] C. O. O'Hern, L. E. Silbert, A. J, Liu, and S. R. Nagel, "Jamming at zero temperature and zero applied stress: the epitome of disorder," *Phys. Rev. Lett.* **68**, 011306 (2003).
[14] N. Xu, J. Blawzdziewicz, and C. S. O'Hern, "Reexamination of random close packing: ways to pack frictionless disks," *Phys. Rev. E* **71**, 061306 (2005).
[15] G.-J. Gao, J. Blawzdziewicz, and C. S. O'Hern, "Studies of the frequency distribution of mechanically stable disk packings," *Phys. Rev. E* **74**, 061304 (2006).
[16] A. Donev, R. Connelly, F. H. Stillinger, and S. Torquato, "Underconstrained jammed packings of hard ellipsoids," *Phys. Rev. E* **75**, 051304 (2007).
[17] W. S. Jodrey and E. M. Tory, "Computer simulation of close random packing of equal spheres," *Phys. Rev. A* **32**, 2347 (1985).
[18] R. J. Speedy, "Random jammed packings of hard discs and spheres," *J. Phys.: Condens. Matter* **10**, 4185 (1998).
[19] A. Zinchenko, "Algorithm for random close packing of spheres with periodic boundary conditions," *J. Comput. Phys.* **114**, 298 (1994).
[20] W. H. Press, B. P. Flannery, S. A. Teukolsky, and W. T. Vetterling, *Numerical Recipes in Fortran 77* (Cambridge University Press, New York, 1986).
[21] T. Majmudar and R. P. Behringer, "Contact force measurements and stress-induced anisotropy in granular materials," *Nature* **435**, 1079 (2005).

[22] S. Henkes, C. S. O'Hern, and B. Chakraborty, "Entropy and temperature of a static granular assembly: an ab initio approach," *Phys. Rev. Lett.* **99**, 038002 (2007).
[23] G. Lois, J. Xie, T. Majmudar, S. Henkes, B. Chakraborty, C. S. O'Hern, and R. Behringer, "Stress correlations in granular materials: An entropic formulation," *Phys. Rev. E* **80**, 060303(R) (2009).
[24] S. Henkes and B. Chakraborty, "Statistical mechanics framework for static granular matter," *Phys. Rev. E* **79**, 061301 (2009).
[25] M. Lundberg, K. Krishan, N. Xu, C. S. O'Hern, and M. Dennin, "Comparison of low amplitude oscillatory shear in experimental and computational studies of model foams," *Phys. Rev. E* **79**, 041505 (2009).
[26] B. Miller, C. O'Hern, and R. P. Behringer, "Stress fluctuations for continuously sheared granular materials," *Phys. Rev. Lett.* **77**, 3100 (1996).
[27] M. P. Allen and D. J. Tildesley, *Computer Simulation of Liquids* (Oxford University Press, New York, 1987).
[28] G.-J. Gao, J. Blawzdziewicz, and C. S. O'Hern, "Geometric families of mechanically stable granular packings," *Phys. Rev. E* **80**, 061303 (2009).
[29] D. J. Lacks and M. J. Osborne, "Energy landscape picture of overaging and rejuvenation in a sheared glass," *Phys. Rev. Lett.* **93**, 255501 (2004).
[30] B. A. Isner and D. J. Lacks, "Generic rugged landscapes under strain and the possibility of rejuvenation in glasses," *Phys. Rev. Lett.* **96**, 025506 (2006).
[31] G. Parisi and F. Zamponi, "Mean field theory of hard sphere glasses and jamming," to appear in *Rev. Mod. Phys.* **82**, 789 (2010).
[32] L. Santen and W. Krauth, "Absence of thermodynamic phase transition in a model glass former," *Nature* **405**, 550 (2000).
[33] M. Mailman, C. F. Schreck, C. S. O'Hern, and B. Chakraborty, "Jamming in systems of ellipse-shapes particles," *Phys. Rev. Lett.* **102**, 255501 (2009).
[34] C. Schreck, J.-K. Yang, H. Noh, H. Cao, and C. S. O'Hern, "Pseudo band gaps in two-dimensional amorphous materials," unpublished 2010.
[35] D. Brown and J. H. R. Clarke, "Constant-stress nonequilibrium molecular dynamics: shearing of the soft-sphere crystal and fluid," *Phys. Rev. A* **34**, 2093 (1986).
[36] G.-J. Gao, J. Blawzdziewicz, C. S. O'Hern, and M. Shattuck, "Experimental demonstration of nonuniform frequency distributions of granular packings," *Phys. Rev. E* **80**, 061304 (2009).
[37] Z. Zeravcic, N. Xu, A. Liu, S. Nagel, and W. van Saarloos, "Excitations of ellipsoid packings near jamming," *Europhys. Lett.* **87**, 26001 (2009).
[38] B. J. Berne and P. Pechukas, "Gaussian model potentials for molecular interactions," *J. Chem. Phys.* **56**, 4213 (1972).
[39] J. G. Gay and B. J. Berne, "Modification of the overlap potential to mimic a linear site–site potential," *J. Comp. Phys.* **74**, 3316 (1981).
[40] D. J. Cleaver, C. M. Care, M. P. Allen, and M. P Neal, "Extension and generalization of the Gay–Berne potential," *Phys. Rev. E* **54**, 559 (1996).
[41] J. W. Perram and M. S. Wertheim, "Statistical mechanics of hard ellipsoids I: overlap algorithm and the contact function," *J. Comp. Phys.* **58**, 409 (1985).
[42] J. W. Perram, J. Rasmussen, and E. Prætgaard, "Ellipsoid contact potential: theory and relation to overlap potentials," *Phys. Rev. E.* **54**, 6565 (1996).
[43] C. Schreck and C. S. O'Hern, "A new jamming critical point controls the glassy dynamics of ellipsoidal particles," unpublished 2010.
[44] S. Torquato and F. H. Stillinger, "Multiplicity of generation, selection, and classification procedures for jammed hard-particle packings", *J. Phys. Chem. B* **105**, 11849 (2001).

[45] C. F. Moukarzel, "Isostatic phase transition and instability in stiff granular materials," *Phys. Rev. Lett.* **81**, 1634 (1998).
[46] R. Blumenfeld, "Stresses in isostatic granular systems and emergence of force chains," *Phys. Rev. Lett.* **93**, 108301 (2004).
[47] H. A. Makse, D. L. Johnson, and L. M. Schwartz, "Packing of compressible granular materials," *Phys. Rev. Lett.* **84**, 4160 (2000).
[48] C. S. O'Hern, S. A. Langer, A. J. Liu, and S. R. Nagel, "Random packing of frictionless particles," *Phys. Rev. Lett.* **88**, 075507 (2002).
[49] W. Kob, C. Donati, S. J. Plimpton, P. H. Poole, and S. C. Glotzer, "Dynamical heterogeneities in a supercooled Lennard–Jones liquid," *Phys. Rev. Lett.* **79**, 2827 (1997).
[50] L. E. Silbert, A. J. Liu, and S. R. Nagel, "Structural signatures of the unjamming transition at zero temperature," *Phys. Rev. E* **73**, 041304 (2006).
[51] A. Donev, F. H. Stillinger, and S. Torquato, "Unexpected density fluctuations in jammed disordered sphere packings," *Phys. Rev. Lett.* **95**, 090604 (2005).
[52] K. E. Daniels and R. P. Behringer, "Hysteresis and competition between disorder and crystallization in sheared and vibrated granular flow," *Phys. Rev. Lett.* **94**, 168001 (2005).
[53] N. Duff and D. J. Lacks, "Shear-induced crystallization in jammed systems," *Phys. Rev. E* **75**, 031501 (2007).
[54] T. M. Truskett, S. Torquato, and P. G. Debenedetti, "Towards a quantification of disorder in materials: distinguishing equilibrium and glassy sphere packings," *Phys. Rev. E* **62**, 993 (2000).
[55] H.-J. Lee, T. Cagin, W. L. Johnson, and W. A. Goddard, "Criteria for formation of metallic glasses: the role of atomic size ratio," *J. Chem. Phys.* **119**, 9858 (2003).
[56] C. De Michele, R. Schilling, and F. Sciortino, "Dynamics of unixial hard ellipsoids," *Phys. Rev. Lett.* **98**, 265702 (2007).
[57] P. J. Steinhardt, D. R. Nelson, and M. Ronchetti, "Bond-orientational order in liquids and glasses," *Phys. Rev. B* **28**, 784 (1983).
[58] L. E. Silbert, A. J. Liu, and S. R. Nagel, "Vibrations and diverging length scales near the unjamming transition," *Phys. Rev. Lett.* **95**, 098301 (2005).
[59] W. G. Ellenbroek, E. Somfai, and M. van Hecke, W. van Saarloos, "Critical scaling in linear response of frictionless granular packings near jamming," *Phys. Rev. Lett.* **97**, 258001 (2006).

3

Soft random solids: particulate gels, compressed emulsions, and hybrid materials

ANTHONY D. DINSMORE

3.1 Introduction

This chapter will review aspects of soft random solids, including particulate gels, compressed emulsions, and other materials having similar basic features. Soft random solids are appealing for a number of reasons – and not just because of their taste (as in foods) or appearance (as in cosmetics). They offer the potential for insight into much broader and quite elegant problems in nonequilibrium thermodynamics such as the dynamics of phase transitions, the origin of the glass transition, and stresses and flows in granular media. They are also central players in a host of industries in the form of paint, ink, concrete, asphalt, dairy foods, and cosmetics. From the range of examples and the techniques involved, it should be apparent that investigation of these materials is a multi-disciplinary process, combining contributions from physicists, chemists, and engineers.

Since the 1980s, studies of soft random solids have benefited enormously from advances in microscopy, computer-aided image analysis, computer simulations, and new methods to synthesize colloidal particles with controlled shape, surface chemistry, and interactions. The chapter's purpose is to summarize areas of recent investigations and point out experimental breakthroughs, remaining questions, and relevant experimental methods. Despite the recent progress, a number of fundamentally interesting and practically relevant questions remain, and it is hoped that this chapter will stimulate further work in these areas.

The chapter's survey of soft, random solids begins with particulate gels in order to establish some useful concepts such as network architecture and topology, fractal scaling and correlation lengths. We will also explore insights into gelation – a nonequilibrium process – that come from principles of equilibrium thermodynamics. The chapter will then turn to another nonequilibrium soft random solid made from a mixture of phase-separating liquids and small particles. These materials

Experimental and Computational Techniques in Soft Condensed Matter Physics, ed. Jeffrey Olafsen.
Published by Cambridge University Press.

have a quasi-two-dimensional surface that confers solidity to the sample and also divides it into two separately continuous regions. These materials are known as "bijels" (for bicontinuous, interfacially jammed emulsion gel); they are solid yet – remarkably – they allow fluids to flow through them in two separate but intertwined compartments [2, 8, 9]. Toward the end of the chapter, we will encounter compressed emulsions, which have a number of interesting structural properties as well as a spontaneous chain-like arrangement of interior stresses in response to applied load. Here, the structures of interest appear not only in the arrangement of the *particles*, but also in the spatial distribution of internal *stresses* or inter-particle contact *forces*.

The chapter focuses on experimental results and methods, including scattering and optical microscopy. Thanks to confocal and related techniques of microscopy and new methods of image analysis, heterogeneous structure can be measured in two or three dimensions with single-particle resolution. (Technical aspects of these techniques are described in Chapter 1.) An equally important but less obvious power of imaging is to quantify forces, free-energies, and elastic moduli. Hence much of the discussion will focus on the results of microscopy. The chapter will also describe how tuning the interactions among colloidal particles or droplets affects gelation. Experiments, simulations, and theoretical models have addressed the problem of how the form of the interaction alters the morphology and dynamics, sometimes in rather striking ways.

Sections 3.2–3.4 focus on particulate gels, providing an overview, then describing their structure in detail, then their viscoelasticity. Section 3.5 describes efforts to understand gelation as a phase-separation process that is arrested before reaching equilibrium. Section 3.6 discusses bijels as an example of controlled, arrested phase separation. Section 3.7 discusses compressed emulsions with emphasis on the emergent stress network. Section 3.8 offers a brief conclusion.

3.2 Overview of particulate gels

The term 'gel' typically refers to an amorphous solid composed of a network-like structure with fairly low concentration, leaving discernible pores or voids (Figure 3.1). This chapter will include particulate gels, which are composed of colloidal particles that aggregate. The term gel can also apply to polymer-based materials and – while the same basic questions and theoretical and experimental techniques might apply – these materials will be left out of this chapter. The definition of colloidal or particulate gel is admittedly imprecise and a discussion of this point will help to motivate some of the fundamental questions.

It is generally assumed that gels are trapped out of equilibrium. That is, even gels composed of spherical particles remain amorphous for months or years, but

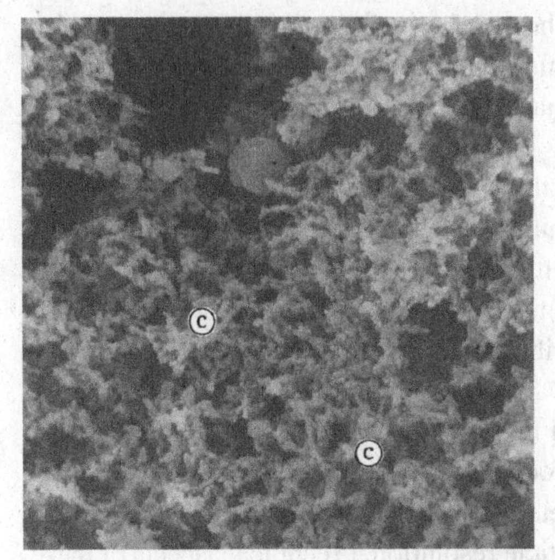

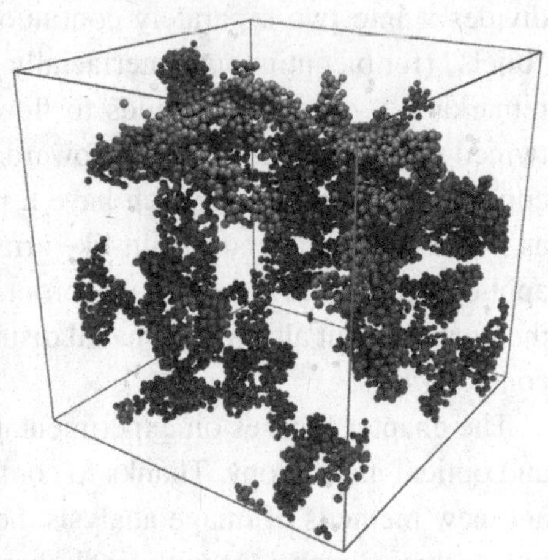

Figure 3.1 Two images of particulate gels, exemplifying the characteristic structure. *Left*: SEM micrograph of yogurt, showing the network of aggregated casein (protein) micelles, two of which are marked with the symbol ©. A single fat globule is visible near the top middle. (Image reprinted from reference [1], Internat. Dairy J. **14**, O. Sandoval-Castilla *et al.*, "Microstructure and texture of yogurt as influenced by fat replacers," (2004), with permission from Elsevier.) *Right*: Computer-generated image of a colloidal gel, reconstructed from particle positions measured by confocal microscopy. (Image from Lu *et al.* [7], reprinted by permission from Macmillan Publishers Ltd, *Nature* **453**, 499, copyright (2008).)

presumably *would become* crystalline given sufficient time (far exceeding the patience of the investigator). In practice, experimentation with spherical, attractive colloidal particles will quickly show that it is rather difficult to form crystals; instead one often finds that increasing the force of attraction between particles leads to a white sediment rather than the hoped-for iridescence. Rather than becoming frustrated by the difficulty of forming crystals, we hope the reader will be inspired to understand how this solidification occurs, why crystals are difficult to obtain, and why the random sediment is by itself a very fascinating and useful material. One can look forward with confidence to progress in understanding gels because, remarkably, the structural and rheological properties of gels have broadly consistent behavior that transcends variation in composition, size, and timescales. Hence, even though gels are formed by a very random process of aggregation of many particles, one can make quite definite predictions of the resulting properties using fundamental concepts.

A particulate gel is typically structurally distinct from a glass. In a glass, the high concentration prevents an individual particle from rearranging unless many other particles also move in order to make space (at least in a hard-sphere glass).

Hence, slow dynamics and solid-like resistance to applied shear stem from the high volume fraction of particles. By contrast, a gel has a low concentration of particles, so that particles have plenty of *space* to rearrange but attractive forces between particles create multiple connections that are not easily broken. Coordinated motion of many particles is needed to break enough bonds to allow rearrangement. Relaxation toward the equilibrium crystal state is enormously slowed. Despite having a concentration as low as 0.01% by volume, a gel is a solid because it extends from one wall of its container to the other. Thus, a shear applied to one wall pulls on bonds and perhaps reduces the network's entropy; this leads to a spring-like restoring force that resists the applied shear.

Whereas a compact solid fills space uniformly, in a gel the particles adopt a combination of one-dimensional, chain-like structures with branches, or roughly two-dimensional, sheet-like structures. The result is a structure having an effective space dimension less than three (i.e., it is fractal). As will be discussed below in some detail, this low-dimensional architecture cleverly allows a network of particles to remain in contact with one another and yet span the length of a macroscopic sample, even at arbitrarily low particle concentrations. By contrast, in a uniform or three-dimensional architecture, particles at low concentration could only fill space by keeping far apart from one another; this material would not be a solid unless there were very long-range interactions. Hence the remarkable and perhaps identifying feature of a gel is the spontaneous emergence of a low-dimensional network. The network also has long structural correlation lengths, which help explain why gels can be so soft (e.g., why yogurt is so much softer than, say, parmesan cheese).

Particulate aggregates, gels, and compressed emulsions are very common in nature and in industry, so that understanding how to control rheological or optical behavior can be quite relevant for food [10], paint [11], and cosmetics [12]. In the case of particulate gels, the particles might be organic or inorganic and nanometers to microns in size. In recent work, mesoscopic aggregates of metallic nanoparticles form a gel that can be molded and then fired to create a shaped metal piece [13,14]. More everyday examples include yogurt, whose viscoelasticity arises from a network of milk protein (casein), which aggregate when rennet is added [1,15,16]. Other proteins form gels too [17–19], as do carbon-black aggregates in motor oil [20] or in ink [21], and cloisite particles in clay [22]. Particulate aggregates (which share structural features of gels) also influence the handling of dry powders such as toner [23] and soot [24, 25]. Very similar architectures are formed by chemical precipitation; this is the route to making aerogels, which are an ultra-low-density solid with a host of useful optical, thermal, acoustic, and other properties [26,27].

3.3 Structure of particulate gels

Particulate gels that arise from random aggregation of Brownian particles share a number of very striking structural properties. At first glance, the emergence of common properties may seem surprising: random aggregation by particles of different sizes, shapes, and inter-particle forces might be expected to yield very distinct structures. On the contrary, gels share many common architectural features over a wide range of length scales, as shown by scattering of neutrons, x-rays, and visible light. (Reviews of scattering experiments are available separately [28]. Reference [29] provides a comprehensive review of colloidal gels.) To start, it will suffice to note that over a wide range of scales (from roughly the particle radius to 1–2 orders of magnitude larger [30, 31]), the scattering follows a power law in the wavevector q, indicating the absence of any characteristic length scale. This scaling arises from the characteristic fractal real-space structure seen in Figures 3.1 and 3.2 and described below.

To understand gelation, it is instructive to consider first a very dilute sample with strongly adhesive particles and no convection or gravity. The standard model describing this limit is known as diffusion-limited cluster aggregation, DLCA [32, 33]. (Its cousin DLA differs by considering the growth of a single cluster, as particles are added to the system one by one [34].) We consider N spherical objects of radius a, distributed at random in a volume V. We define a volume fraction, ϕ, by the fraction of the total volume that is occupied by the particles: $\phi \equiv (4/3)\pi a^3 N/V$. If the particles are smaller than several microns, then gravity can be neglected, at least during early stages of aggregation. Provided that the temperature is uniform and the sample is not stirred, the particles' motions are adequately described by single-particle diffusion. The particles will naturally collide with one another and, if every collision results in a permanent bond, then clusters will appear throughout the sample. Though this process is random, there are nonetheless surprisingly repeatable statistical features. Perhaps the most striking (and famous) feature is the chain-like architecture that is apparent in Figure 3.2. To see how this emerges, the reader can imagine following a single particle as it diffuses toward one of the ramified clusters in Figure 3.2(c). Our particle is unlikely to slip past the outermost tips of the cluster without sticking. In fact, it is more likely to stick to the end of the tip and hence make it longer.

This ramified morphology is not just an unusual shape, it is also a way of packing that is profoundly different from the ball-stacking of our everyday experience. To see this, one can ask how many gelled particles can fit inside a box of lateral dimension L. In ordinary collections of objects, such as oranges in a crate, $N \propto L^3$; for oranges rolling about on the floor, $N \propto L^2$; for beads on a string, $N \propto L$. For the gel shown in Figure 3.2(d), the particles were free to move in 2D (along a water–air

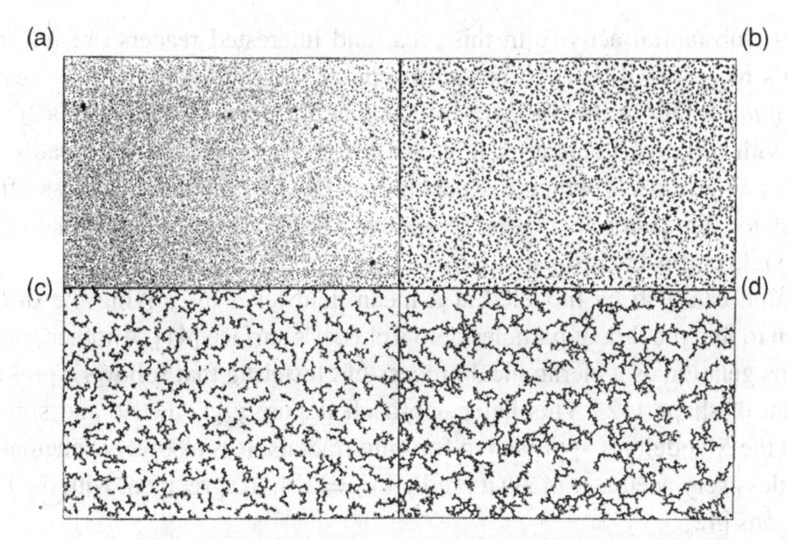

Figure 3.2 Optical micrographs of 1 μm diameter polystyrene spheres trapped at a water–air interface. To initiate gelation, $CaCl_2$ was added to the aqueous subphase at a concentration of 0.73 mol/L. The images were acquired after various delay times: (a) 15 min; (b) 75 min; (c) 105 min; (d) 135 min. The ramified structure characteristic of a gel is clearly visible in (d). Image analysis showed a fractal dimension $d_f = 1.43 \pm 0.04$. (Figure reprinted with permission from D. J. Robinson and J. C. Earnshaw [5]. Copyright (1993) by the American Physical Society.)

interface), but the clusters scaled as $\langle N \rangle \propto L^{d_\mathrm{f}}$, with $d_\mathrm{f} = 1.43 \pm 0.04$ [5]. (The symbols $\langle \cdots \rangle$ indicate the mean of many measurements.) When particles are free in move in 3D, a similar kind of experiment yields $d_\mathrm{f} \approx 1.8$ [35]. (Other experiments will be described below.) The gel structure is something between chain-like (1D) and sheet-like (2D).

The fractal scaling can also be quantified by counting the number of particles in a cluster as a function of its radius of gyration, R_g [33]. The average of many such measurements follows the same scaling as the above case. As a practical matter, the same power-law scaling applies again with a different definition of cluster radius [35]. Alternatively, one can obtain d_f from the power-law decay of the two-point density correlation function measured from images. If one only has access to an image of a slice through the sample, then the decay of the correlation function provides $d_\mathrm{f} - 1$ and the fractal dimension can still be measured accurately [36]. This scaling also appears in the structure factor of the particles and d_f is thus obtained from scattering experiments with x-rays, neutron, or light in the single-scattering limit or even in the presence of some multiple scattering [28,37].

The DLCA model offers a few different routes to predicting the fractal dimension. The most direct method is to simulate DLCA with a computer [38,39]. There has been substantial activity in this area, and interested readers are referred to Meakin's review [40]. For the present purpose, we will simply list the resulting values of d_f for DLCA. In 2D, $d_f = 1.42$, and in 3D, $d_f = 1.78$ [40]. This 2D result agrees with the experimental example shown in Figure 3.2. More broadly, these values show a trend in which the packing of DLCA gels becomes less efficient as the space dimension increases. (Note that DLCA in one-dimensional systems should yield compact clusters with $d_f = 1$.)

Gelation can also be modeled at a mean-field level by solving the diffusion equation to find the flux of particles onto a cluster's surface [41]. Another approach envisions gelation as a hierarchical process of clustering that follows a prescribed evolution of shape [42]. This latter approach is purely geometric and similar in spirit to the Mandelbrot approach to forming fractals. It is also very schematic but nonetheless very accurate: in 2D it predicts $d_f = 1.4$, in agreement with the DLCA simulations [42].

Early measurements of d_f in particulate gels with low concentration were made in the 1980s by Weitz and collaborators [43]. They began with aqueous suspensions of charged gold nanoparticles of radius $a = 7.5$ nm. After adding a relatively large amount of pyridine (to achieve 0.01 mol/L), they found that the particles formed large aggregates within minutes. Electron-microscopy images of clusters showed $d_f = 1.75 \pm 0.05$, and light-scattering showed $d_f = 1.77 \pm 0.05$. When less pyridine was added, the aggregation occurred more slowly and a larger d_f was measured; we will return to this observation below in the context of 'reaction-limited' aggregation. Rapid aggregation of particles composed of other materials (e.g., silica, polystyrene) were found to follow the same aggregation dynamics and fractal scaling [44, 45]. For the two-dimensional aggregation experiment of Figure 3.2, the images showed $d_f = 1.43 \pm 0.04$ [5]. Soot from an annular flame of acetylene burning in air also forms 2D aggregates with $d_f = 1.40 \pm 0.04$ [46]. These and many subsequent measurements of pre-gel clusters and of fully formed gels agree with the DLCA model.

Before exploring gelation under other conditions, we should consider more carefully the consequences of fractal scaling. The first is a structural *correlation length* R_c, which can become very large at low concentrations. To see how this characteristic length emerges, consider the average volume fraction ϕ of particles within a spherical shell of radius $R : \phi(R) \propto \langle N(R) \rangle a^3/R^3$, where $\langle N(R) \rangle$ is the average number of particles inside the shell. From the fractal scaling, we have $\phi(R) \propto (a/R)^{3-d_f}$. Because $\phi(R)$ scales with R raised to a negative power, at large scales the clusters have vanishing volume fractions – they are mostly empty. Clearly, this trend cannot continue indefinitely, since as $R \to \infty$, $\phi(R)$ must approach the sample-averaged volume fraction, ϕ. The upper limit to the range of fractal scaling determines the correlation length, $R_c = a\phi^{-1/(3-d_f)}$ [28, 32]. At larger scales, the sample is a uniform, 3D solid. The correlation lengths can be measured from scattering from dilute particulate gels, where a peak appears at a wavenumber $\sim 1/R_c$ [47]. Real-space images of gels also show a length scale R_c, beyond which the radial distribution function becomes roughly constant.

A second consequence of fractal scaling is that the *topology* or 'connectedness' of gels also follows power-law scaling. To quantify topology, a common approach is to imagine a tiny bug walking along the network and measure the shortest length it must crawl along the gel to move from one point to another. Looking again at Figure 3.2(d), imagine choosing two particles (α and β) a distance r from one another. To crawl from α to β, our bug must travel a minimum pathlength $S_{\alpha\beta}$. From measurements of particle positions (in experiments or simulations), a list of the shortest pathlengths S can be obtained. The result is $\langle S/a \rangle \propto (r/a)^{d_b}$, as long as $r < R_c$. This scaling defines a new dimension, d_b, known as the backbone or chemical dimension [48, 49]. We will see in Section 3.4 that the chemical dimension plays a role in the elasticity of gels. Confocal microscopy of poly(methylmethacrylate) spheres suspended in organic solvent with short-range depletion and a weak long-range repulsion show $d_b = 1.2 \pm 0.1$ [50]. Computer simulations of DLCA in 2D and 3D find $d_b = 1.15 \pm 0.04$ and 1.25 ± 0.05, respectively [49]. The 3D result agrees with the experiment, despite the presence of weak long-range repulsion in the latter. It should be noted that the results for 2D and 3D are quite similar, and in both cases are close to 1. To our tiny bug, the gel looks like a nearly one-dimensional network.

The dynamics of aggregation can also be understood from fairly basic principles. For example, to predict how rapidly aggregates will form, we can first find out how rapidly particles collide with one another. We place one particle at the origin, treat it as a perfectly absorbing surface of radius a, then solve the diffusion equation to find the steady-state concentration, c, around it [51–53]. The boundary conditions require $c = 0$ at the surface and $c \to 3\phi/(4\pi a^3)$ as $r \to \infty$. One can then solve first for the flux of colliding particles at $r = a$, and then for the number of particles hitting the surface per unit time. The inverse of this rate is τ_{agg}, the "doubling time" or characteristic time per diffusive collision: $\tau_{agg} = a^2/(6\phi D) = \pi a^3 \eta/(\phi k_B T)$, where $D = k_B T/(6\pi \eta a)$ is the single-particle diffusion coefficient [51–53]. (It should be noted that the effective diffusion constant describing mean-square separation between two spheres is twice D.) The quantity η is the shear viscosity of the liquid. For μm-sized particles in water, τ_{agg} is on the order of 1 s, but clearly is a strong function of a for a given ϕ.

The mean cluster radius R_g increases with time, t, as $(t \times c)^{1/d_f}$ [43, 51, 54]. Extrapolating this result, one can estimate the time for gelation as the time for the mean cluster size to reach R_c [55]. Even before the gel is fully formed – as clusters appear and grow at random throughout the sample – the statistical distribution of cluster sizes is a universal function of the mean cluster size [56,57]. This remarkable result means that the aggregation dynamics are, to a great extent, captured by just the aggregation time τ_{agg} and the fractal dimension d_f [57]. This result also provides a pathway to a useful thermodynamic descriptor of the system, as will be described in Section 3.5.

The power of the random aggregation model is that its features are observed – albeit with quantitative variations – over a far broader range of parameter space than strictly allowed by the DLCA model. A common example arises in systems where the particle collisions *rarely* lead to bonds (in contrast to DLCA, where bonds always form). The resulting clusters are more compact than in DLCA, but still fractal. In experiments with Ludox spheres in water with acid and salt added to induce aggregation, x-ray scattering showed $d_f = 2.12 \pm 0.05$ [58, 59]. This fractal dimension is noticeably larger than the DLCA prediction. Returning to the experiments with gold nanoparticles described above: when a relatively *small* amount of pyridine was added, the particles aggregated *slowly* because the remaining weak repulsion made bond formation a rare event. In this limit, the experiments showed $d_f = 2.05 \pm 0.05$. Remarkably, the fractal scaling and the value of d_f are reproducible, just as they were in the strong-attraction, fast-aggregation (DLCA) limit [43, 44]. The limit of very improbable bonding is now known as "reaction-limited" cluster aggregation, or RLCA [60,61]. According to computer simulations of RLCA, $d_f = 1.53$ in 2D and 1.99 in 3D [40]. The latter is close to the experimental values [43, 44, 58]. Because of the repulsion, the rate of aggregation is slowed and τ_{agg} is increased by a factor known as the Fuchs stability ratio, W. The ratio can be estimated from an integration over particle separation of the Boltzmann factor divided by a factor that accounts for hydrodynamic friction between two particles [15,53,55,62,63].

Hydrodynamic interactions among particles can also change the aggregate structure. Even when the particles are neutrally buoyant and do not sediment, movement of the particles leads to solvent flow (e.g., see [64, 65]). The flow, in turn, leads to long-range interactions. In equilibrium, these hydrodynamic interactions cancel, but during aggregation they could have a major effect. Using computer simulations of 2D, Tanaka and Araki pointed out that hydrodynamic flows around a dimer approaching another cluster would tend to align it parallel to the direction of motion, hence leading to a ramified structure [66]. By comparison, "free-draining" (nonhydrodynamic) Brownian dynamic simulations of the same particles yielded substantial local crystalline order [66]. This interaction is unavoidable in particulate suspensions, though the magnitude of the effect in 3D might be quite different. A different hydrodynamic effect arises from sedimentation when the particles are not perfectly density-matched to the solvent. Even if the sedimentation of individual particles is very slow (which is readily achieved in experiments), the difficulty is that the sedimentation velocity scales with the square of the cluster radius. Therefore, the aggregates may eventually become large enough to sediment and the resulting flows can lead to more compact clusters (e.g., $d_f = 2.3$ [67]; see also [29,68]).

A finite attraction allows rearrangement of bonds after they form. Although this rearrangement leads to gradual relaxation of the structure, there is still a clearly discernible fractal dimension that is less than 3. Studies of Ludox particles aggregating in water show that d_f is initially either 1.75 ± 0.05 or 2.05 ± 0.05, depending on the sample conditions (and consistent with either DLCA or RLCA). For rapidly aggregating samples, however, d_f increased from 1.75 to 2.05 over a period of minutes or hours, *via* internal restructuring [59]. Similarly, light scattering from polystyrene-particles show that d_f slowly climbs from 1.8 to nearly 2.0 as a result of rearrangement over a period of two weeks [69]; in these experiments the particles were buoyancy-matched to suppress sedimentation. The weaker attraction could also lead to a more compact structure and relaxation even during the initial aggregation process. Monte Carlo simulations in 2D and 3D have demonstrated this effect [70, 71], though full accounting of hydrodynamic effects in 3D might yield a different result.

Experimental exploration of these effects might benefit from a recently developed trick to tune the strength of attraction *in situ*. When surfactant micelles are used to induce depletion, the size and concentration of micelles – and hence the depletion attraction – can vary strongly with temperature. Thus, modest changes in temperature (as little as 1°C) can reversibly induce crystallization, gelation, or melting of the particles [72, 73].

Why are the aggregating particles trapped in a random, nonequilibrium state rather than forming crystals with lower free energy? One answer is that the bonds themselves might be rigid (i.e., they might resist changes of bond angle), perhaps because of faceted or bumpy surfaces. This is the obvious explanation for jagged powders or bumpy or faceted colloidal particles. Even for spheres, rigid bonds can arise as shown by a nice set of measurements with charge-stabilized polystyrene spheres in water. The classic Derjaguin–Landau–Verwey–Overbeek (DLVO) model for colloidal interactions describes a combination of van der Waals attraction at short range and electrostatic/entropic repulsion at long range [54, 74]. Although these interactions should be centro-symmetric (nonrigid), Pantina and Furst showed that single-file strings of aggregated spheres nonetheless resist bending forces applied with an optical trap [75]. Another example of rigid bonds with seemingly centro-symmetric forces is found in colloidal spheres where attraction is induced by addition of nonadsorbing polymer; this example will be described in Section 3.4 [76]. Ultimately the mechanism of the rigidity might be nanometer-scale roughness of the particle surfaces. At the other limit, *liquid droplets* form molecularly

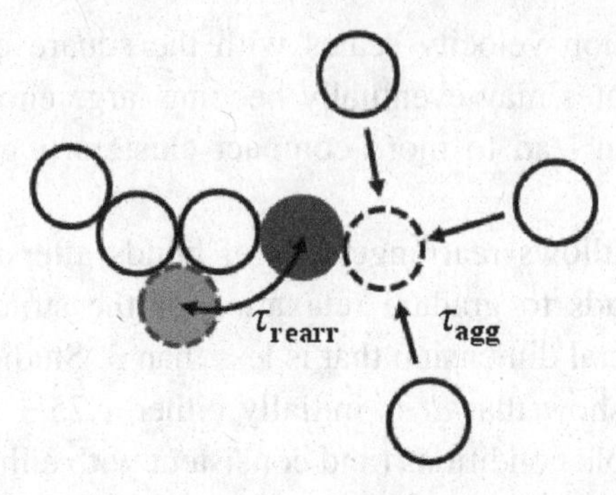

Figure 3.3 Illustration of the competition between bond-angle rearrangement in a cluster and addition of new particles to the cluster. The typical time for the solid gray particle to diffuse along the cluster, make a new bond, and adopt a locally crystalline packing is τ_{rearr}. The typical time for a new particle from suspension, to bind, to the gray particle is τ_{agg}.

smooth surfaces and have no mechanism for bond-bending rigidity (as long as the surfactant layer is fluid). Recent experiments and simulations have explored this interesting regime, now referred to as "slippery DLCA." Liquid droplets do indeed form low-density solids (gels) in the presence of strong attractive forces [77–79]. They also show fractal scaling of the mass with $d_f = 1.8$, similar to the DLCA (rigid-bond) limit [47]. At first glance, this result defies intuition because the droplets can slip past one another and should therefore form a compact structure (such as a crystal). Computer simulations show that the slippery bonds do indeed lead to compact and rigid tetrahedral arrangement of droplets – but only at short lengthscales. The tetrahedra do aggregate and form a hierarchical DLCA-like structure [80].

This still might not fully explain why spheres with isotropic (centro-symmetric) forces do not crystallize. Whereas liquid droplets form short-range order, experiments with solid spheres find none [50, 81]. Some insight comes from comparing time-scales for aggregation and for bond rearrangement. Figure 3.3 shows an illustration of a small cluster in the early stages of gelation. To estimate whether the cluster can crystallize, we consider the motion of a target particle shown in gray. Assuming it is strongly bound to its neighbor with an isotropic bond, the solid gray particle will undergo a random walk along its neighbor's surface until it reaches the position shown by the dashed, light-gray circle, where it forms one additional bond and is then held rigidly in place. The characteristic timescale for this process is labeled τ_{rearr}. On the other hand, if another particle from the surrounding suspension first binds to the gray particle, then its motion is slowed because rearrangement requires multiple particles to move together. The timescale for one new particle to bind to the gray one is τ_{agg} (defined above). In this very simplified approach, therefore, the cluster should crystallize if $\tau_{rearr}/\tau_{agg} \ll 1$, but otherwise remain amorphous. An estimate of $\tau_{\rm rearr}$ comes from the time to diffuse (in 1D) a distance of approximately the particle diameter, hence $\tau_{\rm rearr} = 2a^2/D_{\rm b}$, where D_b is the diffusion constant of a particle bonded to another particle. The condition for crystallization would therefore be $\tau_{\rm rearr}/\tau_{\rm agg} = 12\phi D/D_{\rm b} \ll 1$. When $\phi \ll 1$ it might appear that cluster should always crystallize, but we should take care with $D_{\rm b}$: for *solid* particles $D_{\rm b}$ is smaller than D – potentially far smaller – because of the hydrodynamic friction of the thin layer of fluid between the particles [82, 83]. Hence hydrodynamic friction would make the particles behave as though rigidly bonded. This very qualitative model suggests: (i) that liquid droplets are more likely to crystallize than solid spheres because the hydrodynamic friction is less (though a viscous surfactant layer might also slow rearrangement); (ii) that a strong attraction leads to a smaller separation between particles, smaller $D_{\rm b}$, and slower rearrangement into ordered clusters; (iii) that spherical particles with isotropic interactions in a vacuum would crystallize, at least locally. Computer simulations typically do not account for hydrodynamic interactions of the sort that determine $D_{\rm b}$, and so might not (yet) shed light on this problem. Experiments with confocal microscopy showed that a stronger inter-particle attraction leads to fewer rigidly packed regions [50], which might be interpreted as a reduced tendency to adopt local order; this result supports point (ii).

Finally, a great deal of recent work has gone into changing the overall form of the inter-particle potential and observing how the behavior changes. An interesting example is spheres with a short-range attraction (which ultimately causes aggregation) and a long-range repulsion (typically electrostatic). The combination is sometimes known as SALR [84]. Early experimentation with SALR colloids showed that, in some cases, fractal aggregates appear but reach a steady state (perhaps even equilibrium) with a finite cluster size [18, 50, 85–88]. There are fractal clusters but they do not gel! This structure has been called a 'fluid-cluster phase' – it is literally a fluid of clusters. Similarly finite-sized clusters are found in some protein (lysozyme) solutions [18, 19], in clay [22], and with charged liposomes in water [89]. The fluid-cluster phase is attributed to a balance between aggregation of particles at the tips of these clusters and repulsion from the main bodies of the clusters [90, 91]. It should be noted, however, that some authors concluded that the fluid-cluster phase appears even in samples *without* repulsion, when the attraction is weak [92, 93]. Within the context of the fluid-cluster phase, gelation may be viewed as a glass transition of the clusters [92, 94]. More generally, theory and simulations have shown that SALR can lead to a variety of equilibrium phases, such as a fluid of solid or fluid clusters, lamellar phases and solids with voids in 2D [95–97], and fluid-cluster phases and fluids with voids in 3D [84]. It might be that the range of possible *non*equilibrium states of SALR is correspondingly expanded.

Short-range attraction and long-range repulsion (SALR) by no means represents the full range of possibilities. For example, mixing *attractive* particles with *repulsive* particles provides an additional control of gel structure and dynamics [98]. This is a clever means of tailoring gel properties and future research will likely uncover other useful and illuminating lessons.

3.4 Viscoelasticity of particulate gels

The response of gels to applied stress (i.e., their rheology) is, naturally, a function of their structure. In detail, this is clearly a very difficult problem: even if the inter-particle bonds may be considered harmonic (a common assumption), the gel resembles a complex mix of springs in series and in parallel. Nonetheless, the observation that gel structure has so many common features, despite differences in composition, offers hope of common features in rheology.

Experiments have borne this out. In a series of experiments with carbon-black aggregates, Trappe, Prasad, Weitz, and co-workers applied oscillatory shear strain at a frequency ω while measuring the in-phase and out-of-phase components of stress [3,99]. (In some cases, a controlled stress is applied and the resulting strain is measured.) These measurements provide the real and imaginary parts of the complex shear modulus, which represent the elastic and viscous components G' and G'', respectively. Solid materials, by common definition, have $G' \to$ constant (referred to as the plateau modulus G_p) and G'' decreasing as ω approaches some practical limit near zero. As a caveat, it should be remembered that very slow aging makes the true $\omega = 0$ limit inaccessible to experiments and difficult to interpret. At the other limit, $G'' \to \eta\omega$ as $\omega \to \infty$, reflecting the viscous response of the solvent, characterized by its shear viscosity η. A very important (and satisfying) observation is that although $G'(\omega)$ and $G''(\omega)$ differ for each sample, they exhibit a remarkable scaling. Each sample has a characteristic timescale τ and modulus G_0. Trappe and Weitz showed that a plot of G'/G_0 (and, separately, G''/G_0) as a function of $\omega\tau$ has a common form for all samples studied [99]. Figure 3.4 shows data from gels composed of carbon-black particles in oil. The data curves superimpose across ten decades in timescale and more than six decades in modulus. Moreover, for each sample the two scale factors are related by $\tau \propto \eta/G_0$, where η is the shear viscosity of the solvent [3]. Aside from showing a remarkable universality, the scaling has significant pragmatic utility: it allows a measurement of the plateau modulus by extrapolation even when the experimentally accessible values of ω are too high to show a plateau in G'.

How does the gel's structure determine the plateau (low-frequency) modulus G_p? As noted earlier, the randomness of the structure makes a general theory appear daunting, yet measurements have shown that G_p scales with particle volume

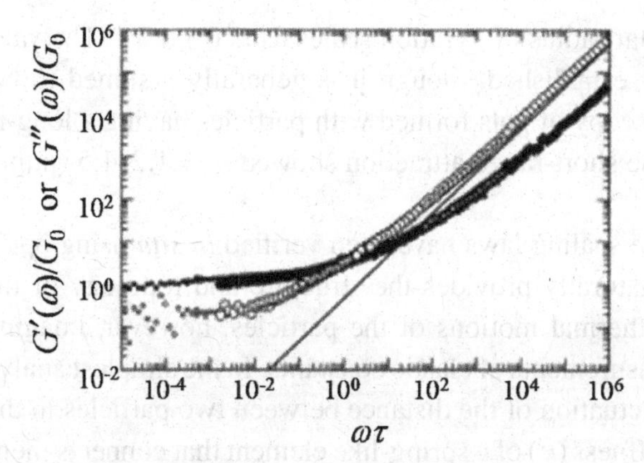

Figure 3.4 Measured shear moduli, G, of particulate gels as functions of oscillatory frequency ω. The gels are subjected to an oscillatory strain and the resulting stress is measured. The in-phase component (G') is the storage modulus, represented by the filled symbols. The out-of-phase component (G'') is the loss modulus, represented by the open symbols. The larger symbols correspond to seven samples with ϕ ranging from 0.056 to 0.149. The small circular symbols correspond to samples with $\phi = 0.14$ and varying inter-particle potentials. For each sample, the ω and G axes are rescaled by two empirical constants τ and G_0. (Plot from V. Prasad *et al.* [3], reproduced by permission of the Royal Society of Chemistry.)

fraction ϕ with an exponent that ranges between 3 and 4. These scaling exponents have been predicted starting from an elegantly simple fractal elastic model proposed by Kantor and Webman [100]. The model uses the fractal scaling of the mass described above. At length scales (r) that are shorter than the fractal correlation length R_c, a region of the gel is viewed as a single elastic fractal beam. If the inter-particle bonds do not resist bending, then the chain is simply a linear collection of springs. The spring constant that corresponds to pulling on the ends of the fractal segment, $\kappa(r)$, would then scale as $\kappa_0/N(r)$, where κ_0 is the single-bond spring constant and N is the number of particles in the chain whose end-to-end (Euclidean) length is r. If, however, the bonds *resist* bending, then the elastic constant of the fractal chains is given by $\kappa(r) = \chi/[N(r)r_\perp^2(r)]$, where χ is the rigidity of a single bond or segment of the chain and $r_\perp$ is the radius of gyration of the chain for rotations about the end-to-end axis. (For later use, it will be useful to note here that computer simulations of the strain-response of DLCA gels show that 70% of the energy is stored in bending [101] and should thus be accounted for with this modulus.) This expression for κ shows why it is easier to compress a u-shaped wire than a straight one: bending the wire increases $r_\perp$. Returning to gels, simulations and experiments show that $N(r) \propto r^{d_b}$, where d_b is known as the "bond" or "chemical" dimension and has the value 1.2 ± 0.1 in imaging experiments [50] and 1.25 in computer simulations of DLCA [49] (see Section 3.3).

The perpendicular radius of gyration scales as $r_{\perp}^2(r) \propto r^a$. The value of a appears not to be firmly established, though it is generally assumed to be 2 for DLCA. Confocal microscopy of gels formed with particles having a long-range repulsion in addition to the short-range attraction showed $a = 1.2$–1.5 (numerically similar to d_b) [50].

Some of these scaling laws have been verified *in situ* using optical microscopy [76]. Imaging naturally provides the structure and topology of the gel network. By visualizing thermal motions of the particles, however, imaging also leads to quantitative measurements of elastic constants. In the simplest analysis, the average mean-square fluctuation of the distance between two particles in the gel, $\langle \Delta r^2 \rangle$, is related to the stiffness (κ) of a spring-like element that connects them. According to the equipartition theorem, $\kappa = k_B T / \langle \Delta r^2 \rangle$. The effective κ can then be compared to the shape of the shortest gel segment that connects the two particles. In gels composed of particles with short-range attraction and weak long-range repulsion, segments of the colloidal gel had spring constants scaling as $\kappa(r) \propto \kappa_0 / N(r)$, indicating a lack of bond-bending rigidity [76]. In this regime, the gel structure was held in place by the multiple connections among gel segments, which frustrated reorganization [76]. In gels formed by particles having a shorter-range and stronger attraction, the scaling changed to $\kappa(r) \propto 1/[N(r) r_{\perp}^2(r)] \sim r^{(-d_b + a)}$. This scaling indicated that rigid bonds had formed, perhaps owing to interlocking of asperities on the particle surfaces. In both cases, the fractal-elastic-beam model agreed with the experiments for segments of the gel whose size was smaller than or comparable to the fractal correlation length, R_c [76].

At length scales greater than R_c, the gel is modeled as a homogeneous solid of springs of size R_c and spring constant $\kappa(R_c)$. This simplified approach very elegantly allows quantitative prediction of the dependence of shear or Young's modulus on volume fraction, particle size, and inter-particle bond stiffness [102–107]. As dimensional analysis will show, if the solid becomes homogeneous at scales larger than R_c, then the modulus of such a structure would then scale as $\kappa(R_c)/R_c$. For *nonrigid* bonds, we thereby arrive at the prediction that $G_p \sim \kappa_0 R_c^{-(1+d_b)}$. Recalling from above that $R_c \sim \phi^{-1/(3-d_f)}$, one predicts that $G_p \propto \phi^{\alpha}$, where $\alpha = -(1 + d_b)/3 - d_f)$. If the fractal dimensions are given by the DLCA model ($d_f = 1.78$ and $d_b = 1.25$, though strictly speaking DLCA assumes rigid bonds), then $\alpha \approx 1.8$. If the bonds are *rigid*, then the modulus has a different scaling: $G_p \sim \gamma R_c^{-(a+d_b+1)}$. Hence, $G_p \propto \phi^{\alpha}$ where $\alpha = -(1 + a + d_b)/(3 - d_f)$. The early models assumed $a = 2$ so that $\alpha = -(3 + d_b)/(3 - d_f) \approx 3.5$ [102, 107]. This scaling-law prediction should apply to shear and Young's moduli of the network. (It should be noted, though, that experimental measurements of Young's modulus will be strongly influenced by the surrounding fluid, even at low frequency.) Finite-element-method numerical calculations of the Young's modulus of DLCA solids show $\alpha = 3.6$, which agrees very nicely with the predicted scaling [108].

Measurements of the ϕ-scaling of the modulus are broadly consistent with the above power-law, though there are variations. For nonspherical particles or in the limit of strong attraction, one expects the rigid-bond result to apply (i.e., $\alpha \approx 3.5$). From measurements of gels composed of colloidal alumina, Shih *et al.* found $G' \propto \phi^\alpha$ with $\alpha = 4.1$–4.2 [103]. For colloidal polystyrene, rheometer experiments gave $\alpha = 4.6 \pm 0.3$ (250-nm radii in water) [105]. Later experiments with gels composed of 10-nm-radius silica spheres in water showed that the shear modulus increased as a power law with time; accounting for this effect, Manley *et al.* found that the time-independent modulus scales with ϕ^α where $\alpha = 3.6$, in excellent agreement with the model [109]. (These highly controlled experiments were also done under microgravity conditions on the International Space Station.) The shear modulus can also be measured by dynamic light scattering in a technique commonly referred to as microrheology [110–112] (see also Chapter 6). In this approach, the scattering effectively measures the strain caused by thermal fluctuations. Studies of 10-nm radius polystyrene spheres suspended in neutrally buoyant mixture of H_2O and D_2O showed power-law scaling of the modulus with $\alpha = 3.9$ [113]. Aerogels show similar scaling of the shear and Young's moduli with exponents in the range of $\alpha = 2.6 - 3.8$ [108]. (Note that in aerogels the solvent is removed by super-critical drying [26] so that the Young's modulus comes only from the particle network.)

For spherical particles with relatively weak and long-range attraction, gels are only formed when ϕ exceeds some minimum value (ϕ_g). This observation contradicts the DLCA model, where gels can form even in the limit where $\phi \rightarrow 0$. Remarkably, similar scaling of the modulus is still observed if ϕ is replaced by $(\phi - \phi_g)$. In rheology measurements of gels formed by functionalized silica spheres in cooled cyclohexane, Grant *et al.* found $G' \propto (\phi - \phi_g)^\alpha$ with $\alpha = 3.0 \pm 0.5$ [114]. Prasad *et al.* studied gels composed of poly(methylmethacrylate) spheres with depletion induced by added polystyrene [3]. In those experiments, the range of the depletion attraction, set by the size of the polymer, altered the scaling. With a relatively short-range attraction, they reported $G' \propto (\phi - \phi_g)^\alpha$ with $\alpha = 3.3 \pm 0.1$, whereas a long-range attraction led to $\alpha = 2.10 \pm 0.05$. Once again, the scaling model agrees with these results if *short*-range attraction causes rigid bonds and *long*-range attraction causes nonrigid bonds (consistent with the confocal microscopy study described earlier [76]). Finally, measurements of carbon-black gels in oil yielded a slightly larger value, $\alpha = 4.1$, but the difference might arise from the relatively large value of d_f (2.2) [99, 115]. In general, the range of results for α probably came from from slightly different values of the exponents that characterized the gel's mass (d_f) or topological properties (d_b or a). For example, it is known from confocal microscopy that when there is a long-range repulsion in addition to the short-range attraction, the particle "chains" become straighter than expected for DLCA, so that $a = 1.2\text{–}1.5 \approx d_{\mathrm{b}}$ [50]. The scaling argument described above would then predict that $\alpha \approx 2.9$. From a pragmatic point of view, this suggests how to tune the modulus by means of ϕ, interaction potential, *and* fractal architecture.

3.5 Gelation and arrested phase-separation dynamics

The power of equilibrium thermodynamics and the absence of a known counterpart in nonequilibrium phenomena inspire us to look to the former for guidance. There are at least two specific ways in which equilibrium thermodynamics can make useful and verifiable predictions relevant to gelation. The first is that equilibrium thermodynamics can predict the regions of parameter space where a particular phase is *unstable* with respect to small fluctuations of density (or some other order parameter). There is now compelling evidence that these unstable regions, known as spinodal regions [116], might well explain *some* of the important aspects of gelation, even as others remain very much open. The second power of equilibrium thermodynamics is to indicate likely sequences of stages in the dynamics of phase transitions. A suspension of spherical particles can adopt many possible structures in equilibrium: gas, liquid, face-centered cubic crystal, body-centered cubic crystal, to name a few. When a sample undergoes a phase transition, it is now well established that the dynamic process might involve several intermediate steps, each characterized by a structurally distinct phase. For example, a supercooled gas that crystallizes can first form liquid droplets, which then crystallize. Or a supercooled gas or liquid might first adopt one crystalline structure and then change to the true equilibrium crystal structure as it grows. According to "Ostwald's Rule of Stages," these intermediate states appear in order of decreasing free energy [117–121]. The Rule has a mostly empirical support and is not quantitative, but it is nonetheless a very useful tool. Recent models and experiments have shown that the above extensions of equilibrium thermodynamics might explain aspects of gelation, as will be described just below. For still other thermodynamic approaches to modeling gelation, including the increasingly successful mode-coupling theory, the reader is referred to published references [122–128], one of which offers a comprehensive review [127].

Before discussing *non*equilibrium dynamics, it will be useful to clarify a few points about the equilibrium states and summarize a few key results relevant to colloids (which are also summarized in a recent review [129]). In a dilute suspension, the particles may be viewed as a thermodynamic gas phase. The term "gas," of course, refers to the random arrangement of the particles, not to the solvent.

It is well established that lowering the temperature or increasing particle–particle attractions in such a system can lead to formation of crystalline or fluid phases, as well as gel and glass phases. When the attraction is of relatively long range (i.e., the range exceeds approximately 14% of the particle diameter), then the equilibrium phase diagram contains regions of gas–liquid, gas–crystal, and liquid–crystal coexistence, with a triple point. [129–132]. There is also a gas–liquid critical point and a region where gas–liquid separation occurs by spinodal decomposition [133]. The topology of this phase diagram resembles that of atomic and molecular systems. However, when the range of attraction falls below approximately 14%, there remains only one amorphous phase in equilibrium; gas–liquid coexistence does not occur in equilibrium. The equilibrium phases are gas and/or crystal [130]. (With very short-range attraction, a second crystal phase with the same structure becomes stable [134, 135].) Although the liquid phase might be "lurking" behind the scenes as a metastable phase, it could nonetheless play a major role in the observed behavior.

Studies of particle gelation in the 1990s showed that the structure evolves during gelation in a way that resembles spinodal decomposition of a molecular gas following a deep quench. By imaging, it is clear that gels form a bicontinuous structure, a characteristic shared by spinodal decomposition [92, 136]. (This bicontinuous structure will be described in more detail below and in Figure 3.5.) To quantify this resemblance, Carpineti and Giglio reported light-scattering data with 19 nm diameter polystyrene spheres in water with very low ϕ (0.003 or less). On addition of a divalent salt, the particles aggregated and eventually formed a gel. The structure factor $S(q)$ formed a peak at a wavenumber q_{m}, then grew in magnitude and shifted toward smaller q as time passed [51, 137]. The results are consistent with some essential pieces of the DLCA model: d_{f} appeared in the scaling of $S(q)$, q_{m} scaled with ϕ in the same way that the fractal correlation length does, and the inferred cluster size grew in time as expected from DLCA (at least at early times) [51]. The authors, however, pointed out a similarity to the scaling behavior seen in molecular fluids undergoing spinodal decomposition. Partway into the gelation process ($t \gtrsim 10^4$ s, much longer than τ_{agg}), the structure factor followed a time-invariant form. That is, plots of $(q_{\mathrm{m}}^{d_{\mathrm{f}}})S(q/q_{\mathrm{m}}(t))$ as a function of q/q_{m} at various times collapsed onto a common curve with $d_{\mathrm{f}} = 1.90 \pm 0.02$. A true mapping to spinodal decomposition would require that the scaling exponent equal the space dimension (i.e., 3), but the proposal sparked many subsequent studies. Later authors, however, expressed reservations about the link to spinodal decomposition. A similar scaling of $S(q)$ was reported in two dimensions in samples with area fractions $\sim$10% by Robinson and Earnshaw, but they also pointed out that similar scaling could arise from nonspinodal processes [5]. Later analysis showed that the resemblance of $S(q)$ to spinodal decomposition may be coincidental, since a peak in $S(q)$ can arise from a competition between two length scales rather than the unstable wavenumber characteristic of spinodal decomposition [78, 138, 139].

More recent experiments have looked at colloidal gels with larger volume fractions: $\phi = 0.10$ by light scattering [140], and $0.045 < \phi < 0.16$ by confocal microscopy [7, 127]. In the latter experiments, the solvent's density was matched to that of the particles so that the growing clusters did not sediment. (The importance of neutral buoyancy is demonstrated by another investigation, which showed that sedimentation can alter the gelation process [85].) The structure factor $S(q)$ evolved *over time* in a manner reminiscent of spinodal decomposition of molecular fluids (at least in some cases). The peak wavenumber, $q_{\mathrm{m}}(t)$, shifted toward smaller values as a power-law in time t, then nearly stopped once the sample became solid. Moreover, the structure factor followed a common form when plotted as $(q_{\mathrm{m}}^3)S(q/q_{\mathrm{m}}(t))$ *vs.* q/q_{m}: this time, the scaling is with the space dimension rather than the fraction dimension [7, 127] (though the exponent reverted to d_{f} with strong attraction in one experiment [140]).

If gelation is indeed triggered by spinodal decomposition, then theoretical predictions of the spinodal region should identify which samples in an experiment form a gel. As a practical limitation, though, phase diagrams have only been predicted for particular forms of the interaction potential. Therefore, one needs a way to map an experimental system to a theoretically tractable system. Lu *et al.* showed how this could be done [7]: for any system having short-range attractions, an "extended principle of corresponding states" proposed by Noro and Frenkel [132] predicts that the phase behavior for different inter-particle potentials can be described with just one parameter, the second virial coefficient, B_2. (Recall that B_2 is the second term in the series expansion of the pressure as a function of the density [141].) By this principle, the magnitude and the range of the potential do not *separately* matter; only the particular combination represented by B_2 does. Lu *et al.* went further and pushed the corresponding-states principle to a *non*equilibrium quantity, the distribution of cluster sizes during gelation. Thus, by matching experimentally measured cluster-size distributions to those computed in simulations, those authors obtain B_2 for their experimental system. For experimentalists, this is potentially a very useful contribution that might allow phase diagrams – and perhaps even nonequilibrium properties – to be plotted in a universal form as a function of B_2, rather than as a function of polymer concentration or size, temperature, or other system-dependent parameters. From this analysis, Lu, Zaccarelli and co-workers concluded that the gel transition in their experiment occurred at a B_2-value that coincided with the spinodal boundary in a theoretical reference model (the Baxter sticky hard-sphere model) [7, 142]. It should be noted, however, that earlier attempts to compare gelation to the predicted spinodal curve showed a discrepancy when $\phi > 0.2$. Further experimental testing of this idea over a broader range in ϕ would be useful.

The notion of a spinodal decomposition may go a long way toward explaining gelation dynamics. Comparisons of experimental scattering data to mode-coupling-theory predictions provides evidence that spinodal decomposition is arrested by the formation of a glass in the more concentrated fluid phase [126]. Still, several questions remain, such as when or at what lengthscale the spinodal decomposition arrests, and why the structure-factor scaling exponent changes from d_f to 3 with increasing ϕ. Even if the early stages are dictated by a thermodynamic instability in the starting gas, one still must account for the fact that the gas–*liquid* coexistence is metastable with respect to gas–*crystal* coexistence. Crystals should eventually appear.

To explain what happens to the crystal phase, one can turn again to concepts of equilibrium thermodynamics, as suggested at the start of this section. (This question, why crystals do not form, was addressed to some extent at the single-particle level in Section 3.3, but here we consider an alternative approach.) Approximately 40 years ago, Cahn described how the equilibrium phase behavior can allow us to predict the dynamics of phase separations. The topic has been very nicely described by Poon and co-workers in the context of colloids [131,141–145]. The analysis starts with a common approach to determining the equilibrium state by plotting of the free-energy density, f, as a function of particle concentration. If two or more phases can coexist, then such a curve has two or more local minima. The region of equilibrium coexistence is found by the "double-tangent" construction. Regions where $f(\phi)$ has negative curvature correspond to the spinodal region, since even small fluctuations in the local f can reduce the total free energy.

When there are three phases in equilibrium, these plots can indicate in what order the phase separation occurs if one starts from an initially homogeneous suspension. From the plot of $f(\phi)$, one can also obtain the difference of free energy between the initial state and any of the equilibrium states [131,144]. From this information, regions of parameter space can be defined where an initially homogeneous fluid will have a lower free-energy barrier by first forming a liquid, then a crystal. In other regions, the free-energy barrier favors the formation of a crystal first, then a liquid.

The "kinetic maps" predicted by this approach compare well to the experimental observations: in suspensions of sterically stabilized poly(methylmethacrylate) spheres ($a = 240$ nm) with polymer added to induce depletion. The polymers had a radius of gyration of 88 nm, so that the range of the depletion attraction was long enough to observe gas–liquid coexistence as well as gas–crystal and liquid–crystal [131]. Poon *et al.* used scattering and low-magnification imaging to measure the phase-separation dynamics. They identified three distinct types of behavior: (i) gas–liquid separation, followed by crystallization in the liquid phase, (ii) crystallization in the liquid phase, followed by gas–liquid separation, and

(iii) crystallization in pockets of the gas phase surrounded by liquid, followed by macroscopic separation. All of these behaviors can be qualitatively, but very elegantly, explained by features of the $f(\phi)$ curve [131].

Even when the liquid phase is not seen in equilibrium (through choice of ϕ or interaction strength, or because the attraction is of too short a range), the kinetic-map approach still applies, and the liquid phase can still play a transient role [131]. The metastable liquid was shown to play a major role in enhancing crystal nucleation by Monte Carlo simulations [146] and by recent experiments using colloids with a tunable depletion attraction [73].

To explain gelation, Poon *et al.* propose that the liquid phase forms a glass at high density, which then arrests further evolution. Consider a homogeneous fluid that in *equilibrium* would separate into coexisting crystal and gas phases. If the kinetic map indicates that the crystal phase should nucleate directly from the liquid (as in case (ii) above), then the sample might crystallize. Crystals could nucleate and grow while the surrounding liquid phase becomes gradually less and less concentrated. Nevertheless, the kinetic map could indicate that gas–liquid separation should occur first (even though liquid is metastable); this would be analogous to case (i) above. In this case, the liquid phase would become more concentrated over time and could vitrify in the same way that uniform suspensions with large ϕ form glass. The dynamics at this stage might be more appropriately thought of as viscoelastic phase separation [147]. If the kinetic-map approach is correct, then it may be possible to suppress gelation by tuning the form of the particle interactions to broaden the type (ii) regime.

3.6 Bicontinuous solid networks from arrested phase separation

Controlling the dynamics of liquid–liquid phase transitions provides an elegant alternative route to random solids with a remarkable structure. Bicontinuous, interfacially jammed emulsion gels ("bijels" [8]) consist of an approximately two-dimensional surface that separates two distinct domains in three dimensions (Figure 3.5) [2,6,8,148–150]. The bijel contains two continuous liquid phases separated by an elastic interface. Each of the two fluid phases can flow, yet the material as a whole is solid because of the network-like architecture of the interface. To put it another way, two microscopic fish that start off on opposite sides of the dividing interface can each swim across the entire sample, yet they are prevented by the membrane from touching one another. As in the particulate gel, a 3D solid can be constructed from an arbitrarily low particle volume fraction ϕ. Once again, the particles are brought into contact with one another (and therefore can be strongly bonded), and yet they permeate the entire sample by forming a low-dimensional network. As in the particulate gel, tunable structural correlation lengths should allow considerable power to tune permeability, elasticity, or other properties for a variety of technological applications.

The bijel is formed by intentional arrest of spinodal decomposition of two (molecular) fluids. As described in the previous section, a supercooled fluid can rapidly separate into two phases by the process of spinodal decomposition if the temperatures and concentrations are appropriately adjusted. In general, spinodal decomposition leads to a scattering peak at high q (i.e., at short length scales), corresponding to a bicontinuous arrangement of the two fluids. The structure coarsens by reducing the interfacial area, which leads to a larger characteristic length and a shift of the scattering peak to smaller q. To make bijels, small particles are added to the starting fluid. The particles are chemically modified to segregate to the interface (where they lower the interfacial energy by removing some of the area of contact between the two fluids [151]). As the material coarsens, the particles are squeezed together until they solidify, or "jam." The remarkable feature of the bicontinuous structure is that it is then trapped; forming droplets or macroscopically separated phases would first require a reduction of the interfacial area. This mechanism has an interesting similarity to the proposed spinodal route to particle gels, but in the bijel case the solidification occurs by jamming in 2D, whereas we have seen that gelation might be explained by solidification or vitrification in 3D.

Chung *et al.* showed that phase separation in polymer blends can also be arrested by addition of particles [148]. They investigated thin films ($\sim$2 μm or less in thickness) composed of a fifty/fifty blend of poly(methylmethacrylate) with a random copolymer, poly(styrene-ran-acrylonitrile). After the samples were heated to 195°C, 30°C above the critical temperature, the polymers phase separated. In this case, the resulting structure arose from a combination of phase separation (perhaps spinodal decomposition) and possibly de-wetting of one phase from the surface. In samples with 20-nm particles added, the structure was trapped with a correlation length as low as 1 μm.

A fully 3D, trapped, bicontinuous structure was formed by Herzig *et al.* in a mixture of 2,6-lutidine and water mixed with partially hydrophobic silica spheres 580 nm in diameter ($\phi = 0.02$) [2]. The water-lutidine mixture had approximately critical composition and was heated to 40°C, 6°C above the critical temperature. Figure 3.5 shows confocal microscopy images of the fluorescent particles, clearly showing that they pack densely at the interface. Fourier transforms of the images (akin to a structure factor) show a stretched exponential behavior, which can be fit to obtain a correlation length of approximately 40 μm. Reconstruction of the images shows that the structure is indeed bicontinuous, even over a sample depth that is several times larger than the correlation length. In addition, it was preserved for several months while the sample temperature was maintained in the two-phase region. Even though the bulk phases are fluid, the bijel can support weight: it is

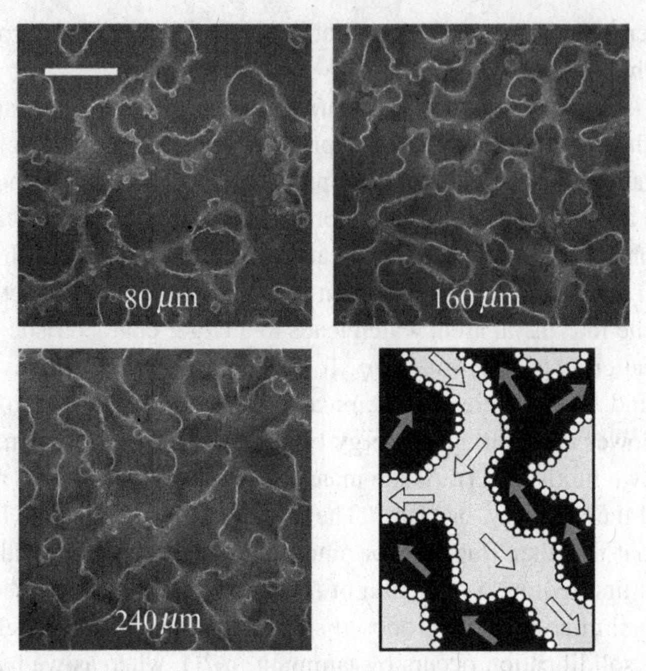

Figure 3.5 Confocal optical micrographs of a "bijel," consisting of two distinct fluid domains separated by a solid interface packed with jammed microparticles. The microparticles are fluorescent. The images are parallel to the (x, y) plane at various heights in the sample (as marked on the images). The structure was formed by spinodal decomposition of the molecular fluid, and further coarsening was prevented by the jammed interface. Scale bar: 100 μm. (Images by E.M. Herzig *et al.* [2]. Reprinted by permission from Macmillan Publishers Ltd *Nature Materials* **6**, 966, copyright (2007).) *Lower-right*: Illustration of the bijel structure, showing two fluid phases (light and dark gray) separated by a densely packed monolayer of particles. (Illustration from M. E. Cates and P. S. Clegg [6], reproduced by permission of the Royal Society of Chemistry.)

a solid [2]. Earlier work by the same group explored thin, quasi-two-dimensional samples of alcohol-alkane mixtures (e.g., ethanol and dodecane, methanol and hexane or heptanes), but these structures were less stable than the ones formed in water–lutidine mixtures [149].

Firoozmand *et al.* recently reported slowed spinodal decomposition (but not quite arrested) in phase-separating mixtures of gelatin, starch, and sugar in water [152]. Further investigation of these biopolymer mixtures might lead to application of the bijel approach to foods, where the trapped nonequilibrium structure should have a significant impact on food texture and appearance. In biopolymer mixtures, there is the interesting possibility of electrostatic interactions among polyelectrolytes and charged particles, which might allow an additional way to solidify the particle-coated interface.

Chung *et al.* and Herzig *et al.* found that the particles' effectiveness at arresting phase separation varied with chemical modification of the interface [2,148]. Computer simulations have raised the possibility that the particles might be more easily ejected from the liquid interface than expected from the binding energy [8,153], though van der Waals and other inter-particle forces were not taken into account. Finding the optimal surface chemistry remains an important practical issue.

The correlation length that appears in the bijel (roughly the characteristic length that is seen in Figure 3.5) can be tuned by the particle volume fraction, ϕ. The correlation length ξ can be roughly predicted by assuming that (i) all of the particles bind to the interface and (ii) the structure can be modeled as a close-packed array of spherical droplets, each fully coated by the particles. This model predicts $\xi = (2\pi/\sqrt{3})(a/\phi)$, where a is the particle radius. The scaling with ϕ was confirmed in experiments [2] and in computer simulations using dissipative particle dynamics [154]. As with the particulate gels, the correlation length could become arbitrarily large in principal. It may therefore be possible to tune elasticity, permeability, and other properties over a wide range.

3.7 Stress and structure in compressed emulsions

Compressed emulsions are a form of soft, random solid that is routinely found in your medicine cabinet, in the form of Vaseline and other skin lotions. Even though these materials contain only liquids, they can nonetheless form solid materials – a scoop of these materials maintains its shape against gravity. So although the individual particles are squishy (easily deformed) and slippery (lacking static friction), they resist shear stress at low frequency. These samples are referred to as compressed emulsions, indicating that some pressure is typically needed to exceed the jamming volume fraction and distinguish them from liquid suspensions of droplets (such as vinaigrette). When the droplets are deformed by shear, their surface area increases and the interfacial tension, γ, causes a restoring force. More precisely, the deformation is resisted by the excess (Laplace) pressure inside each droplet, which is proportional to γ/a. Consequently, the measured shear modulus scales with γ/a [155].

The arrangement of droplets within compressed emulsions is compact. In the language of Section 3.3, $d_f = 3$ and there are negligible structural correlations beyond the scale of a droplet size [4,156–159]. Nonetheless, there are reports of significant variations in 'connectedness' – the number and orientation of contacts among particles – which are described below. Emulsions that have a broad distribution of sizes (i.e., are polydisperse) offer an additional form of randomness: the neighbors around a given particle can vary in size as well as in proximity and angular position. Here, the structural randomness does not arise from thermal fluctuations; indeed thermal motion in compressed emulsions is generally negligibly small, especially when the droplets are several microns in size or larger. A recent study by Clusel *et al.* [160] used 3D images of compressed, polydisperse emulsions to measure the number of neighbors n (defined by a Voronoi-like analysis), and the number of contacts Z (defined as touching droplets). The measurements showed that n and Z both vary from droplet to droplet, but the ratio Z/n has a well-defined average that is independent of droplet size: $\langle Z/n \rangle = 0.41$. The data are accurately described by a statistical model that reduces to the problem of randomly packing a layer of spheres of various sizes around a single target sphere. This elegant single-particle view (or 'granocentric' view, in the words of Clusel *et al.*) includes the enormously simplifying assumptions that the particles are statistically independent of one another aside from well-defined geometric constraints, and that only nearest neighbors need to be considered. While this approach works for a variety of problems, its general validity is not obvious *a priori*, since perturbing a single particle in a dense solid can have long-range effects (see, for example, [161]). The single-particle concept is not new to this field, but the polydisperse emulsion provides an elegant confirmation of its accuracy and demonstrates its potential to address many problems associated with random media.

When stress is externally applied to a compressed emulsion, by means of a weight pressing on the droplets, the *stress* is spontaneously focused into lightning-bolt-like regions of the samples, known as "force chains" [4, 159]. Force chains were first reported in granular media, i.e. in piles of macroscopic solid grains (see [161, 162] and cited references). To measure this effect in emulsions, experiments take advantage of the softness of droplets in a way first used by Brujic *et al.* [156, 157]: when two droplets are pushed together with a force F, they both deform. The magnitude of F can be obtained from the sizes of the droplets, the area of the deformed contact region, and γ. By imaging the contact regions with confocal microscopy, one can measure the positions, orientations, and magnitudes of hundreds or thousands of contact forces in a sample [4, 156, 157, 159]. Figure 3.6 shows a 3D confocal image of droplets coated with a layer of fluorescent nanoparticles. (They assemble on the surface for the same reasons that applied in the bijel case.) Computer analysis of the 3D images led to identification of the contact regions and a list of quantitative forces. As an alternative method to visualize the interfaces, one can dissolve fluorophore (Nile red dye) uniformly in the droplet phase. Nile red dye molecules that reside at the oil-water interface have a fluorescence emission spectrum that is shifted relative to emission from the oil phase [158]. Hence, contrast between interface and droplet interior can be achieved with a color filter [158] without addition of interfacially bound nanoparticles. The contact regions can then be identified from their greater

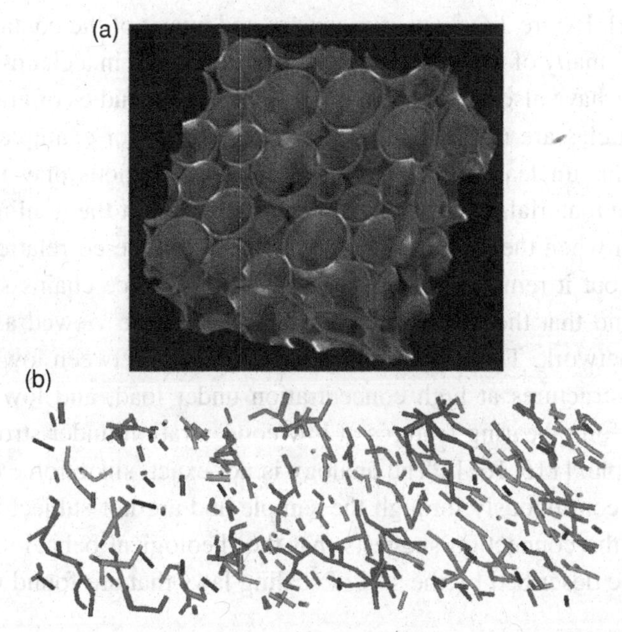

Figure 3.6 Images of contact forces and geometry in a compressed emulsion. (a) 3D confocal-microscope image of droplets coated with fluorescent nanoparticles. The mean droplet diameter was 30 μm. (b) Reconstructed image of contact forces F, looking only at forces greater than $1.6\langle F\rangle$. Each cylindrical section represents an individual contact, with the cylinder radius proportional to $|F|$. Forces acting on interfaces of the same droplet are represented by connected cylinders. The grayscale is arbitrarily chosen to aid visualization. The chain-like arrangement is apparent in the image, where the longest chain is 12 diameters in length. (Images from Zhou [4]). Reprinted with permission from AAAS.)

intensity because of the double layer of interfacially bound fluorophores in those regions [157,158].

From the measurements of individual inter-drop contact forces, it is possible to look for chains or other spatial correlations. The sum of all of the scalar force magnitudes on each droplet has negligible correlations beyond roughly a diameter (unless the sample is sheared, as shown by experiments with photoelastic discs in 2D [163]). The average contact force on a droplet also has negligible correlations [4, 156, 157]. To identify chains, the direction of the force should be considered: when two inter-drop contacts (i, j) are separated by the vector $r\boldsymbol{e}_{ij}$, then correlations of $|\boldsymbol{F}_i \cdot \boldsymbol{e}_{ij}|$ and $|\boldsymbol{F}_j \cdot \boldsymbol{e}_{ij}|$ describe the alignment of forces in addition to correlations of their magnitudes [4]. Here, $\boldsymbol{F}_i$ is the force at a given contact (not the net force on a droplet) and $\boldsymbol{e}_{ij}$ is the unit vector from i to j. More precisely, correlations appear only when one looks selectively at the contact forces that are large compared to the sample average, $\langle F\rangle$. The correlations become rapidly stronger with larger F and correlation lengths of 10–12 diameters were reported for contacts with

$F > 1.6\langle F\rangle$ [4]. Figure 3.6 shows a reconstructed image of the contact forces that exceed $1.6\langle F\rangle$; many of these contacts are clearly aligned in a chain-like fashion.

Force chains have also been reported in experimental studies of granular media, where the particles are macroscopic solid objects (see, for example [162–164]). While it remains unclear what role these high-stress regions play in the overall response of the material, there is evidence that they affect the nonlinear response of an emulsion when the system size approaches the force-correlation length [4]. It is possible (but it remains to be proven) that these force chains support much of the stress and that the aging of these materials can be viewed as remodeling of this stress network. There is an intriguing similarity between low-dimensional force-bearing structures at high concentration under load, and low-dimensional structural and force-bearing features at low concentration under strong attraction (see, for example, [115, 165]). The analogy is not exact, since force chains might not propagate continuously through the sample and are not subject to mass conservation, but the connection suggests that the rheological behavior of the force chains could be described by the sort of scaling laws that are found in particulate gels.

3.8 Conclusion

This chapter provided a summary of some of the exciting advances from the past three decades. Many of these advances were made possible by new methods of imaging, image analysis, chemical synthesis, and new theoretical paradigms. The range of materials that qualify as soft, random solids is very broad. This chapter focused on particulate gels, emulsions, and hybrids of the two (e.g., bijels). Among other things, these examples highlight the potential to fabricate new materials by trapping them in nonequilibrium states in a controllable and reproducible fashion. There continues to be great potential for breakthroughs in fundamental science and invention of new materials.

References

[1] O. Sandoval-Castilla, C. Lobato-Calleros, E. Aguirre-Mandujano, and E. J. Vernon-Carter, "Microstructure and Texture of Yogurt as Influenced by Fat Replacers," *Internat. Dairy J.* **14**, 151 (2004).

[2] E. M. Herzig, K. A. White, A. B. Schofield, W. C. K. Poon, and P. S. Clegg, "Bicontinuous Emulsions Stabilized Solely by Colloidal Particles," *Nature Mater.* **6**, 966 (2007).

[3] V. Prasad, V. Trappe, A. D. Dinsmore, P. N. Segre, L. Cipelletti, and D. A. Weitz, "Universal Features of the Fluid to Solid Transition for Attractive Particles," *Faraday Disc.* **123**, 1 (2003).

[4] J. Zhou, S. Long, Q. Wang, and A. D. Dinsmore, "Measurement of Forces inside a Three-Dimensional Pile of Frictionless Droplets," *Science* **312**, 1631 (2006).
[5] D. J. Robinson and J. C. Earnshaw, "Long-Range Order in 2-Dimensional Fractal Aggregation," *Phys. Rev. Lett.* **71**, 715 (1993).
[6] M. E. Cates and P. S. Clegg, "Bijels: A New Class of Soft Materials," *Soft Matter* **4**, 2132 (2008).
[7] P. J. Lu, E. Zaccarelli, F. Ciulla, A. B. Schofield, F. Sciortino, and D. A. Weitz, "Gelation of Particles with Short-Range Attraction," *Nature* **453**, 499 (2008).
[8] K. Stratford, R. Adhikari, I. Pagonabarraga, J. C. Desplat, and M. E. Cates, "Colloidal Jamming at Interfaces: A Route to Fluid-Bicontinuous Gels," *Science* **309**, 2198 (2005).
[9] A. D. Dinsmore, "Colloids – a Useful Boundary," *Nature Mater.* **6**, 921 (2007).
[10] R. G. M. van der Sman and A. J. van der Goot, "The Science of Food Structuring," *Soft Matter* **5**, 501 (2009).
[11] E. Kostansek, "Controlling Particle Dispersion in Latex Paints Containing Associative Thickeners," *J. Coat. Technol. Res.* **4**, 375 (2007).
[12] C. Gallegos and J. M. Franco, "Rheology of Food, Cosmetics and Pharmaceuticals," *Curr. Opin. Colloid Interface Sci.* **4**, 288 (1999).
[13] R. Klajn, K. J. M. Bishop, M. Fialkowski, M. Paszewski, C. J. Campbell, T. P. Gray, and B. A. Grzybowski, "Plastic and Moldable Metals by Self-Assembly of Sticky Nanoparticle Aggregates," *Science* **316**, 261 (2007).
[14] R. Klajn, T. P. Gray, P. J. Wesson, B. D. Myers, V. P. Dravid, S. K. Smoukov, and B. A. Grzybowski, "Bulk Synthesis and Surface Patterning of Nanoporous Metals and Alloys from Supraspherical Nanoparticle Aggregates," *Adv. Funct. Mater.* **18**, 2763 (2008).
[15] L. G. B. Bremer, B. H. Bijsterbosch, P. Walstra, and T. Vanvliet, "Formation, Properties and Fractal Structure of Particle Gels," *Adv. Colloid Interface Sci.* **46**, 117 (1993).
[16] J. A. Lucey, P. A. Munro, and H. Singh, "Rheological Properties and Microstructure of Acid Milk Gels as Affected by Fat Content and Heat Treatment," *J. Food Sci.* **63**, 660 (1998).
[17] S. Ikeda, E. A. Foegeding, and T. Hagiwara, "Rheological Study on the Fractal Nature of the Protein Gel Structure," *Langmuir* **15**, 8584 (1999).
[18] A. Stradner, H. Sedgwick, F. Cardinaux, W. C. K. Poon, S. U. Egelhaaf, and P. Schurtenberger, "Equilibrium Cluster Formation in Concentrated Protein Solutions and Colloids," *Nature* **432**, 492 (2004).
[19] H. Sedgwick, K. Kroy, A. Salonen, M. B. Robertson, S. U. Egelhaaf, and W. C. K. Poon, "Non-Equilibrium Behavior of Sticky Colloidal Particles: Beads, Clusters and Gels," *Eur. Phys. J. E* **16**, 77 (2005).
[20] P. Bezot, C. Hessebezot, B. Rousset, and C. Diraison, "Effect of Polymers on the Aggregation Kinetics and Fractal Structure of Carbon-Black Suspensions in an Aliphatic Solvent – a Static and Dynamic Light-Scattering Study," *Colloid Surf. A-Physicochem. Eng. Asp.* **97**, 53 (1995).
[21] A. Kyrlidis and A. Shim, in *Nip24/Digital Fabrication 2008: 24th International Conference on Digital Printing Technologies, Technical Program and Proceedings*, p. 72 (2008).
[22] A. Shalkevich, A. Stradner, S. K. Bhat, F. Muller, and P. Schurtenberger, "Cluster, Glass, and Gel Formation and Viscoelastic Phase Separation in Aqueous Clay Suspensions," *Langmuir* **23**, 3570 (2007).

[23] A. Castellanos, J. M. Valverde, and M. A. S. Quintanilla, "Physics of Compaction of Fine Cohesive Particles," *Phys. Rev. Lett.* **94**, 075501 (2005).

[24] C. M. Sorensen, C. Oh, P. W. Schmidt, and T. P. Rieker, "Scaling Description of the Structure Factor of Fractal Soot Composites," *Phys. Rev. E* **58**, 4666 (1998).

[25] C. M. Sorensen, B. Hageman, T. J. Rush, H. Huang, and C. Oh, "Aerogelation in a Flame Soot Aerosol," *Phys. Rev. Lett.* **80**, 1782 (1998).

[26] A. S. Dorcheh and M. H. Abbasi, "Silica Aerogel; Synthesis, Properties and Characterization," *J. Mater. Process. Technol.* **199**, 10 (2008).

[27] J. Fricke and A. Emmerling, "Aerogels – Preparation, Properties, Applications," *Struct. Bond.* **77**, 37 (1992).

[28] C. M. Sorensen, "Light Scattering by Fractal Aggregates: A Review," *Aerosol Sci. Technol.* **35**, 648 (2001).

[29] W. C. K. Poon and M. D. Haw, "Mesoscopic Structure Formation in Colloidal Aggregation and Gelation," *Adv. Colloid Interface Sci.* **73**, 71 (1997).

[30] D. Asnaghi, M. Carpineti, M. Giglio, and A. Vailati, "Light-Scattering-Studies of Aggregation Phenomena," *Physica A* **213**, 148 (1995).

[31] S. Manley, L. Cipelletti, V. Trappe, A. E. Bailey, R. J. Christianson, U. Gasser, V. Prasad, P. N. Segre, M. P. Doherty, S. Sankaran, A. L. Jankovsky, B. Shiley, J. Bowen, J. Eggers, C. Kurta, T. Lorik, and D. A. Weitz, "Limits to Gelation in Colloidal Aggregation," *Phys. Rev. Lett.* **93**, 108302 (2004).

[32] P. Meakin, "Formation of Fractal Clusters and Networks by Irreversible Diffusion-Limited Aggregation," *Phys. Rev. Lett.* **51**, 1119 (1983).

[33] M. Kolb, R. Botet, and R. Jullien, "Scaling of Kinetically Growing Clusters," *Phys. Rev. Lett.* **51**, 1123 (1983).

[34] T. A. Witten, Jr. and L. M. Sander, "Diffusion-Limited Aggregation: A Kinetic Critical Phenomenon," *Phys. Rev. Lett.* **47**, 1400 (1981).

[35] D. A. Weitz and M. Oliveria, "Fractal Structures Formed by Kinetic Aggregation of Aqueous Gold Colloids," *Phys. Rev. Lett.* **52**, 1433 (1984).

[36] A. Thill, M. Wagner, and J. Y. Bottero, "Confocal Scanning Laser Microscopy as a Tool for the Determination of 3d Floc Structure," *J. Colloid Interface Sci.* **220**, 465 (1999).

[37] M. Lattuada, H. Wu, and M. Morbidelli, "Estimation of Fractal Dimension of Colloidal Gels in the Presence of Multiple Scattering," *Phys. Rev. E* **64**, 061404 (2001).

[38] R. Jullien, M. Kolb, and R. Botet, "Aggregation by Kinetic Clustering of Clusters D > 2," *J. Phys. Lett.* **45**, L211 (1984).

[39] M. D. Haw, W. C. K. Poon, and P. N. Pusey, "Structure and Arrangement of Clusters in Cluster Aggregation," *Phys. Rev. E* **56**, 1918 (1997).

[40] P. Meakin, "A Historical Introduction to Computer Models for Fractal Aggregates," *J. Sol-Gel Sci. Technol.* **15**, 97 (1999).

[41] M. Muthukumar, "Mean-Field Theory for Diffusion-Limited Cluster Formation," *Phys. Rev. Lett.* **50**, 839 (1983).

[42] C. M. Sorensen and C. Oh, "Divine Proportion Shape Preservation and the Fractal Nature of Cluster–Cluster Aggregates," *Phys. Rev. E* **58**, 7545 (1998).

[43] D. A. Weitz, J. S. Huang, M. Y. Lin, and J. Sung, "Limits of the Fractal Dimension for Irreversible Kinetic Aggregation of Gold Colloids," *Phys. Rev. Lett.* **54**, 1416 (1985).

[44] M. Y. Lin, H. M. Lindsay, D. A. Weitz, R. C. Ball, R. Klein, and P. Meakin, "Universality in Colloid Aggregation," *Nature* **339**, 360 (1989).

[45] M. Y. Lin, H. M. Lindsay, D. A. Weitz, R. Klein, R. C. Ball, and P. Meakin, "Universal Diffusion-Limited Colloid Aggregation," *J. Phys.-Condens. Matter* **2**, 3093 (1990).
[46] C. M. Sorensen and W. B. Hageman, "Two-Dimensional Soot," *Langmuir* **17**, 5431 (2001).
[47] J. Bibette, T. G. Mason, H. Gang, and D. A. Weitz, "Kinetically Induced Ordering in Gelation of Emulsions," *Phys. Rev. Lett.* **69**, 981 (1992).
[48] S. Havlin and R. Nossal, "Topological Properties of Percolation Clusters," *J. Phys. A: Math. Gen.* **17**, L427 (1984).
[49] P. Meakin, I. Majid, S. Havlin, and H. E. Stanley, "Topological Properties of Diffusion Limited Aggregation and Cluster–Cluster Aggregation," *J. Phys. A: Math. Gen.* **17**, L975 (1984).
[50] A. D. Dinsmore and D. A. Weitz, "Direct Imaging of Three-Dimensional Structure and Topology of Colloidal Gels," *J.Phys. – Condens. Matter* **14**, 7581 (2002).
[51] M. Carpineti and M. Giglio, "Aggregation Phenomena," *Adv. Colloid Interface Sci.* **46**, 73 (1993).
[52] L. G. B. Bremer, P. Walstra, and T. Vanvliet, "Estimations of the Aggregation Time of Various Colloidal Systems," *Colloid Surf. A-Physicochem. Eng. Asp.* **99**, 121 (1995).
[53] A. S. Kyriakidis, S. G. Yiantsios, and A. J. Karabelas, "A Study of Colloidal Particle Brownian Aggregation by Light Scattering Techniques," *J. Colloid Interface Sci.* **195**, 299 (1997).
[54] D. A. Weitz, J. S. Huang, M. Y. Lin, and J. Sung, "Dynamics of Diffusion-Limited Kinetic Aggregation," *Phys. Rev. Lett.* **53**, 1657 (1984).
[55] W. B. Russel, D. A. Saville, and W. Schowalter, *Colloidal Dispersions* (Cambridge University Press, 1989).
[56] D. A. Weitz and M. Y. Lin, "Dynamic Scaling of Cluster-Mass Distributions in Kinetic Colloid Aggregation," *Phys. Rev. Lett.* **57**, 2037 (1986).
[57] M. Broide and R. Cohen, "Experimental Evidence of Dynamic Scaling in Colloidal Aggregation," *Phys. Rev. Lett.* **64**, 2026 (1990).
[58] D. W. Schaefer, J. E. Martin, P. Wiltzius, and D. S. Cannell, "Fractal Geometry of Colloidal Aggregates," *Phys. Rev. Lett.* **52**, 2371 (1984).
[59] C. Aubert and D. S. Canell, "Restructuring of Colloidal Silica Aggregates," *Phys. Rev. Lett.* **56**, 738 (1986).
[60] M. Kolb and R. Jullien, "Chemically Limited Versus Diffusion Limited Aggregation," *J. Physique Lett.* **45**, L977 (1984).
[61] P. Meakin and F. Family, "Structure and Kinetics of Reaction-Limited Aggregation," *Phys. Rev. A* **38**, 2110 (1988).
[62] C. S. Wen and L. Z. Zhang, "Rate of Coagulation of Brownian Particles in the Presence of a Double Layer," *J. Colloid Interface Sci.* **188**, 372 (1997).
[63] A. Vaccaro, J. Sefcik, and M. Morbidelli, "Characterization of Colloidal Polymer Particles through Stability Ratio Measurements," *Polymer* **46**, 1157 (2005).
[64] J. C. Crocker, "Measurement of the Hydrodynamic Corrections to the Brownian Motion of Two Colloidal Spheres," *J. Chem. Phys.* **106**, 2837 (1997).
[65] S. K. Sainis, V. Germain, and E. R. Dufresne, "Statistics of Particle Trajectories at Short Time Intervals Reveal fN-Scale Colloidal Forces," *Phys. Rev. Lett.* **99**, 018303 (2007).
[66] H. Tanaka and T. Araki, "Simulation Method of Colloidal Suspensions with Hydrodynamic Interactions: Fluid Particle Dynamics," *Phys. Rev. Lett.* **85**, 1338 (2000).

[67] C. Allan, M. Cloitre, and M. Wafra, "Aggregation and Sedimentation in Colloidal Suspensions," *Phys. Rev. Lett.* **74**, 1478 (1995).
[68] P. Dimon, S. K. Sinha, D. A. Weitz, C. R. Safinya, G. S. Smith, W. A. Varady, and H. M. Lindsay, "Structure of Aggregated Gold Colloids," *Phys. Rev. Lett.* **57**, 595 (1986).
[69] L. Cipelletti, S. Manley, R. C. Ball, and D. A. Weitz, "Universal Aging Features in Restructuring of Fractal Colloidal Gels," *Phys. Rev. Lett.* **84**, 2275 (2000).
[70] J. M. Jin, K. Parbhakar, L. H. Dao, and K. H. Lee, "Gel Formation by Reversible Cluster–Cluster Aggregation," *Phys. Rev. E* **54**, 997 (1996).
[71] S. D. Orrite, S. Stoll, and P. Schurtenberger, "Off-Lattice Monte Carlo Simulations of Irreversible and Reversible Aggregation Processes," *Soft Matter* **1**, 364 (2005).
[72] J. R. Savage, D. W. Blair, A. J. Levine, R. A. Guyer, and A. D. Dinsmore, "Imaging the Sublimation Dynamics of Colloidal Crystallites," *Science* **314**, 795 (2006).
[73] J. R. Savage and A. D. Dinsmore, "Experimental Evidence for Two-Step Nucleation in Colloidal Crystallization," *Phys. Rev. Lett.* **102**, 198302 (2009).
[74] J. N. Israelachvili, *Intermolecular and Surface Forces* (Academic Press, Amsterdam, 2003).
[75] J. P. Pantina and E. M. Furst, "Elasticity and Critical Bending Moment of Model Colloidal Aggregates," *Phys. Rev. Lett.* **94**, 138301 (2005).
[76] A. D. Dinsmore, V. Prasad, I. Y. Wong, and D. A. Weitz, "Microscopic Structure and Elasticity of Weakly Aggregated Colloidal Gels," *Phys. Rev. Lett.* **96**, 185502 (2006).
[77] J. Bibette, T. G. Mason, G. Hu, D. A. Weitz, and P. Poulin, "Structure of Adhesive Emulsions," *Langmuir* **9**, 3352 (1993).
[78] J. N. Wilking, S. M. Graves, C. B. Chang, K. Meleson, M. Y. Lin, and T. G. Mason, "Dense Cluster Formation During Aggregation and Gelation of Attractive Slippery Nanoemulsion Droplets," *Phys. Rev. Lett.* **96**, 015501 (2006).
[79] R. Penfold, A. D. Watson, A. R. Mackie, and D. J. Hibberd, "Quantitative Imaging of Aggregated Emulsions," *Langmuir* **22**, 2005 (2006).
[80] C. R. Seager and T. G. Mason, "Slippery Diffusion-Limited Aggregation," *Phys. Rev. E* **75**, 011406 (2007).
[81] E. H. A. de Hoog, W. K. Kegel, A. van Blaaderen, and H. N. W. Lekkerkerker, "Direct Observation of Crystallization and Aggregation in a Phase-Separating Colloid-Polymer Suspension," *Phys. Rev. E* **6402**, 021407 (2001).
[82] H. Brenner, "The Slow Motion of a Sphere through a Viscous Fluid Towards a Plane Surface," *Chem. Eng. Sci.* **16**, 242 (1961).
[83] J. Y. Walz and L. Suresh, "Study of Sedimentation of a Single Particle toward a Flat Plate," *J. Chem. Phys.* **103**, 10714 (1995).
[84] A. J. Archer and N. B. Wilding, "Phase Behavior of a Fluid with Competing Attractive and Repulsive Interactions," *Phys. Rev. E* **76**, 031501 (2007).
[85] H. Sedgwick, S. U. Egelhaaf, and W. C. K. Poon, "Clusters and Gels in Systems of Sticky Particles," *J. Phys.-Condens. Matter* **16**, S4913 (2004).
[86] A. I. Campbell, V. J. Anderson, J. S. van Duijneveldt, and P. Bartlett, "Dynamical Arrest in Attractive Colloids: The Effect of Long-Range Repulsion," *Phys. Rev. Lett.* **94**, 208301 (2005).
[87] R. Sanchez and P. Bartlett, "Equilibrium Cluster Formation and Gelation," *J. Phys.-Condens. Matter* **17**, S3551 (2005).
[88] T. Ohtsuka, C. P. Royall, and H. Tanaka, "Local Structure and Dynamics in Colloidal Fluids and Gels," *EPL* **84**, 46002 (2008).

[89] C. Haro-Perez, L. F. Rojas-Ochoa, R. Castaneda-Priego, M. Quesada-Perez, J. Callejas-Fernandez, R. Hidalgo-Alvarez, and V. Trappe, "Dynamic Arrest in Charged Colloidal Systems Exhibiting Large-Scale Structural Heterogeneities," *Phys. Rev. Lett.* **102**, 018301 (2009).
[90] J. Groenewold and W. K. Kegel, "Anomalously Large Equilibrium Clusters of Colloids," *J. Phys. Chem. B* **105**, 11702 (2001).
[91] J. Groenewold and W. K. Kegel, "Colloidal Cluster Phases, Gelation and Nuclear Matter," *J. Phys.-Condens. Matter* **16**, S4877 (2004).
[92] P. N. Segre, V. Prasad, A. B. Schofield, and D. A. Weitz, "Glasslike Kinetic Arrest at the Colloidal-Gelation Transition," *Phys. Rev. Lett.* **86**, 6042 (2001).
[93] P. J. Lu, J. C. Conrad, H. M. Wyss, A. B. Schofield, and D. A. Weitz, "Fluids of Clusters in Attractive Colloids," *Phys. Rev. Lett.* **96**, 028306 (2006).
[94] F. Sciortino, S. Mossa, E. Zaccarelli, and P. Tartaglia, "Equilibrium Cluster Phases and Low-Density Arrested Disordered States: The Role of Short-Range Attraction and Long-Range Repulsion," *Phys. Rev. Lett.* **93**, 055701 (2004).
[95] M. A. Glaser, G. M. Grason, R. D. Kamien, A. Kosmrlj, C. D. Santangelo, and P. Ziherl, "Soft Spheres Make More Mesophases," *EPL* **78**, 46004 (2007).
[96] A. J. Archer, "Two-Dimensional Fluid with Competing Interactions Exhibiting Microphase Separation: Theory for Bulk and Interfacial Properties," *Phys. Rev. E* **78**, 031402 (2008).
[97] A. J. Archer, C. Ionescu, D. Pini, and L. Reatto, "Theory for the Phase Behaviour of a Colloidal Fluid with Competing Interactions," *J. Phys.-Condens. Matter* **20**, 415106 (2008).
[98] A. Mohraz, E. R. Weeks, and J. A. Lewis, "Structure and Dynamics of Biphasic Colloidal Mixtures," *Phys. Rev. E* **77**, 060403 (2008).
[99] V. Trappe and D. A. Weitz, "Scaling of the Viscoelasticity of Weakly Attractive Particles," *Phys. Rev. Lett.* **85**, 449 (2000).
[100] Y. Kantor and I. Webman, "Elastic Properties of Random Percolating Systems," *Phys. Rev. Lett.* **52**, 1891 (1984).
[101] H. S. Ma, J. H. Prevost, and G. W. Scherer, "Elasticity of DLCA Model Gels with Loops," *Int. J. Solids Struc.* **39**, 4605 (2002).
[102] R. Buscall, P. D. A. Mills, J. W. Goodwin, and D. W. Lawson, "Scaling Behavior of the Rheology of Aggregate Networks Formed from Colloidal Particles," *J. Chem. Soc. – Faraday Trans. I* **84**, 4249 (1988).
[103] W.-H. Shih, W. Y. Shih, S.-I. Kim, J. Lu, and I. A. Aksay, "Scaling Behavior of the Elastic Properties of Colloidal Gels," *Phys. Rev. A* **42**, 4772 (1990).
[104] L. G. B. Bremer and T. Vanvliet, "The Modulus of Particle Networks with Stretched Strands," *Rheologica Acta* **30**, 98 (1991).
[105] R. de Rooij, D. Vandenende, M. H. G. Duits, and J. Mellema, "Elasticity of Weakly Aggregating Polystyrene Latex Dispersions," *Phys. Rev. E* **49**, 3038 (1994).
[106] A. A. Potanin and W. B. Russel, "Fractal Model of Consolidation of Weakly Aggregated Colloidal Dispersions," *Phys. Rev. E* **53**, 3702 (1996).
[107] W. Wolthers, D. van den Ende, V. Breedveld, M. H. G. Duits, A. A. Potanin, R. H. W. Wientjes, and J. Mellema, "Linear Viscoleastic Behavior of Aggregated Colloidal Dispersions," *Phys. Rev. E* **56**, 5726 (1997).
[108] H. S. Ma, A. P. Roberts, J. H. Prevost, R. Jullien, and G. W. Scherer, "Mechanical Structure–Property Relationship of Aerogels," *J. Non-Cryst. Solids* **277**, 127 (2000).
[109] S. Manley, B. Davidovitch, N. R. Davies, L. Cipelletti, A. E. Bailey, R. J. Christianson, U. Gasser, V. Prasad, P. N. Segre, M. P. Doherty, S. Sankaran,

A. L. Jankovsky, B. Shiley, J. Bowen, J. Eggers, C. Kurta, T. Lorik, and D. A. Weitz, "Time-Dependent Strength of Colloidal Gels," *Phys. Rev. Lett.* **95**, 048302 (2005).

[110] T. G. Mason and D. A. Weitz, "Optical Measurements of Frequency-Dependent Linear Viscoelastic Moduli of Complex Fluids," *Phys. Rev. Lett.* **74**, 1250 (1995).

[111] J. C. Crocker, M. T. Valentine, E. R. Weeks, T. Gisler, P. D. Kaplan, A. G. Yodh, and D. A. Weitz, "Two-Point Microrheology of Inhomogeneous Soft Materials," *Phys. Rev. Lett.* **85**, 888 (2000).

[112] P. Cicuta and A. M. Donald, "Microrheology: A Review of the Method and Applications," *Soft Matter* **3**, 1449 (2007).

[113] A. H. Krall and D. A. Weitz, "Internal Dynamics and Elasticity of Fractal Colloidal Gels," *Phys. Rev. Lett.* **80**, 778 (1998).

[114] M. C. Grant and W. B. Russel, "Volume-Fraction Dependence of Elastic Moduli and Transition Temperatures for Colloidal Silica Gels," *Phys. Rev. E* **47**, 2606 (1993).

[115] V. Trappe, V. Prasad, L. Cipelletti, P. N. Segre, and D. A. Weitz, "Jamming Phase Diagram for Attractive Particles," *Nature* **411**, 772 (2001).

[116] J. W. Cahn, "Spinodal Decomposition (the 1967 Institute of Metals Lecture)," *Trans. Metal. Soc. AIME* **242**, 166 (1967).

[117] J. Nyvlt, "The Ostwald Rule of Stages," *Crystal Research and Technology* **30**, 443 (1995).

[118] P. R. ten Wolde and D. Frenkel, "Homogeneous Nucleation and the Ostwald Step Rule," *PCCP Phys. Chem. Chem. Phys.* **1**, 2191 (1999).

[119] H. Schmalzried, "On the Equilibration of Solid Phases. Some Thoughts on Ostwalds Contributions," *Z. Phys. Chemie* **217**, 1281 (2003).

[120] S. J. L. Billinge, "How Do Your Crystals Grow?," *Nat. Phys.* **5**, 13 (2009).

[121] S. Y. Chung, Y. M. Kim, J. G. Kim, and Y. J. Kim, "Multiphase Transformation and Ostwald's Rule of Stages During Crystallization of a Metal Phosphate," *Nat. Phys.* **5**, 68 (2009).

[122] J. Bergenholtz, M. Fuchs, and T. Voigtmann, "Colloidal Gelation and Non-Ergodicity Transitions," *J. Phys.-Condens. Matter* **12**, 6575 (2000).

[123] K. N. Pham, A. M. Puertas, J. Bergenholtz, S. U. Egelhaaf, A. Moussaid, P. N. Pusey, A. B. Schofield, M. E. Cates, M. Fuchs, and W. C. K. Poon, "Multiple Glassy States in a Simple Model System," *Science* **296**, 104 (2002).

[124] V. Trappe and P. Sandkuhler, "Colloidal Gels – Low-Density Disordered Solid-Like States," *Curr. Opin. Colloid Interface Sci.* **8**, 494 (2004).

[125] M. E. Cates, M. Fuchs, K. Kroy, W. C. K. Poon, and A. M. Puertas, "Theory and Simulation of Gelation, Arrest and Yielding in Attracting Colloids," *J. Phys.-Condens. Matter* **16**, S4861 (2004).

[126] S. Manley, H. M. Wyss, K. Miyazaki, J. C. Conrad, V. Trappe, L. J. Kaufman, D. R. Reichman, and D. A. Weitz, "Glasslike Arrest in Spinodal Decomposition as a Route to Colloidal Gelation," *Phys. Rev. Lett.* **95**, 238302 (2005).

[127] E. Zaccarelli, "Colloidal Gels: Equilibrium and Non-Equilibrium Routes," *J. Phys.-Condens. Matter* **19**, 323101 (2007).

[128] D. C. Viehman and K. S. Schweizer, "Theory of Gelation, Vitrification, and Activated Barrier Hopping in Mixtures of Hard and Sticky Spheres," *J. Chem. Phys.* **128**, 084509 (2008).

[129] V. J. Anderson and H. N. W. Lekkerkerker, "Insights into Phase Transition Kinetics from Colloid Science," *Nature* **416**, 811 (2002).

[130] S. M. Ilett, A. Orrock, W. C. K. Poon, and P. N. Pusey, "Phase Behavior of a Model Colloid-Polymer Mixture," *Phys. Rev. E* **51**, 1344 (1995).

[131] W. C. K. Poon, F. Renth, R. M. L. Evans, D. J. Fairhurst, M. E. Cates, and P. N. Pusey, "Colloid-Polymer Mixtures at Triple Coexistence: Kinetic Maps from Free-Energy Landscapes," *Phys. Rev. Lett.* **83**, 1239 (1999).

[132] M. G. Noro and D. Frenkel, "Extended Corresponding-States Behavior for Particles with Variable Range Attractions," *J. Chem. Phys.* **113**, 2941 (2000).

[133] A. E. Bailey, W. C. K. Poon, R. J. Christianson, A. B. Schofield, U. Gasser, V. Prasad, S. Manley, P. N. Segre, L. Cipelletti, W. V. Meyer, M. P. Doherty, S. Sankaran, A. L. Jankovsky, W. L. Shiley, J. P. Bowen, J. C. Eggers, C. Kurta, T. Lorik, P. N. Pusey, and D. A. Weitz, "Spinodal Decomposition in a Model Colloid-Polymer Mixture in Microgravity," *Phys. Rev. Lett.* **99**, 205701 (2007).

[134] P. Bolhuis and D. Frenkel, "Prediction of an Expanded-to-Condensed Transition in Colloidal Crystals," *Phys. Rev. Lett.* **72**, 2211 (1994).

[135] D. Frenkel, "Colloidal Encounters: A Matter of Attraction," *Science* **314**, 768 (2006).

[136] N. A. M. Verhaegh, D. Asnaghi, H. N. W. Lekkerkerker, M. Giglio, and L. Cipelletti, "Transient Gelation by Spinodal Decomposition in Colloid-Polymer Mixtures," *Physica A* **242**, 104 (1997).

[137] M. Carpineti and M. Giglio, "Spinodal-Type Dynamics in Fractal Aggregation of Colloidal Clusters," *Phys. Rev. Lett.* **68**, 3327 (1992).

[138] M. D. Haw, M. Sievwright, W. C. K. Poon, and P. N. Pusey, "Structure and Characteristic Length Scales in Cluster–Cluster Aggregation Simulation," *Physica A* **217**, 231 (1995).

[139] H. Huang, C. Oh, and C. M. Sorensen, "Structure Factor Scaling in Aggregating Systems," *Phys. Rev. E* **57**, 875 (1998).

[140] W. C. K. Poon, A. D. Pirie, and P. N. Pusey, "Gelation in Colloid–Polymer Mixtures," *Faraday Discussions* **101**, 65 (1995).

[141] R. K. Pathria, *Statistical Mechanics* (Elsevier, Oxford, 1994).

[142] E. Zaccarelli, P. J. Lu, F. Ciulla, D. A. Weitz, and F. Sciortino, "Gelation as Arrested Phase Separation in Short-Ranged Attractive Colloid–Polymer Mixtures," *J. Phys.-Condens. Matter* **20**, 494242 (2008).

[143] R. M. L. Evans and W. C. K. Poon, "Diffusive Evolution of Stable and Metastable Phases: 2. Theory of Nonequilibrium Behavior in Colloid-Polymer Mixtures," *Phys. Rev. E* **56**, 5748 (1997).

[144] W. C. K. Poon, F. Renth, and R. M. L. Evans, "Kinetics from Free-Energy Landscapes – How to Turn Phase Diagrams into Kinetic Maps," *J. Phys.-Condens. Matter* **12**, A269 (2000).

[145] R. M. L. Evans, W. C. K. Poon, and F. Renth, "Classification of Ordering Kinetics in Three-Phase Systems," *Phys. Rev. E* **64**, 031403 (2001).

[146] P. R. ten Wolde and D. Frenkel, "Enhancement of Protein Crystal Nucleation by Critical Density Fluctuations," *Science* **277**, 1975 (1997).

[147] H. Tanaka, T. Koyama, and T. Araki, "Network Formation in Viscoelastic Phase Separation," *J. Phys.-Condens. Matter* **15**, S387 (2003).

[148] H. Chung, K. Ohno, T. Fukuda, and R. J. Composto, "Self-Regulated Structures in Nanocomposites by Directed Nanoparticle Assembly," *Nano Lett.* **5**, 1878 (2005).

[149] P. S. Clegg, E. M. Herzig, A. B. Schofield, S. U. Egelhaaf, T. S. Horozov, B. P. Binks, M. E. Cates, and W. C. K. Poon, "Emulsification of Partially Miscible

Liquids Using Colloidal Particles: Nonspherical and Extended Domain Structures," *Langmuir* **23**, 5984 (2007).

[150] P. S. Clegg, "Fluid-Bicontinuous Gels Stabilized by Interfacial Colloids: Low and High Molecular Weight Fluids," *J. Phys.-Condens. Matter* **20**, S3433 (2008).

[151] P. Pieranski, "Two-Dimensional Interfacial Colloidal Crystals," *Phys. Rev. Lett.* **45**, 569 (1980).

[152] H. Firoozmand, B. S. Murray, and E. Dickinson, "Interfacial Structuring in a Phase-Separating Mixed Biopolymer Solution Containing Colloidal Particles," *Langmuir* **25**, 1300 (2009).

[153] E. Kim, K. Stratford, R. Adhikari, and M. E. Cates, "Arrest of Fluid Demixing by Nanoparticles: A Computer Simulation Study," *Langmuir* **24**, 6549 (2008).

[154] M. J. A. Hore and M. Laradji, "Microphase Separation Induced by Interfacial Segregation of Isotropic, Spherical Nanoparticles," *J. Chem. Phys.* **126**, 244903 (2007).

[155] T. G. Mason, J. Bibette, and D. A. Weitz, "Elasticity of Compressed Emulsions," *Phys. Rev. Lett.* **75**, 2051 (1995).

[156] J. Brujic, S. F. Edwards, D. V. Grinev, I. Hopkinson, D. Brujic, and H. A. Makse, "3d Bulk Measurements of the Force Distribution in a Compressed Emulsion System," *Faraday Discuss.* **123**, 207 (2003).

[157] J. Brujic, S. F. Edwards, I. Hopkinson, and H. A. Makse, "Measuring the Distribution of Interdroplet Forces in a Compressed Emulsion System," *Physica A* **327**, 201 (2003).

[158] J. Brujic, C. M. Song, P. Wang, C. Briscoe, G. Marty, and H. A. Makse, "Measuring the Coordination Number and Entropy of a 3d Jammed Emulsion Packing by Confocal Microscopy," *Phys. Rev. Lett.* **98**, 248001 (2007).

[159] J. Zhou, S. Long, Q. Wang, and A. D. Dinsmore, "Correction to Zhou *et al.*," *Science* **314**, 254 (2006).

[160] M. Clusel, E. I. Corwin, A. O. N. Siemens, and J. Brujic, "A 'Granocentric' Model for Random Packing of Jammed Emulsions," *Nature* **460**, 611 (2009).

[161] H. M. Jaeger, S. R. Nagel, and R. P. Behringer, "Granular Solids, Liquids, and Gases," *Rev. Mod. Phys.* **68**, 1259 (1996).

[162] C. H. Liu, S. R. Nagel, D. A. Schecter, S. N. Coppersmith, S. Majumdar, O. Narayan, and T. A. Witten, "Force Fluctuations in Bead Packs," *Science* **269**, 513 (1995).

[163] T. S. Majmudar and R. P. Behringer, "Contact Force Measurements and Stress-Induced Anisotropy in Granular Materials," *Nature* **435**, 1079 (2005).

[164] J. F. Geng, D. Howell, E. Longhi, R. P. Behringer, G. Reydellet, L. Vanel, E. Clement, and S. Luding, "Footprints in Sand: The Response of a Granular Material to Local Perturbations," *Phys. Rev. Lett.* **87**, 035506 (2001).

[165] A. J. Liu and S. R. Nagel, "Nonlinear Dynamics – Jamming Is Not Just Cool Any More," *Nature* **396**, 21 (1998).

4

Langmuir monolayers

MICHAEL DENNIN

4.1 Introduction

Langmuir monolayers have proven themselves to be a powerful experimental system for the study of a range of issues in soft condensed matter [1–4]. Essentially, Langmuir monolayers are a single layer of insoluble molecules at the air–water interface. As such, they form an almost ideal two-dimensional system. This offers the opportunity to study fundamental questions in the phase behavior and material properties of two-dimensional systems. Secondly, they are readily transferred from the surface to a solid substrate. This has been the motivation for studying these systems for a range of technological applications. Finally, they are a natural system for the study of biological questions given that much of biology relies on processes at interfaces, such as cellular membranes and membranes of cellular components.

The experimental techniques associated with Langmuir monolayers can loosely be divided into two classes: formation of the monolayer and characterization of the monolayer. In this chapter, we will briefly review important issues in the formation of Langmuir monolayers. The focus will be mostly on the general concepts and issues, with some specific recipes given for illustrative purposes. However, it should be recognized that many groups specializing in Langmuir monolayers have refined and developed specific methods that are optimized for their system. So, some care needs to be taken in applying any given specific recipe. In the area of characterization, we will briefly review standard surface pressure characterization and provide a survey of commercial tools, including x-ray and neutron scattering techniques. However, the bulk of the chapter will focus on optical and mechanical characterization of Langmuir monolayers. This choice reflects the relative ease with which custom optical and mechanical systems can be constructed for relatively low cost as well as the corresponding usefulness of such systems.

Experimental and Computational Techniques in Soft Condensed Matter Physics, ed. Jeffrey Olafsen.
Published by Cambridge University Press.

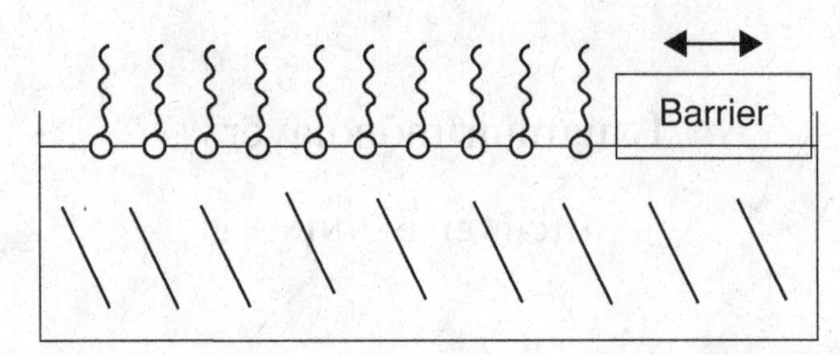

Figure 4.1 Schematic representation of a "classic" Langmuir monolayer on a water subphase. The molecules at the air–water interface are represented by the circular headgroups that are hydrophilic and the carbon chains that are hydrophobic. The rectangle labeled "barrier" represents a movable element (as indicated by the arrows) of the apparatus that is used to compress and expand the monolayer at the water surface. Typically, the barrier is a Teflon bar that is pressed onto the edges of the trough so it seals the surface, but it is still free to move.

4.2 Monolayer basics

There are two main classes of monolayers: soluble and insoluble. By raising the issue of solubility, we need to briefly comment on the issue of the *subphase*. Subphase refers to the fluid that the monolayer sits on top of. Though we generally will be considering monolayers confined to an air–water interface, there is nothing particularly special about using water. In fact, many important classes of experiments, especially those of a biological nature, require that an appropriate buffer solution or other fluid substrate is used (e.g. [5]). So, even when we refer to water, keep in mind that everything we discuss applies to a general fluid–air interface. The situation is illustrated schematically in Figure 4.1, where the molecules are shown as having a circular "headgroup" and a long chain. This represents the generic molecular monolayer. However, both molecular [1,2] and nanoparticle [6] systems are actively studied using similar techniques.

Monolayers composed of soluble particles are generally referred to as Gibbs monolayers [7]. A key feature of these monolayers is that the concentration of particles on the surface is set by the equilibrium between particles in a solution and on the surface. These monolayers will not be explicitly discussed in this chapter, though many of the same techniques can be used with them [7]. We will focus on monolayers of insoluble particles, generally referred to as Langmuir monolayers [1,2].

A key feature of Langmuir monolayers is the fact that the concentration of particles is determined by the number of particles on the air–water interface and the area of the interface. We are specifically using the word particles because Langmuir monolayers can be molecular in composition or composed of nanoparticles. Much of the early work has focused on simple long chain hydrocarbons, such as fatty

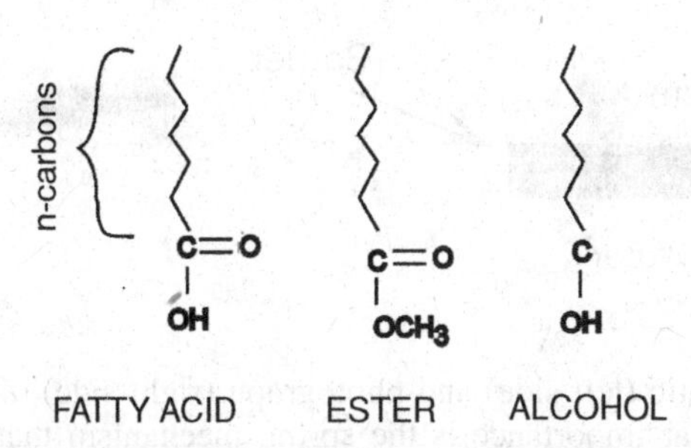

Figure 4.2 Schematic representation of three classic systems used as monolayers. The difference between the molecules is in the headgroup, for which the chemical composition is shown explicitly. The carbon chain in each case is indicated by the straight lines. Typical chain lengths for insoluble monolayers are in the range from 16 to 24 carbons.

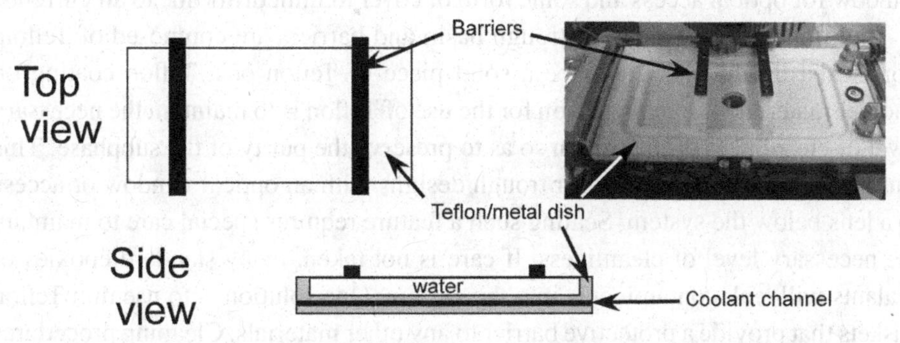

Figure 4.3 Schematic representation of the top a Langmuir trough (left side) and corresponding photograph (right side). The top view shows two barriers that can be moved laterally to change the area/molecule of the system. Additionally, there is a schematic side view that shows the basin for holding the subphase and an optional extra channel used for temperature control.

acids, esters, and alcohols, examples of which are given in Figure 4.2 [2]. Also, phospholipids have been heavily studied because of their prevalence in biological membranes [4]. More recently, there has been a growing interest in monolayers of nanoparticles, largely because of the technological motivations.

4.2.1 Langmuir trough

The main element of any Langmuir monolayer experiment is the *trough.* A schematic representation and photograph of a trough are shown in Figure 4.3. The photograph is of a custom-made trough, but many commercial troughs are available. (Two main producers of commercial troughs are Nima Technology and KSV Instruments Ltd.) The trough is the container for the water, or subphase, and the basic design is rather standard. The trough is composed of a basin that holds

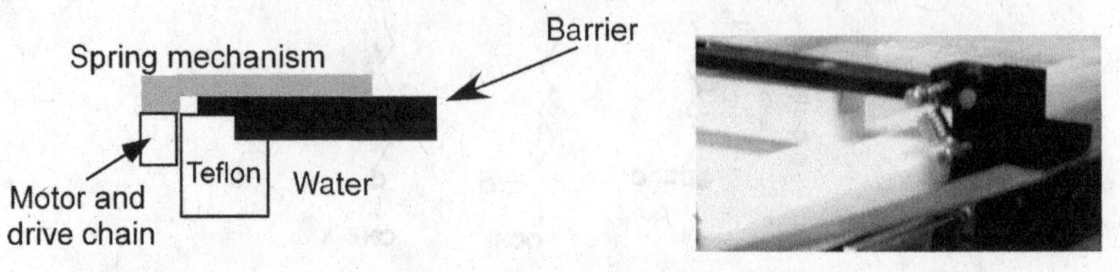

Figure 4.4 Schematic (left side) and photograph (right side) of a typical barrier design. Of particular importance is the spring mechanism that helps provide a good seal between the barrier and the trough.

the subphase, a barrier or barriers for compressing the monolayer, and a method for controlling the temperature of the subphase. Additionally, the trough may involve a window for optical access and some form of cover to limit drifts due to air currents.

Traditionally, the Langmuir trough basin and barriers are composed of Teflon. For the basin, this can either be a solid piece of Teflon or a Teflon coating on another material. The main reason for the use of Teflon is to maintain the necessary level of cleanliness of the trough so as to preserve the purity of the subphase. This can be a particular challenge for trough designs with an optical window or access to a lens below the system. Sealing such a feature requires special care to maintain the necessary level of cleanliness. If care is not taken, many standard epoxies or sealants will leak contaminants into the system. One solution is to use thin Teflon gaskets that provide a protective barrier to any other materials. Cleaning procedures are sufficiently important that they will be given their own section.

Another consideration in the design of the experiment is the nature of the barriers. At least one movable barrier is critical for a well-controlled monolayer experiment because it allows for systematic changes in the surface concentration of the monolayer. There are three basic choices: single barriers, double barriers, and centrosymmetric barriers. Most troughs have the basic rectangular design illustrated in Figure 4.3. In this case, there is either a single barrier for compressing the monolayer or two barriers. Generally, two barriers are used to compress the monolayer symmetrically from both sides, so the material in the center is not moving. Figure 4.4 provides an illustration of the two elements of the barrier design that lead to a good seal between the barrier and the trough basin: a step structure to the barrier and a top-loaded force. The use of a step structure in the design allows for the barrier to extend below the water surface. The use of an additional force on the barrier helps provide a good seal. Often, this force is provided by a spring mechanism. Finally, to achieve motion of the barrier, it needs to be attached to a motor by a drive chain.

The geometry of rectangular troughs is such that shear stresses are induced in the monolayer during compression. Depending on the experiment, this may or may not

Figure 4.5 Image of a centrosymmetric trough. This is a version in which the outer barrier is composed of individual fingers [12].

be relevant. In contrast, centro-symmetric troughs have been designed to provide essentially uniform compression of the monolayer [8–11]. An image of such a trough is shown in Figure 4.5. There have been two basic designs for the barriers for such troughs: an elastic band supported by "fingers" [8], or Teflon pieces that form an iris, also supported on individual fingers [12]. It should be noted that some experiments have used a conically shaped basin and controlled the concentration by changing the subphase level [9]. This provides another method of obtaining symmetric compression/expansion of the monolayer without any moving parts.

One of the most common classes of experiments using Langmuir monolayers are phase studies. The range of phases (solid, liquid, liquid crystalline, gas, etc.) will be discussed in a later section. However, for most phase studies, a method of controlling the temperature of the subphase is critical. There are two basic designs: water-controlled and Peltier device-based methods. A Peltier device is solid-state-based mechanism that uses thermo-electric cooling/heating, and it can typically be attached to the bottom of the trough. Some troughs combine the two methods. For a water cooling systems, channels are made in the trough base and temperature-controlled water is flowed through the channels. Often, with such a design the Teflon basin is placed in a secondary metal (usually aluminum) base through which the water is flowed. This increases the heat transfer between the water and the trough. The only disadvantage of a water-controlled system is that one is limited to heating and cooling in the range for which water is still a liquid. For most experiments of biological relevance, this is not a severe limitation.

The most common alternative to water cooling is the use of Pelletier devices to actively cool or heat the trough. A Pelletier device works by converting a voltage to a temperature difference. Generally, such a system is used when one is interested in temperature ranges that are difficult to achieve with purely water-controlled systems. As a Pelletier device generally needs a good heat sink if it is used in a cooling mode, it is often combined with a water-based cooling system.

One of the challenges in achieving precision temperature control in Langmuir monolayer systems is the air–water interface. Because of this, the temperature of the water subphase is controlled indirectly through the temperature of the base, and the temperature of the air above the trough is generally not controlled. This makes it necessary to monitor the temperature of the suphase directly. An excellent option for this is Teflon-coated thermistors, which are available commercially. This both protects the thermistor from the water and meets the necessary cleanliness requirements.

A final step in temperature control is to add a cover to the system. This serves two purposes. First, it provides an additional insulating barrier that helps to maintain constant temperature of the subphase. And for situations in which the subphase is heated above room temperature, heating the cover can help maintain a constant temperature. Also, the cover helps to minimize air flows. The morphology of low-density phases are particularly sensitive to air-currents, so, when making optical measurements, a cover is extremely useful. There are a number of challenges associated with covering the monolayer. The first has to do with the relation between the cover and the barriers. Generally, the cover has to be very close to the air–water interface if a microscope objective is going to be able to focus on the system. This precludes placing the cover above the barrier. Therefore, generally systems with a cover require a single barrier. For systems in which the optical access is from below, one has more options on the cover design.

4.2.2 Cleaning procedures

As mentioned, an important feature of any Langmuir monolayer experiment is the cleaning protocols for the trough. It is beyond the scope of this book to cover all possible detailed cleaning procedures, in part because this can depend on the materials being studied. For instance, many proteins are notoriously "sticky" and require a more rigorous cleaning protocol. But, also over the years, many groups have developed there own variations of cleaning, which generally work equally well. However, we will give some general principles and a few specific examples. The basic issue is to recognize the types of contaminants you wish to avoid, and design a cleaning procedure to eliminate those. Two common problems are surfactant contamination and ionic contamination.

When cleaning the trough, one has to be aware of a number of other issues – avoiding scratching the trough and possible contamination by the cleaning agent itself. The first issue is important because so many troughs have a composite design in which the Teflon dish is contained in another material, or even simply a coating on another material. In this design, if the Teflon is scratched in the process of cleaning, its integrity can be compromised. This is especially important in situations where the Teflon is in a metal container, as the metal can provide ionic contaminates to the subphase. Therefore, any scrubbing of the trough should involve the use of Teflon forceps to hold soft tissue paper.

To avoid contamination by the solvents, it is a good idea to use a cycle of solvents. The general principle is to use an excellent polar solvent to remove any surfactant contaminants followed by a solvent for the first cleaning agent that is also soluble in water. The final stage is to clean with pure water. This principle is highlighted by a common combination: chloroform followed by ethanol followed by pure water. Chloroform is a powerful solvent for organics, but relatively insoluble in water. Ethanol provides a bridge between the chloroform and water, as it is soluble with both. Therefore, a cleaning cycle of chloroform, ethanol, and pure water provides a mechanism for removing organics and ending with pure water.

The above suggestion does not generally deal with ionic contaminates. Often, one uses lipids with various charged head groups (fatty acids, for example). In this case, the presence of ionic contaminants is a potential problem. Often, simply soaking the trough in a pH-2 acid solution and rinsing with pure water is sufficient to clean the trough. This technique also serves as a good test for any "metallic leaks." If the trough is Teflon within a metal dish, any holes in the Teflon will reveal themselves in an acid bath. Such holes, even if extremely small, can be a source of future ionic contaminants. Occasionally, if the ionic contaminant is a specific ion, a chelating agent is required. Chelating agents are chemicals that have a high affinity for specific ionic species.

A final stage of cleaning usually involves the monolayer itself. In general, even with the best cleaning procedures, a small amount of contaminant will be present in the water if the trough is new or has not been used recently. This will show up when one compresses the barriers on a pure water surface. If a change in the surface pressure is measured (measuring surface pressure is discussed in the next section), there is material at the surface. This can be "vacuumed" off with an aspirator. An aspirator is basically a needle on the end of a hose attached to a vacuum system. To protect the vaccum system, the hose is generally connected through a flask for collection of water. This is illustrated in Figure 4.6.

The basic vacuuming procedure is as follows. The barriers are compressed to minimize the area between them. At this point, one carefully removes a small amount of the water by passing the aspirator needle back and forth across the surface. The system is expanded and compressed again. This process is repeated until no measurable change in surface pressure is observed upon compression. If

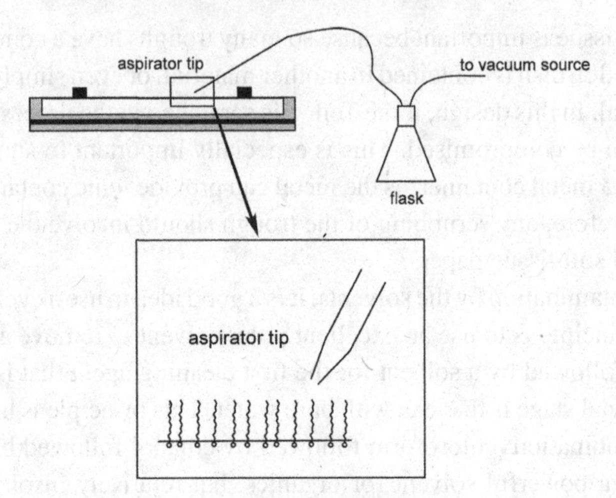

Figure 4.6 Schematic of a system for aspirating the air–water interface with a close up of the aspirator tip near the monolayer surface.

necessary, a small amount of pure water is added to replace the volume of water removed by the vacuuming process. As with any interaction with the system, the needle itself must be clean. Two main styles of needles are used: Teflon and glass. A glass aspirator is relatively straightforward to make using a standard disposable glass pipet. It is useful to heat the edge of the pipet to smooth it. This helps avoid scratching the Teflon in the trough if contact is accidentally made. In either the glass or Teflon case, the cleaning of the needle is similar to the cleaning of the trough.

A final comment on general lab cleaning is necessary. Though in many regards a clean room would be ideal, surprisingly it does not seem to be necessary for running careful Langmuir monolayer experiments. However, maintaining a generally clean lab space is extremely helpful. This includes regular mopping of floors and a general cleaning of the area around the trough. There also needs to be consideration of the ventilation system when placing the trough. Having the trough too close to an air-vent can cause problems.

4.2.3 Solution preparation

In order to make accurate measurements of monolayers, it is useful to know the amount of material placed on the surface. Therefore, solution preparation is an important step in any monolayer experiment. Generally, a known weight of material (and hence number of molecules) is added to a known volume of solvent. When a fixed volume of this solution is added to the water surface, it is known how many molecules have been added.

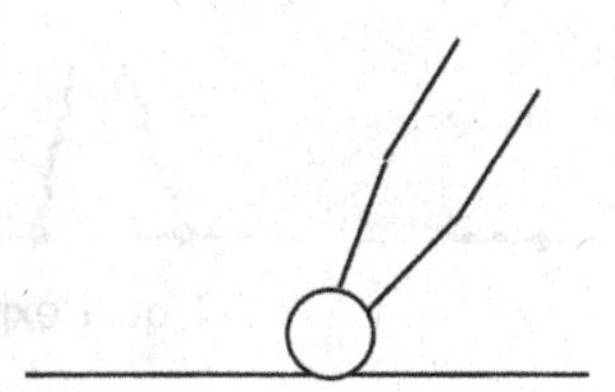

Figure 4.7 Schematic representation of a drop of solvent containing the monolayer molecules being placed on the water surface.

Arguably the most important step in the preparation of a monolayer solution is the selection of a solvent. The basic principle is to use a solvent that is insoluble in water and that evaporates rapidly. This allows for placing the monolayer directly on the surface of water with a high degree of confidence in the number of molecules that are present. In general, this makes chloroform an ideal choice, but often one needs to be prepared to use mixtures of chloroform and alcohols, or other solvents all together.

As with the trough, solution preparation requires great care to maintain a clean solution without additional contamination. The glassware involved should be cleaned with the same care as the trough; in this case, finishing the cleaning with the solvent that will be used to form the solution. One does need to be careful with any syringe/needle combination that is used for handling of the solution. These often contain epoxies that are dissolved by the solvents used to make the monolayer solution.

Finally, placing the monolayer on the water surface requires some practice. Ideally, one places the material a single drop at a time (see Figure 4.7). The drop should be touched to the water surface and it will visibly spread when this occurs. One places the drops in different positions around the trough, to minimize any accidental aggregation of the molecules or particles. Dripping the drops from above the surface runs the risk that the drops will go below the surface and not deposit the material on the surface.

4.3 Phase characterization

The characterization of phase transitions in monolayers relies on three main techniques: isotherms, optical studies, and x-ray scattering. In this section, we review the basic principles and techniques in measuring isotherms and optical studies. (However, we include a number of references on x-ray studies [1,13–18].) Before discussing the techniques, we will outline briefly the nomenclature and general characteristics of the main phases in Langmuir monolayers. We should refer to one of the excellent reviews of this topic for more details [1,2].

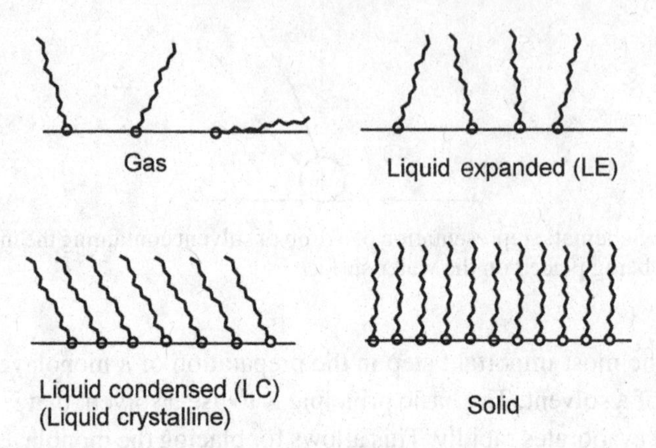

Figure 4.8 Schematic representation of the relative spacing of the headgroups and tilt of the molecular chains for the four main categories of Langmuir monolayer phases: gas, liquid expanded (LE), liquid condensed (LC), and solid.

For surfactant monolayers, there are two aspects to the observed phases: the orientation of the headgroups (or of the molecules in the air–water interface plane) and the orientation of the hydrocarbon chains. There are four basic phase categories: gas, liquid expanded (isotropic liquid), liquid condensed (liquid crystalline), and solid. These are shown schematically in Figure 4.8. The key element is the spacing of the molecules, or the area per molecule. At the lowest densities, the chains are completely free to move and the system is in a gas phase. At higher densities, the chains are still free to move, but remain essentially off the water surface on average. This is the liquid expanded (LE) phase, which is essentially an isotropic liquid. At higher densities, the molecules are still free to rotate, and maintain enough mobility so that they are still in an essentially liquid phase, but now the chains take on orientational order. They now possess a fixed angle (on average) relative to the vertical, and point in the same azimuthal direction. There are a wide number of such liquid condensed phases, depending on the nature of the headgroup orientation and the direction of alignment of the tilt. For example, the L_2 and L_2' phases are distinguished by a tilt toward the nearest or next-nearest neighbor, as illustrated in Figure 4.9. Finally, at sufficiently high densities, the molecules are all vertical and the system is in a solid phase. As with the LC phases, there exist a number of different solid phases.

In addition to the chain ordering, the headgroups can exhibit different degrees of ordering, such as hexatic and herringbone [1, 2]. A common situation for LC phases and nanoparticle systems is for the headgroups to have some degree of hexatic ordering. This is shown schematically in Figure 4.10. A hexatic phase is characterized by local hexagonal ordering of the molecules with no long-range positional order of the molecules [1, 2]. However, there is long-range orientational

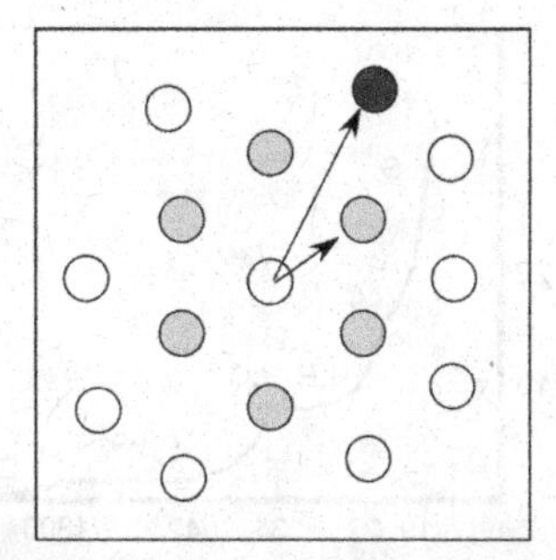

Figure 4.9 Schematic representation of two options for the direction of tilt alignment: nearest neighbor (the gray circles) and next nearest neighbor (one of which is highlighted in solid black).

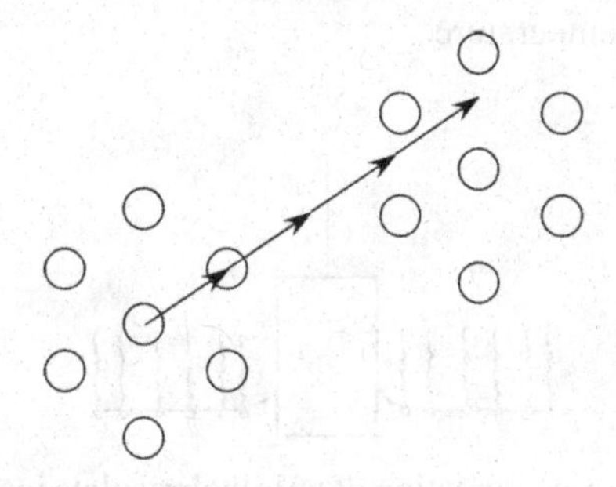

Figure 4.10 Schematic representation of ordering in a hexatic phase.

order to the hexagonal structure. This is indicated in Figure 4.10 by showing two local hexagonal structures that are orientated in the same direction, but the molecules are not a fixed lattice spacing apart (as indicated by the arrows). For naonparticle systems, there is generally no issue of the chain ordering, just the in-plane ordering of the molecules. For such systems, one is generally interested in gas, liquid, glassy, hexatic, and crystalline phases.

4.3.1 Isotherms

Isotherm measurements are arguably the most basic characterization of Langmuir monolayers, and should be part of any study. Because monolayers are essentially two-dimensional systems, the basic thermodynamic variables are the area per molecule (two-dimensional density), temperature, and the surface pressure (Π). The surface pressure is directly related to surface tension, and is defined as the difference between the surface tension of the pure interfaces (γ_w) and the surface tension with the surfactant present (γ): $\Pi = \gamma_w - \gamma$. Recall that surface tension is the force per length acting on a line through the surface. Another way to think of

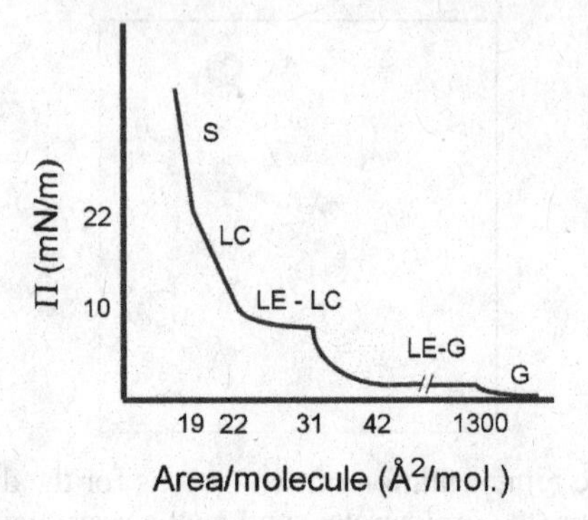

Figure 4.11 Schematic representation of an isotherm for which a number of transitions occur. In this case, a single LC phase is assumed to exist. In many systems, multiple LC phases will exist. The area/molecule scale is typical of single chain fatty acids near room temperature.

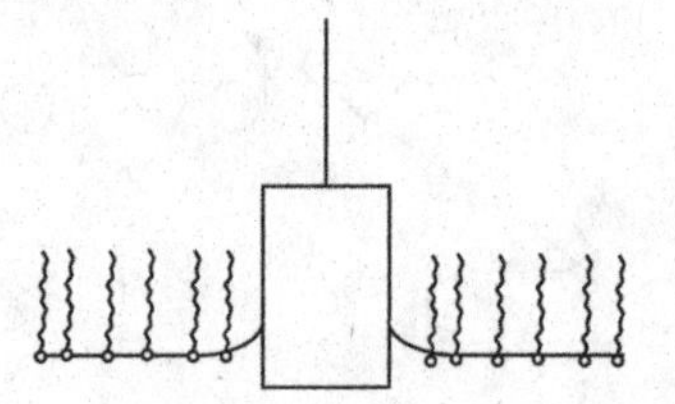

Figure 4.12 Schematic representation of a Wilhelmy plate inserted into the surface of the subphase.

surface tension is the energy per area to create a piece of surface. Therefore, surfactants always lower surface tension. Therefore, this definition of surface pressure has important properties. For the pure surface, the surface pressure is zero as expected. Because surfactants lower the surface tension, the surface pressure increases as the density of the molecules on the surface increases. A very schematic isotherm is shown in Figure 4.11.

Because the surface pressure is defined in terms of the surface tension, there are a number of straightforward methods for measuring it. These all use the same basic principle. A probe is inserted in the surface and measures the force per length on the probe. Perhaps the most common technique for Langmuir monolayers is the Wilhelmy plate. In this case, a thin piece of paper (often filter paper) or a thin metal plate of a known width is touched to the surface (schematically shown in Figure 4.12). The total force on the plate is given by the surface tension force, the weight, and the buoyancy force. The buoyancy force is present because a certain fraction of the plate is submerged. In terms of the weight, for plates made from filter paper, some liquid is absorbed by the paper, changing its weight, so one must make sure this is accounted for. Surface tension itself is not a force. It is the product of the surface tension (which for pure water at room temperature is 72 mN/m) and the perimeter of the plate in contact with the water that is the magnitude of the force due to surface tension. Since one is interested in changes in surface tension, the weight and the buoyancy terms can by zeroed out ahead of time. This is accomplished by placing the plate in contact with the surface before adding any surfactant and waiting a sufficient time for the weight due to water absorption to equilibrate. After zeroing the system, one needs to be aware of evaporation of the subphase altering the relative level of the Wilhemy plate and the water. This is generally a very small effect and only relevant when highly accurate phase measurements are of interest.

The measurement of an isotherm combines the ability to measure surface pressure with the compression/expansion of the monolayer. As mentioned, because Langmuir monolayers are insoluble, they can be compressed with barriers. If this is done at constant temperature while monitoring the surface pressure, one obtains the surface pressure as a function of molecular density at constant temperature: an isotherm. Features in the isotherm are an indication of phase transitions. As with three-dimensional systems, a plateau represents a first-order phase transition with the coexistence of two phases. A kink (or change in slope) in the isotherm represents a second-order phase transition (again, see Figure 4.11 for a schematic representation of these features).

It is worth reviewing some common *nomenclature*. First, the surface pressure in the gas phase is typically indistinguishable from zero pressure. The area per molecule at which the pressure first rises above zero is often called the *lift-off point* and is a measure of the transition to the liquid phase. This point plays a critical role in calibrating a given system as it is a good measure of the amount of material actually on the surface. Also, if one performs cycles of isotherms, changes in the value of the lift-off point imply that there has been a loss of material upon compression either due to collapse or leakage around barriers. Finally, one generally reports isotherms in terms of the *area per molecule*. Occasionally, when only relative changes are of interest, one will report results in terms of the *area of the trough* if the number of molecules is not known with sufficient precision to be useful.

One of the challenges in monolayer experiments is that isotherms are very sensitive to small amounts of contaminants. This can be a problem if the goal is an accurate measurement of the phase behavior based on the isotherm. However, it can be an advantage when using standard systems. The quality of the isotherms can provide a direct indication of the purity of the sample. There are a number of useful resources that provide isotherms for standard systems that can be used to check the cleanliness of our samples and trough. Another challenge for isotherms is barrier leakage. Often, a particular set of barriers works well for relatively low surface pressures. However, upon compression of the monolayer, material leaks around the barriers. A clear signature of this is significant changes in the area per molecule at which transitions are measured during subsequent compressions, especially the lift-off pressure.

An important parameter for any measurement of an isotherm is the rate of compression/expansion. There are a number of aspects of the kinetic response of the monolayer as it relates to the compression speed. Clearly, if the compression speed is too fast, especially through a phase transition region, one can quench the system into a quasi-static state that does not represent the true phase of the system. One would expect that going at the slowest possible speed is always ideal. However, monolayers also possess an equilibrium spreading pressure. This is the surface pressure at which the monolayer coexists with the three-dimensional phase of the material. Typically, this value is at or below surface pressures of interest to any given study. Therefore, many of the high-pressure phases are, strictly speaking, only metastable phases. Generally, the nucleation barrier to forming the three-dimensional phase is sufficiently high that, for all practical purposes, the high-pressure phases are stable. But, one needs to be aware of the possibility of the nucleation of the three-dimensional phase even at relatively low pressures if one is compressing the barriers over sufficiently long time scales.

4.3.2 Optical measures

Some of the most interesting issues in monolayer experiments involve the coexistence of two different phases and/or ordering of the molecules [3, 19–37]. Because of this, optical experiments have proven very useful. We will briefly review two standard techniques: Brewster angle microscopy [23, 38] and fluorescence microscopy [29, 37, 39]. For both of these methods, commercial systems are readily available, and the techniques are well developed. However, one can also construct a custom apparatus, depending on the needs of our particular application.

The most straightforward optical technique is fluorescence microscopy. Here a small amount of fluorescent probe is added to the monolayer (typically a few molar percent). This technique is most useful in situations where one has coexisting phases. In general, there are three distinct regimes probed by fluorescence microscopy. At the lowest densities (gas phase), the molecular probe is free to flop around on the surface, and this results in substantial quenching of the fluorescence. So, the gas (G) phase is usually dark. At the higher densities of the isotropic liquid phase, the probe molecules are supported in a relatively upright position, and one has the maximum fluorescence. Therefore, the liquid expanded (LE) phase is bright. Finally, liquid condensed (LC) phases (ones in which the molecules are closely packed and achieve orientational order) generally expel the probe, as it cannot fit in the packed geometry. Therefore, these phases are again dark. Therefore, straightforward fluorescence is best for observing either G–L coexistence or LE–LC coexistence. At the final stages of collapse, the fluorescence of the bulk phase can become substantial again, so fluorescence can be useful in collapse studies.

For studies of the liquid condensed phases, where the orientation of the molecular chains matters, a variation on simple fluorescence is possible [32, 33]. In this case, a significantly smaller amount of probe is used (usually 0.1–0.01% probe). This allows a reasonable fraction of the probe to remain in the condensed phase. The orientation of the surrounding molecule orients the fluorescent probe so that the fluorescent signal is proportional to the polarization of the incoming light. This allows for direct measurement of textures due to the alignment of the tilt-azimuth orientation. However, the low probe concentrations require a powerful light source, and one generally needs a laser for illumination of the monolayer. It should be noted that this technique played a crucial role in connecting Brewster Angle Microscope studies (with no probe) and fluorescent studies, establishing the fact that fluorescent probes have a minimal impact on the phase behavior.

There are two main issues with fluorescence microscopy: impact of the fluorescent impurity and bleaching of the probe. The first can be minimized by reducing the concentration of the probe. This generally requires using more powerful light sources, and/or intensified CCD cameras for detection. An intensified CCD camera has a separate element that enhances the signal before it strikes the CCD element in the camera. Also, if one is more interested in the morphology of the phases than precise measurements of transition points, the impurities have minimal effect. This has been confirmed with the alternate optical technique of Brewster angle microscopy [23, 32, 38].

Bleaching of the probe refers to the fact that the fluorescence often decays in time as the probe is exposed to light. As this generally results in the degradation of the image, it is often not desirable. Probes which last longer are generally preferred. However, bleaching can be turned into a positive feature for certain questions, especially those involving diffusion.

The basic structure of a fluorescent microscope is shown schematically in Figure 4.13. The main components are the light source (usually a mercury lamp or a laser), a beam splitter, and additional filters. The key physics is that the probe is excited at one wavelength and emits light at a different wavelength. To increase the signal, one uses both a filter and a beam splitter that serves as an additional filter by reflecting the excitation wavelength and passing the fluorescent signal. In a commercial system, these components are all integrated and the filter/beam splitter combinations generally come as integrated cubes that can be easily switched in and out. For custom applications, it is relatively straightforward to put together the

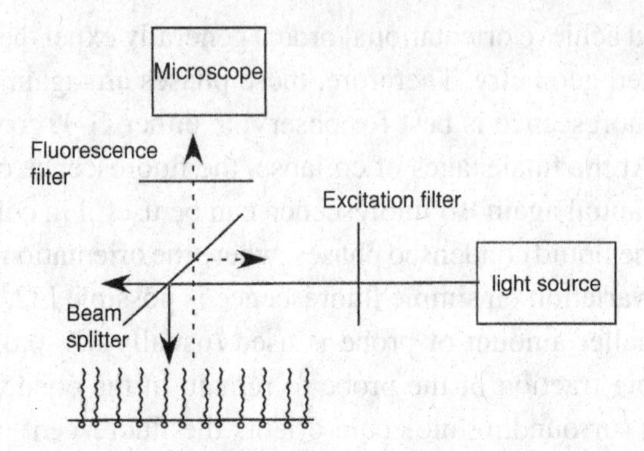

Figure 4.13 Schematic of a fluorescence microscope setup. The light source is filtered to the excitation wavelength of the probe. A beam splitter is used that also reflects in the general range of the excitation wavelength and passes light in the range of the fluorescence. The fluorescent signal is additionally filtered to improve image quality. In the case of polarized fluorescence, an additional polarizer is added to the system.

necessary components. A key element of any fluorescent microscope application is a long working distant objective lens. This is often the most expensive component of the system, but it is critical to maintain the necessary distance between the objective and the water surface, especially if a cover is needed as part of the temperature control.

The Brewster angle microscope (BAM) takes advantage of the minimum in reflection of p-polarized light at the Brewster angle for the air–water interface (or any interface of interest). The presence of the monolayer changes the index of refraction of the interface, which changes the Brewster angle. Therefore, when one sends in p-polarized light at the Brewster angle for the pure water surface, there is no reflection. Addition of the monolayer changes the reflectivity and produces a signal. If there is spatial variation in the properties of the monolayer, those will be detected as variations in the intensity in reflected light. Both variations in the tilt-azimuth and the density of the molecules produce variations in the intensity of the reflected light. The end result is images in which the intensity variations reflect domains of different density or alignment of the tilt-azimuth. An additional polarizer located in the reflected beam can be used as an analyzer. Rotation of the analyzer provides information about the variations of the tilt-azimuth. A BAM has proven to be a powerful tool for studying the textures and morphology of tilted LC domains in monolayers [23, 32, 37, 38, 40].

One can purchase commercial BAMs for a variety of applications. However, if necessary, custom BAMs can be constructed that vary in price. A number of review articles address the construction of a BAM, but it is worth highlighting a few key issues. First, the most common light source is a laser. Because of the coherence of a

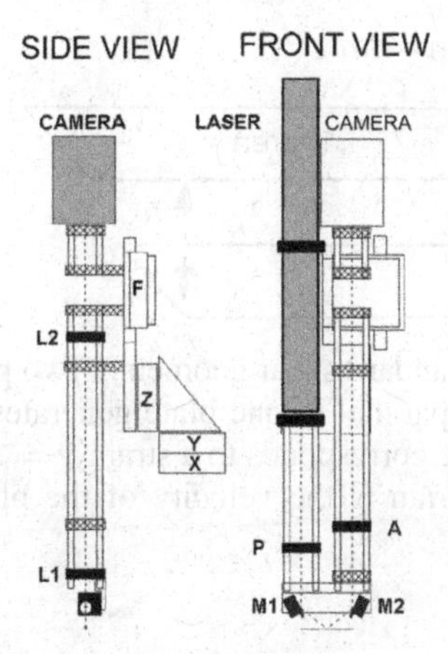

Figure 4.14 A schematic of a custom Brewster angle microscope. Shown in the figure are mirrors (M1 and M2) for reflecting the light at the Brewster angle, lenses (L1 and L2) for forming the image, and polarizers (P and A) for polarizing the light and using as an analyzer.

laser beam, one can end up with reduced image quality due to interference fringes from various optical elements in the BAM. Space requirement will determine the design of the optical path. A system with parallel optical paths that uses mirrors to achieve the Brewster angle is illustrated in Figure 4.14 [40]. Camera sensitivity will dictate the power of laser that is required for reasonable image quality. Finally, an important feature of BAM images is distortion of the image due to the angle of incidence of the light. This is discussed in detail in the original papers on the subject [23, 38].

4.4 Mechanical properties

From the perspective of soft-condensed matter, one of the most interesting questions is the mechanical response of a system. Soft systems are generally special because they exhibit both solid-like and fluid-like behavior. Langmuir monolayers are no different, and the study of flow in monolayers has a long history [9–12, 24, 27, 41–72]. The basic goal of any flow study is the ability to probe both the viscous and elastic properties of the monolayer. As this area is extremely rich, the discussion will be very brief and the reader should refer to the extensive references. The techniques can be loosely divided into macroscopic rheometers and microrheology, or local, techniques. We will provide a brief list of techniques in each of these categories, but the main goal of this section will be to provide a sufficient introduction that the references are useful. To aid in the introduction, we will present two common experimental methods in more detail: Couette

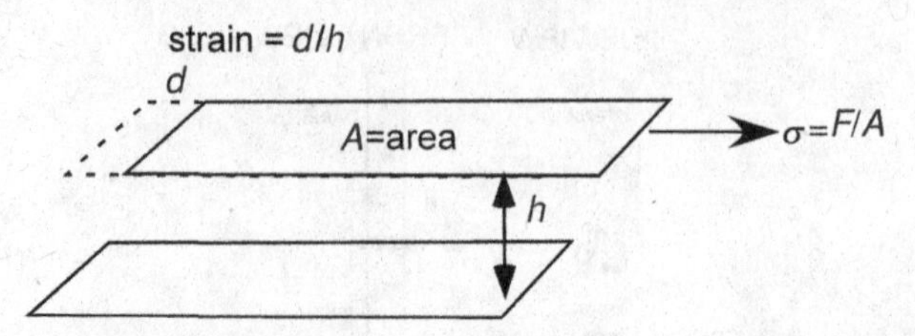

Figure 4.15 Sketch of a standard shear geometry. Two plates of area A confine a material. A force applied parallel to one plate generates a stress $\sigma = F/A$. The displacement d of the plate corresponds to a strain $\gamma = d/h$, where h is the plate separation. The rate of strain is the velocity of the plate divided by the plate separation: $\dot{\gamma} = v/h$.

viscometer [10] and magnetic needle rheometers [44] (for which a commercial setup is available).

Most macroscopic rheometers that have been developed to measure rheological properties of Langmuir monolayers are modeled after standard three-dimensional rheometers in some fashion. Among the more common techniques are various forms of channel viscometers [71,73] and knife-edge torsion pendulum [46,49,74]. Another technique that has been used is light-scattering [52,75]. A powerful method for which a commercial system is available based on the motion of a sphere or a needle floating on the surface [44]. Finally, there exist techniques based on a standard Couette viscometer [10].

The first use of local techniques in Langmuir monolayers was based on relations between translational and rotational mobilities [70,76]. Experiments have included measuring the electric field driven motion of liquid crystalline domains [54], measuring the diffusion of liquid crystalline domains [61], and using liquid crystalline domains as particles for measuring velocity profiles [73]. More recently, there has been work that combines the rotational and translation motion of liquid crystalline domains in the liquid expanded phase [77], and, in the liquid crystalline phase region, Brewster angle microscopy has been used to study the motion of the domains [24,27,56–60,63,68].

As most phases of Langmuir monolayers are viscous or viscoelastic, the general goal is to measure a steady state viscosity or a frequency dependent shear modulus. Figure 4.15 illustrates the general geometry for shear measurements. Recall that for a fluid, the shear stress is proportional to the rate of strain (gradient in the velocity), and the proportionality constant is the viscosity. For a solid, the shear stress (where stress is force per area) is proportional to the shear strain (dimensionless displacement). For a material with both a solid-like and a fluid-like response, one considers the combined response: $\sigma(\omega) = G'\gamma(\omega) + \eta\dot{\gamma}(\omega)$, where $\sigma(\omega)$ is the frequency dependent shear stress, $\gamma(\omega)$ is the shear strain, and $\dot{\gamma}(\omega)$ is the rate of strain. Here

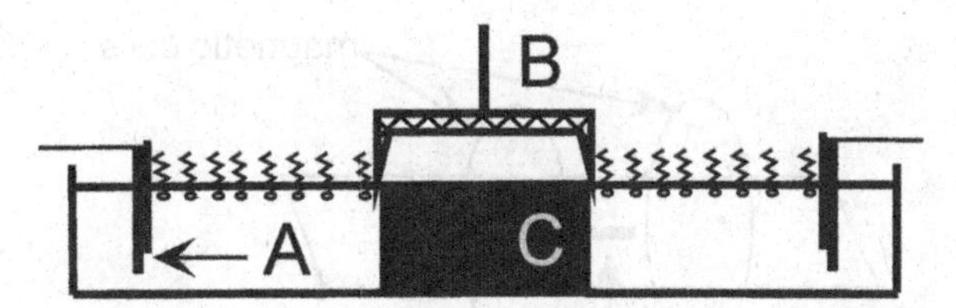

Figure 4.16 Schematic of the Couette trough. Shown here are two of the "fingers" (A) that support the elastic band, the fixed inner cylinder (C), and the knife-edge rotor (B).

G' is the elastic shear modulus and η is the shear viscosity. For the special case of harmonic strain, $\gamma(\omega) = Ae^{i\omega t} \Rightarrow \dot{\gamma}(\omega) = i\omega Ae^{i\omega t} = i\omega\gamma(\omega)$, one discusses the properties of the system in terms of the complex shear modulus $G^* = G' + iG''$, and $\sigma(\omega) = G^*\gamma(\omega)$. Therefore, the two basic experiments are to move a boundary (or probe) with a steady velocity and measure the ratio of stress and rate of strain to determine steady state viscosity (η) or to oscillate a boundary of the sample (or probe placed in the sample) with a fixed force (or displacement) and the resulting displacement (or force) is measured. In this case, the combination of the ratio of the force and displacement and the phase shift between the two determines the complex shear modulus.

Figure 4.16 illustrates a classic geometry for measuring material properties in three-dimensional materials as it can be adapted to Langmuir monolayers: a Couette viscometer [9, 10, 51]. The main idea is to confine the material between two concentric cylinders and either apply a constant rate of strain by rotating the outer cylinder and measuring stress on the inner cylinder (steady state viscosity measurement), or to oscillate the inner cylinder (complex shear modulus measurement). The challenge in adapting this method to Langmuir monolayers is to deal with the connection to the water subphase. There are two elements to this. First, the trough itself has three main components: (A) an outer barrier; (B) and "inner cylinder," which consists of a disk with a knife edge that is just in contact with the monolayer and supported by a torsion wire; and (C) a fixed inner cylinder in the water subphase. The fixed cylinder is critical when the system is used to generate flow in the monolayer, as it ensures the water and the monolayer have essentially the same velocity profile. This minimizes dissipation between the monolayer and the water [10]. For measurements of the complex shear modulus, it is important to account for flow in the water subphase. The details of this are given in reference [9], and are discussed in reference [51].

An alternative method for measuring the complex shear modulus uses a small magnetic needle floating on the water surface. This is illustrated schematically in Figure 4.17, and a detailed discussion of this method is given in reference [44]. Here the basic idea is to either move the needle at its terminal velocity in a constant

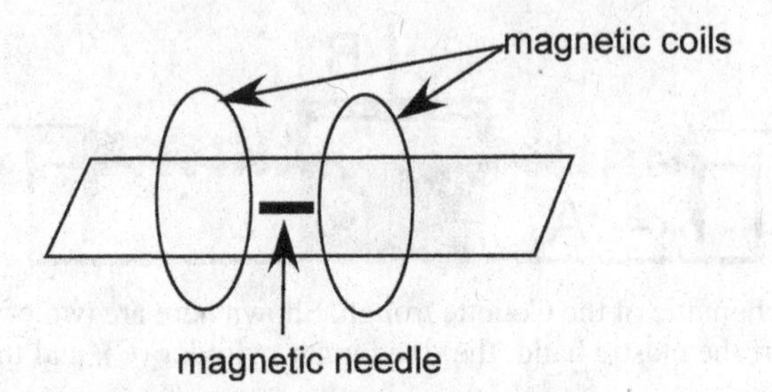

Figure 4.17 A schematic representation of a magnetic needle viscometer. The two key elements are coils to provide the magnetic field and a small magnetic needle floating on the surface of the water.

magnetic field (viscosity measurement) or to oscillate the needle in a magnetic field. Again, the main issue is the need to understand the interaction of the needle with the pure fluid. However, appropriate calibration of the system allows for determination of both the complex shear modulus and the steady state viscosity.

4.5 Summary

This chapter reviews the most basic of techniques for studying monolayers. The perspective has been to provide a starting point for any study that allows for the preparation of the monolayer and the measurement of its most basic characteristics: phase behavior and macroscopic mechanical properties. It should be noted that a range of techniques exist that go well beyond the basic methods reviewed here. These include the already mentioned x-ray studies, but also surface potential measurements, spectroscopic techniques, advance optical techniques, and additional rheological methods. Many of these techniques are best implemented with commercial equipment, details of which are made available by the companies.

References

[1] V. M. Kaganer, H. Mohwald, and P. Dutta, "Structure and phase transitions in Langmuir monolayers," *Rev. Mod. Phys.* **71**, 779 (1999).
[2] C. M. Knobler and R. C. Desai, "Phase transitions in monolayers," *Annual Review of Physical Chemistry* **43**, 207 (1992).
[3] H. M. McConnell, "Structures and transitions in lipid monolayers at the air–water interface," *Annual Review of Physical Chemistry* **42**, 171 (1991).
[4] H. Mohwald, "Phospholipid and phospholipid-protein monolayers at the air/water interface," *Annual Review of Physical Chemistry* **41**, 441 (1990).
[5] A. Rosengarth, A. Wintergalen, H. J. Galla, H. J. Hinz, and V. Gerke, "Ca2+−independent interaction of annexin I with phospholipid monolayers," *Febs Letters* **438**, 279 (1998).

[6] J. H. Fendler, "Nanoparticles at air/water interfaces," *Current Opinion in Colloid and Interface Science* **1**, 202 (1996).
[7] V. Melzer, D. Vollhardt, G. Brezesinski, and H. Mohwald, "Similarities in the phase properties of Gibbs and Langmuir monolayers," *The Journal of Physical Chemistry B* **102**, 591 (1998).
[8] T. M. Bohanon, J. M. Mikrut, B. M. Abraham, J. B. Ketterson, S. Jacobson, L. S. Flosenzier, J. M. Torkelson, and P. Dutta, "Apparatus with an elastic barrier for radial compression of liquid supported monolayers," *Review of Scientific Instruments* **63**, 1822 (1992).
[9] B. M. Abraham, K. Miyano, S. Q. Xu, and J. B. Ketterson, "Centro-symmetric technique for measuring shear modulus, viscosity, and surface tension of spread monolayers," *Review of Scientific Instruments* **54**, 213 (1983).
[10] R. S. Ghaskadvi and M. Dennin, "A two-dimensional Couette viscometer for Langmuir monolayers," *Review of Scientific Instruments* **69**, 3568 (1998).
[11] R. S. Ghaskadvi, J. B. Ketterson, R. C. MacDonald, and P. Dutta, "Apparatus to measure the shear modulus of Langmuir monolayers as functions of strain amplitude and frequency," *Review of Scientific Instruments* **68**, 1792 (1997).
[12] R. S. Ghaskadvi, S. Carr, and M. Dennin, "Effect of subphase Ca++ ions on the viscoelastic properties of Langmuir monolayers," *Journal of Chemical Physics* **111**, 3675 (1999).
[13] J. Als-Nielsen, D. Jacquemain, K. Kjaer, F. Leveiller, M. Lahav, and L. Leiserowitz, "Principles and applications of grazing incidence X-ray and neutron scattering from ordered molecular monolayers at the air–water interface," *Physics Reports* **246**, 252 (1994).
[14] P. Dutta, J. B. Peng, B. Lin, J. B. Ketterson, M. Prakash, P. Georgopoulos, and S. Ehrlich, "X-ray diffraction studies of organic monolayers on the surface of water," *Physical Review Letters* **58**, 2228 (1987).
[15] K. Kjaer, J. Als-Nielsen, C. A. Helm, L. A. Laxhuber, and H. Möhwald, "Ordering in lipid monolayers studied by synchrotron X-ray diffraction and fluorescence microscopy," *Physical Review Letters* **58**, 2224 (1987).
[16] J. Kmetko, A. Datta, G. Evmenenko, M. K. Durbin, A. G. Richter, and P. Dutta, "Ordering in the subphase of a Langmuir monolayer: x-ray diffraction and anomalous scattering studies," *Langmuir* **17**, 4967 (2001).
[17] M. C. Shih, T. M. Bohanon, J. M. Mikrut, P. Zschack, and P. Dutta, "X-ray diffraction study of heneicosanol monolayers on the surface of water," *Journal of Chemical Physics* **97**, 4485 (1992).
[18] M. C. Shih, T. M. Bohanon, J. M. Mikrut, P. Zschack, and P. Dutta, "X-ray-diffraction study of the superliquid region of the phase diagram of a Langmuir monolayer," *Physical Review A* **45**, 5734 (1992).
[19] B. Fischer, E. Teer, and C. M. Knobler, "Optical Measurements of the phase diagram of Langmuir monolayers of fatty acid–alcohol Mixtures," *Journal of Chemical Physics* **103**, 2365 (1995).
[20] B. Fischer, M. W. Tsao, J. Ruizgarcia, T. M. Fischer, D. K. Schwartz, and C. M. Knobler, "Observation of a change from splay to bend orientation at a phase transition in a Langmuir monolayer," *Journal of Physical Chemistry* **98**, 7430 (1994).
[21] T. M. Fischer, R. F. Bruinsma, and C. M. Knobler, "Textures in surfactant monolayers," *Physical Review E* **50**, 413 (1994).
[22] J. Garnaes, D. K. Schwartz, R. Viswanathan, and J. A. N. Zasadzinski, "Domain boundaries and buckling superstructures in Langmuir–Blodgett films," *Nature* **357**, 54 (1992).

[23] D. Hönig and D. Möbius, "Direct visualization of monolayers at the air–water interface by Brewster angle microscopy," *Journal of Physical Chemistry* **95**, 4590 (1991).
[24] J. Ignes-Mullol and D. K. Schwartz, "Molecular orientation in Langmuir monolayers under shear," *Langmuir* **17**, 3017 (2001).
[25] M. L. Kurnaz and D. K. Schwartz, "Morphology of microphase separation in arachidic acid/cadmium arachidate Langmuir–Blodgett multilayers," *Journal of Physical Chemistry* **100**, 11113 (1996).
[26] B. Lin, M. C. Shih, T. M. Bohanon, G. E. Ice, and P. Dutta, "Phase diagram of a lipid monolayer on the surface of water," *Physical Review Letters* **65**, 191 (1990).
[27] T. Maruyama, G. Fuller, C. Frank, and C. Robertson, "Flow-induced molecular orientation of a Langmuir film," *Science* **274**, 233 (1996).
[28] K. Miyano, B. M. Abraham, J. B. Ketterson, and S. Q. Xu, "The phases of insoluble monolayer: comparison between the surface pressure-molecular are diagram and shear modulus measurements," *Journal Chemical Physics* **78**, 4776 (1983).
[29] B. G. Moore, C. M. Knobler, S. Akamatsu, and F. Rondelez, "Phase diagram of Langmuir monolayers of pentadecanoic acid: quantitative comparison of surface pressure and fluorescence microscopy results," *Journal of Physical Chemistry* **94**, 4588 (1990).
[30] G. A. Overbeck and D. Möbius, "A new phase in the generalized phase diagram of monolayer films of long-chain fatty acids," *Journal of Physical Chemistry* **97**, 7999 (1993).
[31] X. Qiu, J. Ruiz-Garcia, C. M. Knobler, K. J. Stine, and J. V. Selinger, "Direct observation of domain structure in condensed monolayer phases," *Physical Review Letters* **67**, 703 (1991).
[32] S. Riviere, S. Hénon, J. Meunier, D. K. Schwartz, M. W. Tsao, and C. M. Knobler, "Textures and phase transitions in Langmuir monolayers of fatty acids – a comparative Brewster angle microscope and polarized fluorescence microscope study," *Journal of Chemical Physics* **101**, 10045 (1994).
[33] D. K. Schwartz and C. M. Knobler, "Direct observations of transitions between condensed Langmuir monolayer phases by polarized fluorescence microscopy," *Journal of Physical Chemistry* **97**, 8849 (1993).
[34] D. K. Schwartz, M. W. Tsao, and C. M. Knobler, "Domain morphology in a two-dimensional anisotropic mesophase – cusps and boojum textures in a Langmuir monolayer," *Journal of Chemical Physics* **101**, 8258 (1994).
[35] M. C. Shih, T. M. Bohanon, and J. M. Mikrut, "Pressure and Ph dependence of the structure of a fatty acid monolayer with calcium ions in the subphase," *Journal of Chemical Physics* **96**, 1556 (1992).
[36] K. J. Stine, S. A. Rauseo, B. G. Moore, J. A. Wise, and C. M. Knobler, "Evolution of foam structures in Langmuir monolayers of pentadecanoic acid," *Phys. Rev. A* **41**, 6884 (1990).
[37] E. Teer, C. M. Knobler, C. Lautz, S. Wurlitzer, J. Kildae, and T. M. Fischer, "Optical measurements of the phase diagrams of Langmuir monolayers of fatty acid, ester, and alcohol mixtures by Brewster-angle microscopy," *Journal of Chemical Physics* **106**, 1913 (1997).
[38] S. Hénon and J. Meunier, "Microscope at the Brewster angle: direct observation of first-order phase transitions in monolayers," *Review of Scientific Instruments* **62**, 936 (1991).
[39] M. Lösche, E. Sackmann, and H. Möhwald, "A fluorescence microscopic study concerning the phase diagrams of phospholipids," *Ber. Bunsenges. Phys. Chem.* **87**, 848 (1983).

[40] G. Marshall, M. Dennin, and C. M. Knobler, "A compact Brewster-angle microscope for use in Langmuir–Blodgett deposition," *Review of Scientific Instruments* **69**, 3699 (1998).
[41] C. Barentin, P. Muller, C. Ybert, J. F. Joanny, and J. M. di Meglio, "Shear viscosity of polymer and surfactant monolayers," *European Physical Journal E* **2**, 153 (2000).
[42] C. Barentin, C. Ybert, J. M. Di Meglio, and J. F. Joanny, "Surface shear viscosity of Gibbs and Langmuir monolayers," *Journal of Fluid Mechanics* **397**, 331 (1999).
[43] H. Bercegol and J. Meunier, "Young's modulus of a solid two-dimensional Langmuir monolayer," *Nature* **356**, 226 (1992).
[44] C. F. Brooks, G. G. Fuller, C. W. Frank, and C. R. Robertson, "An interfacial stress rheometer to study rheological transitions in monolayers at the air–water interface," *Langmuir* **15**, 2450 (1999).
[45] M. R. Buhaenko, J. W. Goodwin, and R. M. Richardson, "Surface rheology of spread monolayers," *Thin Solid Films* **159**, 171 (1988).
[46] S. S. Feng, R. C. MacDonald, and B. M. Abraham, "An exact solution to the viscoelastic damping of a torsion pendulum by a surface film," *Langmuir* **7**, 572 (1992).
[47] M. B. Forstner, J. Kas, and D. Martin, "Single lipid diffusion in Langmuir monolayers," *Langmuir* **17**, 567 (2001).
[48] M. C. Friedenberg, G. G. Fuller, C. W. Frank, and C. R. Robertson, "Direct visualization of flow-induced anisotropy in a fatty acid monolayer," *Langmuir* **12**, 1594 (1996).
[49] R. S. Ghaskadvi, T. M. Bohanon, P. Dutta, and J. B. Ketterson, "Shear response of Langmuir monolayers of heneicosanoic (C-21) Acid studied using a torsion pendulum," *Physical Review E* **54**, 1770 (1996).
[50] R. S. Ghaskadvi and M. Dennin, "Alternate measurement of the viscosity peak in heneicosanoic acid monolayers," *Langmuir* **16**, 10553 (2000).
[51] R. S. Ghaskadvi, J. B. Ketterson, and P. Dutta, "Nonlinear shear response and anomalous pressure dependence of viscosity in a Langmuir monolayer," *Langmuir* **13**, 5137 (1997).
[52] S. Hård and H. Löfgren, "Elasticity and viscosity measurements of monomolecular propyl strearate films at air–water interfaces using laser light scattering techniques," *Journal Colloids and Interface Science* **60**, 529 (1976).
[53] W. D. Harkins and J. G. Kirkwood, "The viscosity of monolayers: theory of the surface slit viscometer," *Journal of Chemical Physics* **6**, 53 (1938).
[54] W. M. Heckl, A. Miller, and H. Möhwald, "Electric-field-induced domain movement in phospholipid monolayers," *Thin Solid Films* **158**, 125 (1988).
[55] A. H. Hirsa, J. M. Lopez, and R. Miraghaie, "Measurement and computation of hydrodynamic coupling at an air/water interface with an insoluble monolayer," *Journal of Fluid Mechanics* **443**, 271 (2001).
[56] J. Ignes-Mullol and D. K. Schwartz, "Alignment of hexatic Langmuir monolayers under shear," *Physical Review Letters* **85**, 1476 (2000).
[57] J. Ignes-Mullol and D. K. Schwartz, "Shear-induced molecular precession in a hexatic Langmuir monolayer," *Nature* **410**, 348 (2001).
[58] A. Ivanova, M. L. Kurnaz, and D. K. Schwartz, "Temperature and flow rate dependence of the velocity profile during channel flow of a Langmuir monolayer," *Langmuir* **15**, 4622 (1999).
[59] A. T. Ivanova, J. Ignes-Mullol, and D. K. Schwartz, "Microrheology of a sheared Langmuir monolayer: elastic recovery and interdomain slippage," *Langmuir* **17**, 3406 (2001).

[60] A. T. Ivanova and D. K. Schwartz, "Transient behavior of the velocity profile in channel flow of a Langmuir monolayer," *Langmuir* **16**, 9433 (2000).
[61] J. F. Klingler and H. M. McConnell, "Brownian motion and fluid mechanics of lipid monolayer domains," *Journal of Physical Chemistry* **97**, 6096 (1993).
[62] J. Krägel, S. Siegel, R. Miller, M. Born, and K.-H. Schano, "Measurement of interfacial shear rheological properties: an automated apparatus," *Colloids and Surfaces A* **91**, 169 (1994).
[63] M. L. Kurnaz and D. K. Schwartz, "Channel flow in a Langmuir monolayer: unusual velocity profiles in a liquid-crystalline mesophase," *Physical Review E* **56**, 3378 (1997).
[64] M. L. Kurnaz and D. K. Schwartz, "A technique for direct observation of particle motion under shear in a Langmuir monolayer," *Journal of Rheology* **41**, 1173 (1997).
[65] J. M. Lopez and A. H. Hirsa, "Surfactant-influenced gas–liquid interfaces: Nonlinear equation of state and finite surface viscosities," *Journal of Colloid and Interface Science* **229**, 575 (2000).
[66] J. M. Lopez and J. Chen, "Hydrodynamic coupling between a viscoelastic gas/liquid interface and a swirling vortex," *Journal Fluids Engineering* **120**, 655 (1998).
[67] J. M. Lopez and A. Hirsa, "Direct determination of the dependence of the surface shear and dilational viscosities on the thermodynamic state of the interface: theoretical foundations," *Journal of Colloid and Interface Science* **206**, 231 (1998).
[68] T. Maruyama, M. Friedenberg, G. G. Fuller, C. W. Frank, C. R. Robertson, A. Ferencz, and G. Wegner, "In-situ studies of flow-induced phenomena in Langmuir monolayers," *Thin Solid Films* **273**, 76 (1996).
[69] M. Sacchetti, H. Yu, and G. Zografi, "A canal surface viscometer for the in-plane steady shear viscosity of monolayers at the air–water interface," *Review of Scientific Instruments* **64**, 1941 (1993).
[70] P. G. Saffman and M. Delbruck, "Brownian motion in biological membranes," *Proc. Nat. Acad. Sci. USA* **72**; 3111 (1975).
[71] H. A. Stone, "Fluid motion of monomolecular films in a channel flow geometry," *Physics of Fluids* **7**, 2931 (1995).
[72] M. Twardos, M. Dennin, and G. Brezesinski, "Characterization of anomalous flow and phase behavior in a Langmuir monolayer of 2-hydroxy-tetracosanoic acid," *J. Phys. Chem. B* **44**, 110 (2006).
[73] D. K. Schwartz, C. M. Knobler, and R. Bruinsma, "Direct observation of Langmuir monolayer flow through a channel," *Physical Review Letters* **73**, 2841 (1994).
[74] R. J. Mannheimer and R. A. Burton, "A theoretical estimation of viscous-interaction effects with a torsional (knife-edge) surface viscometer," *Journal Colloid Interface Science* **32**, 73 (1970).
[75] D. Langevin, "Light scattering study of monolayer viscoelasticity," *Journal Colloid Interface Science* **80**, 412 (1981).
[76] B. D. Hughes, B. A. Pailthorpe, L. R. White, and W. H. Sawyer, "Extraction of membrane microviscosity from translational and rotational diffusion coefficients," *Biophys. J.* **37**, 673 (1982).
[77] P. Steffen, P. Heinig, S. Wurlitzer, Z. Khattari, and T. M. Fischer, "The translational and rotational drag on Langmuir monolayer domains," *Journal of Chemical Physics* **115**, 994 (2001).

5

Computer modeling of granular rheology

LEONARDO E. SILBERT

5.1 Introduction

Granular materials are ubiquitous throughout nature. From the beauty of sand dunes and the rings of Saturn to the destructive power of snow avalanches and mudslides, the flow of ice floes, and the manner in which plate tectonics determine much of the morphology of the Earth [1–4]. These phenomena arise from the interplay between structural and dynamical properties that result in the collective behavior of a vast number of smaller, distinct entities that we call "grains." From a technological point of view, granular materials play a dominant role in numerous industries, such as mining, agriculture, civil engineering, pharmaceuticals manufacturing, and ceramic component design. Even apparently the most mundane of activities from coffee bean bag filling at the grocery store to pouring salt onto our dinner plates at night involve various aspects of granular matter mechanics that continue to puzzle us. It is estimated that particulate media are second only to water as the most manipulated material for human usage [4], amounting to trillions of dollars per annum in the US alone. The importance of granular materials to our daily lives cannot be overstated.

On the face of it, the study of macroscopic, millimeter-sized sand grains might be perceived as the purview of various engineering disciplines [5]. How is it then that granular materials remain at the forefront of contemporary research projects and are of particular interest to physicists? The answer to this question lies in the fact that granular materials are inherently nonequilibrium systems displaying a wealth of fascinating complex phenomena [6–8]. Take even the simplest granular system: a static packing of grains such as a sandpile. The grains remain stuck indefinitely in their original configuration simply because thermal fluctuations at a (room) temperature T do not provide enough energy to allow the grains

Experimental and Computational Techniques in Soft Condensed Matter Physics, ed. Jeffrey Olafsen. Published by Cambridge University Press.

to explore their local environment: $k_B T/mgd \approx 10^{-12}$, for grains of typical size $d \approx 1$ mm and mass density $\rho \approx 2600$ kgm^{-3} [1], in the Earth's gravitational field $g = 9.8$ ms^{-2}, and $k_B = 1.38 \times 10^{-23}$ JK^{-1} is Boltzmann's constant. However, it has long been recognized that providing an external forcing to a granular material, through shearing or shaking, plays an analogous role to thermal temperature in liquids [9]. Thus, when a pile of grains is sufficiently agitated, the grains can be made to flow like a fluid. Hence, through controlled excitations, a grain pile can be made to mimic a liquid or brought to rest into a static or *jammed* state. Indeed, the study of packings of ball bearings [10] was motivated by the belief that the random arrangements of the ball bearings in a container structurally resemble the instantaneous configurations of the molecules in a liquid. It turns out jammed sphere packings share much in common with molecular glasses [11–13], as discussed in more detail in Chapter 2.

Thus, granular materials exhibit many features that are common to a wide range of physical systems that exist in nonequilibrium and/or athermal amorphous states, such as glasses, foams, and colloidal suspensions. Consequently, there is a common belief that the study of granular matter phenomena can provide further insight into the properties of systems of academic, environmental, and technological interest.

The focus of this chapter is the study of the flow properties of a collection of grains using computer simulation methods. Understanding the rheology of granular materials is of direct relevance to many areas of industry and also includes many natural phenomena that affect our everyday lives, from avalanches to grain transport and storage [2, 3].

5.2 Rheology of granular media

Granular flows are usually categorized into three regimes. The principle parameters that delineate these regimes are the packing fraction ϕ and the flow rate. The packing fraction measures the ratio of the volume occupied by the particles to the total volume of the system. For N monodisperse spheres of diameter d, contained in a volume V:

$$\phi = \frac{N\pi d^3}{6V}. \tag{5.1}$$

It is worth pointing out several reference values here for packings of hard grains. The face-centered cubic ordered array of osculating spheres has a packing fraction $\phi_{\text{fcc}} = \pi/\sqrt{18} \approx 0.74$. However, granular materials have a tendency to form disordered piles that are stable over a range of packing fractions, $\phi_{\text{rlp}} < \phi < \phi_{\text{rcp}}$. Here, the limits refer to random close packing (random loose packing) with corresponding values $\phi_{\text{rcp}} \approx 0.64$ ($\phi_{\text{rlp}} \approx 0.55$) [1, 11, 14–19]. The specific value of the packing fraction that a static packing takes is sensitive to preparation protocol and the material properties such as the frictional properties of the grains [1, 19–23]. The following criteria serve as an approximate rule to determine the nature of the flow: dilute flows correspond to low densities where the particles experience only occasional collisions, whereas in dense flows the particles largely remain in contact with their nearest neighbors.

Quasi static flows When the packing fraction is in the region, $\phi \gtrsim \phi_{\mathrm{rlp}}$, flow proceeds through the slow deformation of a quasi-static, continuous medium creeping via plastic deformations. In such cases, the flow properties become velocity independent [5], whence the shear stress is proportional to the strain only, and is independent of the rate of shearing in the quasi-static limit.

Dilute rapid flows At the other extreme, $\phi << \phi_{\mathrm{rcp}}$, the particles are far enough apart so they experience only instantaneous two-body, or *binary collisions*, whereby the typical collision time of contact between two colliding particles is short compared to the mean free time [24]. These systems can then be treated using the framework of Boltzmann kinetic theory applied to a gas of inelastic particles. Although the particles are transported with the mean *hydrodynamic* flow, they also possess large velocity fluctuations commonly referred to as the *granular temperature*, analogous to thermal temperature in equilibrium fluids. The goal then is to determine macroscopic behavior at the continuum level through hydrodynamic fields such as the density, velocity, and granular temperature, to develop constitutive relations that describe granular matter rheology on a similar footing as the Navier–Stokes equations for regular liquids.

Dense granular flows An intermediate class of flows exists with features from both of the regimes described above. These flows are dense, yet, at the macroscopic scale, the material continuously deforms. Dense granular flows generally tend to exhibit a range of highly complex rheological features, such as large spatial and temporal inhomogeneities [25–27], large-scale structural rearrangements that sensitively depend on boundary conditions and the external forcing [28, 29], and even multiple flow profiles within the same system [31, 32]. Despite the wide range of behaviors, several flow geometries admit steady state flow regimes, whereby power dissipated through inelastic and frictional collisions is balanced by the external forcing.

Dense granular flows are the class of flows that are the subject of this chapter. Before we discuss this further below, we introduce the traditional models used to describe granular particles and collisions between grains to be used in computer simulations of granular materials.

5.3 Force model

Although simulation schemes to study liquids and gases have been around for over 50 years, computational studies of granular materials are more recent. Current granular dynamics simulation models are based on pioneering work on the discrete element method of Cundall and Strack [33] and Walton [34]. The main distinctions between the algorithms describing granular materials (*granular dynamics*) and other simulated systems such as Lennard–Jones particles (molecular dynamics [35, 36]) stem from the fact that granular materials lose energy in collisions, and collisions only occur on contact between two particles. The description of inelastic and frictional collisions is quite well known [37, 38] and can be written down and coded straightforwardly [39, 40]. We use spring–dashpot interactions to model the forces in the directions normal and tangential to the mutual contact surface between interacting particles [41,42]. Consider two particles $\{i, j\}$ of masses $\{m_i, m_j\}$, with diameters $\{d_i, d_j\}$, at positions $\{\mathbf{r}_i, \mathbf{r}_j\}$, with translational velocities $\{\mathbf{v}_i, \mathbf{v}_j\}$ and angular velocities $\{\boldsymbol{\omega}_i, \boldsymbol{\omega}_j\}$. The two particles are considered to be in contact if they experience an overlap; that is, their center–center separation, $\mathbf{r}_{ij} = \mathbf{r}_i - \mathbf{r}_j$, is less than the sum of their radii, $|\mathbf{r}_{ij}| < d = \frac{1}{2}(d_i + d_j)$. Then, the normal compression of the contact δ_{ij}, relative normal velocity $\mathbf{v}_{n_{ij}}$, and relative tangential velocity $\mathbf{v}_{t_{ij}}$, for the two contacting particles are given by:

$$\delta_{ij} = d - |\mathbf{r}_{ij}|, \tag{5.2}$$

$$\mathbf{v}_{n_{ij}} = (\mathbf{v}_{ij} \cdot \mathbf{n}_{ij})\mathbf{n}_{ij}, \tag{5.3}$$

$$\mathbf{v}_{t_{ij}} = \mathbf{v}_{ij} - \mathbf{v}_{n_{ij}} - \left(\frac{d_i}{2}\boldsymbol{\omega}_i + \frac{d_j}{2}\boldsymbol{\omega}_j\right) \times \mathbf{r}_{ij}, \tag{5.4}$$

where $\mathbf{n}_{ij} = \mathbf{r}_{ij}/r_{ij}$, is the unit normal vector along their line of centers, with $r_{ij} = |\mathbf{r}_{ij}|$, and their relative velocity, $\mathbf{v}_{ij} = \mathbf{v}_i - \mathbf{v}_j$.

The normal and tangential components of the contact force $\mathbf{F}_{ij} = \mathbf{F}_{n_{ij}} + \mathbf{F}_{t_{ij}}$, are given by:

$$\mathbf{F}_{n_{ij}} = f(\delta/d)(k_n \delta_{ij} \mathbf{n}_{ij} - m_{\text{eff}} \gamma_n \mathbf{v}_{n_{ij}}), \tag{5.5}$$

$$\mathbf{F}_{t_{ij}} = f(\delta/d)(-k_t \boldsymbol{\Delta}\mathbf{s}_{t_{ij}} - m_{\text{eff}} \gamma_t \mathbf{v}_{t_{ij}}), \tag{5.6}$$

where the effective mass, $m_{\text{eff}} = \frac{m_i m_j}{m_i + m_j}$, and the particle parameters $k_{n,t}$ and $\gamma_{n,t}$ are the elastic and dissipative constants in the directions normal (n) and tangential (t) to the mutual contact surface. These determine the stiffness of the individual particles and how much energy is lost during a collision. The prefactor in Equations (5.5) and (5.6), $f(x)$, is unity for Hookean contacts (the force goes linearly with compression) and $\sqrt{x}$ for Hertzian contacts (the force goes as the compression to the 3/2 power) [38]. Note that for the study presented here, the particles are considered to be noncohesive; therefore, the forces are compressive and act only on contact.

The elastic tangential displacement, $\Delta\mathbf{s}_{t_{ij}}$, is set to zero upon formation of a contact and tracks the displacement of the contact point for the duration of the contact:

$$\Delta\mathbf{s}_{t_{ij}} = \int_0^{t'} \mathbf{v}_{t_{ij}} dt'', \tag{5.7}$$

where the integration occurs over the lifetime, t', of the contact. The tangential displacement obeys the equation:

$$\Delta\mathbf{s}_{t_{ij}}^{new} = \Delta\mathbf{s}_{t_{ij}} - (\Delta\mathbf{s}_{t_{ij}}^{old} \cdot \mathbf{n}_{ij})\mathbf{n}_{ij} \tag{5.8}$$

to ensure that $\Delta\mathbf{s}_{t_{ij}}$ always lies in the local tangent plane of contact surface. The static yield criterion, characterized by the local particle friction coefficient μ, is modeled by truncating the magnitude of $\Delta\mathbf{s}_{t_{ij}}$ as necessary to satisfy the local Coulomb yield criterion, $|\mathbf{F}_{t_{ij}}| \leq |\mu\mathbf{F}_{n_{ij}}|$. The surfaces between two particles in contact are treated as "stuck" while $F_{t_{ij}} < \mu F_{n_{ij}}$, and as "slipping" while the yield criterion is satisfied.

Energy dissipation occurs during collisions due to the velocity dependent forces in Equations (5.5) and (5.6). Additionally, the work done in creating tangential elastic strains, $\Delta\mathbf{s}_t$, at an established interparticle contact, is lost with either interparticle slip at the Coulomb criterion or particle separation. For Hookean contacts, the coefficient of restitution parameterizes the dissipative nature of the interparticle collisions. In terms of the parameters introduced in Equation (5.5) for Hookean contacts:

$$e_n = e^{(-\gamma_n t_{col}/2)}, \tag{5.9}$$

where t_{col} is the duration of interparticle contact during a collision:

$$t_{col} = \frac{\pi}{\sqrt{(2k_n/m - \gamma_n^2/4)}}. \tag{5.10}$$

A similar expression can be written down for the tangential restitution coefficient. Note, for velocity dependent forces, the coefficient of restitution is not constant and goes to zero as the relative velocity goes to zero. These expressions also carry over for interactions between the moving particles and any confining boundaries that define the simulation geometry. If the boundary is composed of particles fixed in position with the same properties as the flowing particles, the only term that changes in the force equations is the value of the effective mass, $m_{\text{eff}} \to m_i$, for a rigid, infinite boundary.

In a gravitational field $\mathbf{g}$, the total forces and torques on particle i are obtained by summing over all particles j that are in contact with i:

$$\mathbf{F}_i^{tot} = m_i\mathbf{g} + \sum_j \left(\mathbf{F}_{n_{ij}} + \mathbf{F}_{t_{ij}}\right) \tag{5.11}$$

$$\boldsymbol{\tau}_i^{tot} = -\frac{1}{2}\sum_j \mathbf{r}_{ij} \times \mathbf{F}_{t_{ij}}. \tag{5.12}$$

The translational, $\mathbf{a}_i$, and rotational, $\boldsymbol{\alpha}_i$, accelerations of particle i are determined by Newton's second law for the forces and torques:

$$\mathbf{F}_i = m_i\mathbf{a}_i \tag{5.13}$$

$$\boldsymbol{\tau}_i = I_i\boldsymbol{\alpha}_i, \tag{5.14}$$

where I is the moment of inertia so that $I_i = \frac{2}{5}m_i d_i^2$ for spheres.

The components of the stress tensor $\sigma_{\alpha\beta}$ within a given sampling volume V are computed as the sum over all particles i within V of the contact stresses arising from the contact forces and the streaming stress due to velocity fluctuations:

$$\sigma_{\alpha\beta} = \frac{1}{V}\sum_i \left[\sum_{j\neq i} \frac{1}{2} r_{ij}^{\alpha} F_{ij}^{\beta} + m_i(v_i^{\alpha} - \bar{v}^{\alpha})(v_i^{\beta} - \bar{v}^{\beta})\right], \tag{5.15}$$

where $F_{ij}^{\beta} = F_{n_{ij}}^{\beta} + F_{t_{ij}}^{\beta}$, and $\bar{\mathbf{v}}$ is the time-averaged velocity of the particles within the sampling volume V.

The equations of motion for the translational and rotational degrees of freedom can be integrated with either a Gear predictor–corrector or a velocity–Verlet scheme [35]. The value of the spring constant should be large enough to avoid grain interpenetration, yet not so large as to require an unreasonably small simulation time step δt, since an accurate simulation typically requires $\delta t \sim t_{col}/50$. For a typical particle stiffness of $k_n = 2 \times 10^5 mg/d$, the time step is $\delta t = 1 \times 10^{-4}\tau$, where $\tau = \sqrt{d/g}$. The time step must be decreased by an order of magnitude for every two orders of magnitude increase in the value of k_n.

To study steady state problems and erase any history of the preparation procedure of the initial conditions, the simulation must be run until standard measures become time independent. For example, one might want to monitor the time evolution of the average kinetic energy per particle as such a measure. When this quantity becomes approximately constant in time and exhibits only fluctuations about a well-defined mean value, then the system is considered to have reached the steady state regime. Once in the steady state, production simulations can be run to compute physical properties of interest, such as the stress tensor and other kinematic and

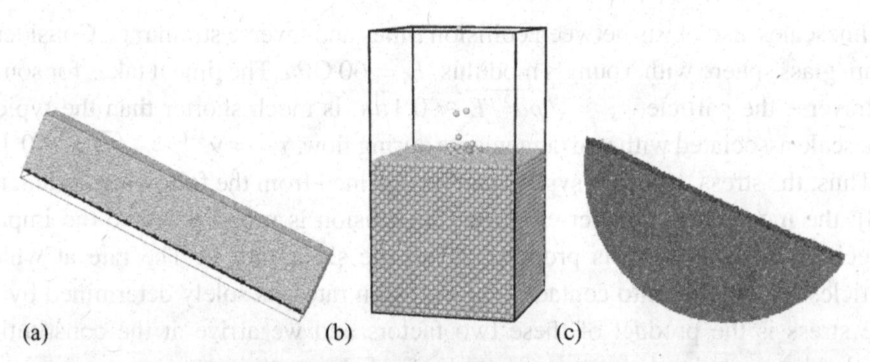

Figure 5.1 Three visualizations of granular dynamic simulations. (a) Inclined plane flow, $N = 160\,000$ [42]. (b) Impact of a single particle onto a quasi-ordered bed containing $N = 16\,000$. (c) Rotating drum, $N = 80\,000$. The black frame is a guide to the size of the computational domain [43].

dynamic information. Several examples of active research in granular dynamic are represented in Figure 5.1.

5.4 Gravity-driven inclined plane granular flows

5.4.1 Constitutive relations for inclined plane flows

A particularly simple class of flow reminiscent of avalanches is that of gravity-driven, dense granular flow down a rough and bumpy inclined plane, or chute flows. This geometry is the archetypal granular flow through which one can study the relation between the stress state and the dynamics and structure, i.e. the constitutive relation of the granular material. Indeed, a number of recent, well-controlled experiments [46–49] and large-scale simulations [41, 42, 50, 51] have motivated numerous theories that capture some of the features of inclined plane flows [52–60]. However, it remains a challenge to reconcile the various observations and results on the nature of particle collisions and contacts, and the mode of stress transmission in dense flows in general [61–65]. Here we provide a view as to how simulations can help us understand various aspects of granular rheology as applied to inclined plane flows, though it should be understood that several issues and controversies persist as open avenues of contemporary research.

To date, the most generally accepted treatment of dense granular rheology is still the physical picture put forward by Bagnold over 50 years ago [66, 67]. Bagnold described a mechanism of momentum transfer between particles in adjacent flowing layers undergoing binary collisions. More generally, it can be understood that in the (infinitely) hard sphere limit, the inverse strain rate is the only relevant time scale in the problem [67]. Provided that the particles are sufficiently stiff, a large separation in timescales also exists between collision times and inverse strain rate. Consider a 1 mm glass sphere with Young's modulus, $E = 60$ GPa. The time it takes for sound to traverse the particle, $\tau_s \approx \sqrt{\rho d^2/E} \approx 0.1\mu s$, is much shorter than the typical timescale associated with rearrangements during flow, $\tau_f \approx \dot{\gamma}^{-1} \approx \sqrt{d/g} \approx 0.1$ s.

Thus, the stress, σ, in the system can be obtained from the following arguments [68]: the momentum transferred during a collision is proportional to the impact velocity, which, in turn, is proportional to the shear rate $\dot{\gamma}$. The rate at which particles are brought into contact – the collision rate – is solely determined by $\dot{\gamma}$. The stress is the product of these two factors and we arrive at the constitutive relation between the shear stress, σ, and rate of shear, strain, $\dot{\gamma}$:

$$\sigma = A_{Bag}^2 \dot{\gamma}^2, \tag{5.16}$$

which is often referred to as Bagnold scaling. The prefactor A_{Bag}, which we call the Bagnold coefficient, is independent of $\dot{\gamma}$, but depends on properties such as the packing fraction, which we will investigate below.

In steady state flow, the Cauchy equations of motion for the stress tensor are:

$$\frac{\partial \sigma_{zz}}{\partial z} = \rho g \cos\theta \tag{5.17}$$

$$\frac{\partial \sigma_{xz}}{\partial z} = \rho g \sin\theta. \tag{5.18}$$

For a given tilt angle, θ, and constant density, $\rho = 6m\phi/\pi d^3$, these reduce to:

$$\sigma_{zz} = (h - z)\rho g \cos\theta, \tag{5.19}$$

$$\sigma_{xz} = (h - z)\rho g \sin\theta, \tag{5.20}$$

where h is the height of the flowing pile. Combining these with the Bagnold rheology equation leads to the following velocity profile in the direction down the incline:

$$v_x(z) = A_{Bag} h^{3/2} \left(\frac{2}{3}\sqrt{\rho g \sin\theta}\right) \left[1 - \left(\frac{h - z}{h}\right)^{3/2}\right], \tag{5.21}$$

which we test below.

Along with the stress and shear rate, the granular temperature is another variable that is used to provide closure to granular kinetic theory. This is defined in an analogous manner to thermal fluids as the mean square velocity fluctuations:

$$T_{gran} = \left(m < (\mathbf{v} - \bar{\mathbf{v}})^2 > + I < (\boldsymbol{\omega} - \bar{\boldsymbol{\omega}})^2 >\right)/3, \tag{5.22}$$

where $\bar{\mathbf{v}}$ and $\bar{\omega}$ are the translational and rotational averaged velocities, and $< \cdots >$ represents a microscopic average over the particles and an ensemble average over the steady state regime. From transport theory and energy balance requirements [69], Bagnold scaling is recovered provided:

$$T_{gran} \propto \dot{\gamma}^2. \tag{5.23}$$

Although the Bagnold equation, Equation (5.16), provides a suitable description of granular rheology in several geometries [46, 50, 64], it is still conceptually difficult to reason why this should be the case in the dense flow regime due to the fact that its origins lie in kinetic theory. Such theories are inherently local in nature in that they depend on the local neighborhood of the particles so that the stress at some point depends on the local value of the shear rate and other fields at this point. This is at odds with several observations made on granular materials that point toward a more nonlocal approach to take into account mechanisms of energy dissipation and stress transmission that occur over length scales much larger than the size of the particle.

Despite these concerns, recent simulation studies examining the role of contact times in the inclined plane geometry [70] have demonstrated that for sufficiently stiff particles Bagnold scaling does describe the bulk rheology. Whereas, when the particles become softer or cohesive forces are present [71, 72], then Bagnold scaling breaks down. This feature of the rheology can be attributed to the formation of long-lasting particle contacts [70], which transmit stresses elastically due to the finite compliance of the particles [68] or through correlated force networks [68, 73–75]. These ideas are consistent with the contributions to the stress tensor, which are dominated by contact forces. The Reynolds stress associated with the kinetic energy of the particles plays a subdominant role [42]. This latter criterion is often used to differentiate between different flowing regimes. In the language of Campbell [68, 76], here we investigate the region of the *inertial–elastic-inertial* regime, using large-scale simulations of gravity driven, dense granular flows down a rough and bumpy inclined plane.

5.4.2 Simulations of inclined plane flows

In the simulations, the imposition of chute flow is achieved by rotating the gravity vector $\mathbf{g}$ in the flow-shear plane by a tilt angle θ away from the $-\mathbf{z}$ direction. In other words, the higher end of the incline is to the left and the grains flow down the slope from left to right, as indicated in Figure 5.1(a). The supporting inclined base is composed of particles that are fixed in position in a random configuration having the same material properties as the flowing grains in the bulk. This particular

Table 5.1 *Parameters used in the simulation studies of the steady state, inclined plane flows of monodisperse spheres.*

Number of particles	N	8000–160000
Force law	$f(x)$	1 (Hookean)
Normal stiffness	k_n	2×10^5 mg/d
Tangential stiffness	k_t	$\frac{2}{7}k_n$
Timestep	δt	$1 \times 10^{-4}\tau$
Normal damping coefficient	γ_n	$50.0\sqrt{g/d}$
Tangential damping coefficient	γ_t	$\gamma_n/2$
Normal restitution coefficient	ϵ	0.88
Friction coefficient	μ	0.5
Angle of repose	θ_r	19.4°
Intermittency angle	θ_i	20.5°
Maximum angle	θ_{max}	26.5°

boundary condition is implemented to reduce slip [28]. The area of the base is $A = L_x L_y$, where L_x and L_y are the dimensions of the simulation cell in the x and y directions respectively. Unless otherwise indicated, most of the studies presented here have $N = 8000$ particles, $L_x = 20d$, and $L_y = 10d$. With these dimensions, we see no system size effects. The direction parallel to the incline is the flow direction and defined to be the x-axis. The direction normal to the plane of the incline is the shear direction and defined to be the z-axis. The third orthonormal y-axis is perpendicular to the xz-plane and in rheological studies is known as the vorticity direction. The parameters chosen for this study are summarized in Table 5.1. We choose simulation parameters that have previously been shown [50] to quantitatively match experimental data [46].

For a typical simulation of granular flow down an inclined plane as described below, a serial algorithm running on a single AMD 2.6 GHz central processing unit (cpu) core can perform 10 million timesteps in 1152 minutes (0.8 days) for a three-dimensional (3D) system containing 8000 spheres. Significant improvement of simulation times can be achieved using a parallelized version of the same code using a standard message-passing interface library such as MPI. The parallel code partitions the simulation domain into smaller 3D subblocks, designating a single cpu to each of these domains, thus reducing the number of calculations per cpu. Nowadays, it is relatively inexpensive to build/purchase a Beowulf Linux computer cluster. Linux clusters are composed of a head node controlling data communications and handling between the compute nodes. Each node typically possesses two or four dual or quad core processors, that is each node can contain between four and 16 cores. Even with standard gigabit communication connections between the head and compute nodes, for the same system mentioned above, the parallel algorithm performs 10 million timesteps in 286 minutes (0.2 days) using one node consisting of two dual-core processors. This represents a linear scaling in compute time over the single cpu run.

This efficiency decreases somewhat when using more than one node due to intra-node communication. On implementing the parallel algorithm, there is a trade-off between sharing the simulation load over several processors and the communication times between the different nodes connecting the different simulation domains. As a rule of thumb, higher parallel efficiencies are obtained for 1000–2000 particles per core. The data presented below were generated using both serial and parallel algorithms. The parallel algorithm used here is known as LAMMPS and was originally developed [44] at Sandia National Laboratories. It is now an open source software [45]. Additional information for particle dynamics algorithms in various programming languages can be found in the following references [35, 36, 40].

5.4.3 Steady flows

Inclined plane flows exhibit several phases. These have been extensively discussed in terms of an effective flow phase diagram [42, 46, 50]. To initiate flow from a static pile of grains, the inclination angle must be increased beyond a threshold angle. This starting angle, θ_{start}, essentially encodes the preparation history of the pile. Once flow has been initiated, a steady state regime exists over a range of inclinations: $\theta_r < \theta < \theta_{max}$. The lower bound, θ_r, is often referred to as the angle of repose. In general, $\theta_r \leq \theta_{start}$, expressing the strong hysteretic nature of granular materials. If the inclination angle for a flowing system is reduced below θ_r, flow ceases. The upper bound, $\theta_{max} > \theta_{start}$, represents the maximum angle at which steady state flows exist. Beyond this angle, the flow accelerates without bound and never achieves a well-defined steady state. Within the steady state regime, yet another phase exists, which is characterized by strong intermittency [26]. The onset of this steady state, intermittent regime occurs at an angle $\theta_i \approx \theta_{start}$, and lies close to, but above, $\theta_r : \theta_i \gtrsim \theta_r$. We focus here on the continuously deforming, steady state regime: $\theta_i < \theta < \theta_{max}$. For the parameters used in this study, see Table 5.1, we restrict the range of inclination angles to the continuous steady state regime: $20.5^\circ < \theta \leq 26^\circ$.

In Figure 5.2, time-depth profiles of various properties of the flow at $\theta = 24^\circ$ indicate that steady state has been established. (1) *Packing fraction*. The packing fraction remains approximately constant throughout the bulk, away from the bottom surface and the saltating top layers. This observation satisfies the assumption leading to the derivation of the Bagnold equation and is a strong motivating factor in accepting Bagnold scaling to describe inclined plane flow. (2) *Coordination number*. The average number of contacts per grain exhibits a mild depth dependence as

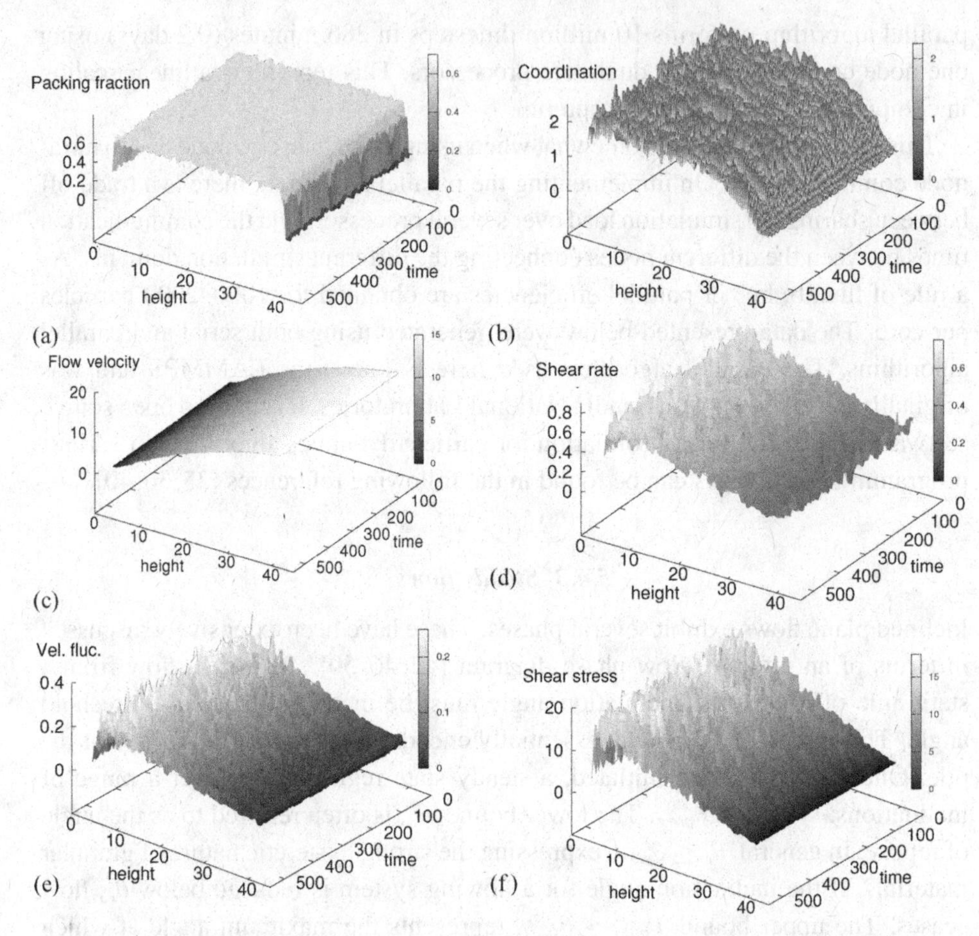

Figure 5.2 Inclined plane flow with $N = 8000$ at $\theta = 24°$. Time-depth profiles of the (a) packing fraction, (b) coordination number, (c) velocity in the direction of flow, (d) shear rate, (e) x-component of the velocity fluctuations, and (f) shear stress.

the overburden of weight from the above layers push particles into contact deeper into the pile. (3) *Flow velocity*. The velocity in the direction of flow (x-component of the velocity vector) is a convex function with height, and as shown below agrees quite well with the Bagnold form, Equation (5.21). (4) *Shear rate*. The derivative of the velocity with respect to height gives the rate of shear strain. The shear rate is maximum close to the bottom boundary and decreases with depth. There is a slight upturn near the top surface in the saltating layer where the particles exist in a thin 'gas' layer. (5) *Velocity fluctuations*. The x-component of the translational velocity fluctuations are obtained from the second term on the right-hand side of Equation (5.15), setting $\alpha = \beta = x$. These also show a maximum in the vicinity of the bottom boundary. (6) *Stress tensor*. The normal and shear (shown) components

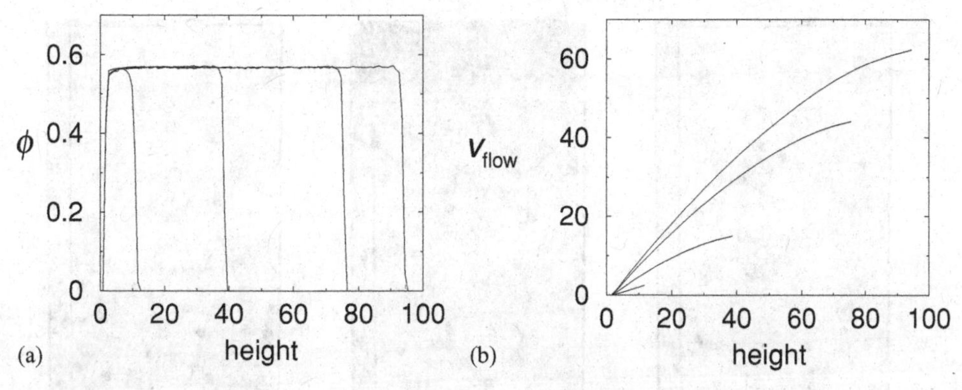

Figure 5.3 Inclined plane flow at $\theta = 24°$ for different system sizes, $2000 \leq N \leq 20\,000$ and base areas $100d^2 \leq A \leq 400d^2$. Depth profiles of the (a) packing fraction ϕ, and (b) velocity in the direction of flow $v_{\rm flow}$ [42].

of the stress tensor all decrease linearly with increasing height through the pile. This is to be expected in the absence of side walls due to the hydrostatic head effect of gravity acting on the pile.

A further test of Bagnold scaling can be obtained through the study of different system sizes at the same inclination angle. In Figure 5.3, systems of varying flow heights at $\theta = 24°$ are compared. The remarkable property shown is the constancy of the packing fraction independent of the height of the flowing system. Thus, despite the increasing weight of the pile on the bottom boundary, the system adapts to achieve the same flow characteristics that scale according to the Bagnold relation, Equation (5.21). The precise way the system achieves such a balance is through an increase in the generation of "granular temperature" with increasing system size. This highlights the unusual and complex interplay that granular materials display during flow.

5.4.4 Microscopics

We can use simulations to examine in more detail the microscopic behavior of the flow at the particle level. Representative, instantaneous snapshots of the flow are shown in Figures 5.4 and 5.5 which display some of these properties. The simulation snapshots of Figure 5.4 show the particle coordination number, and the magnitudes of the translational velocity in the direction of flow and the rotational velocity. These particle-scale data indicate that at the particle level there are noticeable fluctuations.

Another quantity that is of interest in granular matter is the distribution of forces between particles in contact. The snapshots of Figure 5.5 show how the normal and tangential contact forces are distributed at an instant in time during the flow.

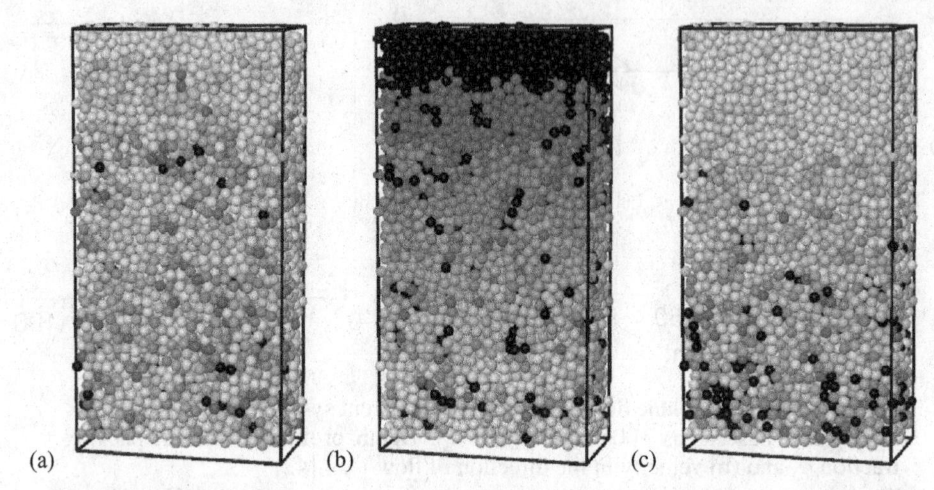

Figure 5.4 Instantaneous snapshots of different particle properties at $\theta = 24°$ for $N = 8000$. (a) Local coordination number, (b) translational velocity, and (c) rotational velocity. Darker shading indicates larger magnitudes. Flow is from left to right. The inclination and the fixed base particles have been removed for clarity.

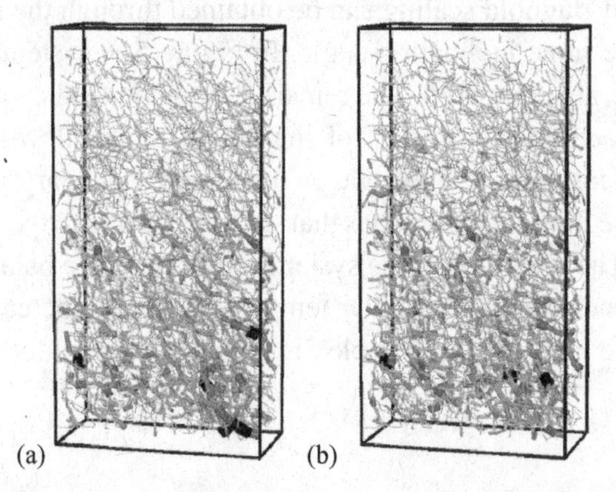

Figure 5.5 Instantaneous snapshot of the forces acting between $N = 8000$ particles in contact at $\theta = 24°$. The lines represent the forces between contacting particles. Particles not shown for clarity. (a) Normal forces perpendicular to the contact surface. (b) Tangential forces in the contact plane. Darker shading and thicker lines indicate larger magnitude of force.

As expected, the influence of gravity leads to the majority of the largest forces to be found near the bottom of the flowing pile due to the overburden of particles above. One common theme of interest is the notion of *force chains*. These are considered to be extended, approximately linear, clusters of particles that bear a significant fraction of the force or stress in the system. There is little evidence for the presence of force chains in the system. There are some extended structures of

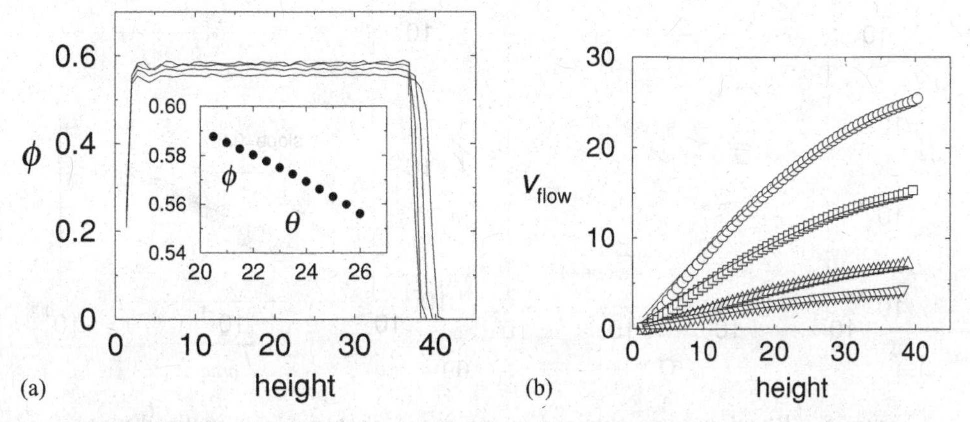

Figure 5.6 Time averaged depth profiles for $N = 8000$ at different inclination angles; $\theta = 26°,\ 24°,\ 22°,\ 21°$. (a) Packing fraction ϕ. Bulk packing fraction decreases with increasing angle. (b) Velocity in the direction of flow v_{flow}. Velocity increases with angle. Lines are fits using Equation (5.21) [42].

forces larger than the average at the bottom of the pile (bottom right of Figure 5.5 (left)), but these extend over a length scale of only a few particle diameters. These force chains are distinct entities from the overall pervading force network of all the forces of varying magnitudes that exist throughout the system. Clearly, this network of forces controls the rheology, but how strongly the presence or lack of force chains influence this remains an open debate [50, 76, 77].

5.4.5 Rheology

We have investigated some of the macroscopic and microscopic features of the inclined plane flow at one angle. Now we turn to the flow behavior as a function of inclination angle. In this geometry, the inclination angle controls the flow rate. From computing the flow data over a range of angles we can therefore extract the bulk rheological behavior and test this against the Bagnold scaling of Equation (5.16).

Before we do this, we wish to determine whether varying the inclination angle leads to any dramatic qualitative or quantitative changes to the flow. As shown in Figure 5.6(a), we find that the steady state regime is characterized by a constant bulk packing fraction that depends on the angle (see inset to Figure 5.6(a)). Increasing the inclination leads to an increase in the velocity down the incline, i.e. flow rate, as shown in Figure 5.6(b), which, in turn, leads to a decreasing bulk density due to a slight dilation of the pile. In Figure 5.6(b) the lines are fit to the Bagnold velocity profile of Equation (5.21). We find that the velocities agree quite well with the

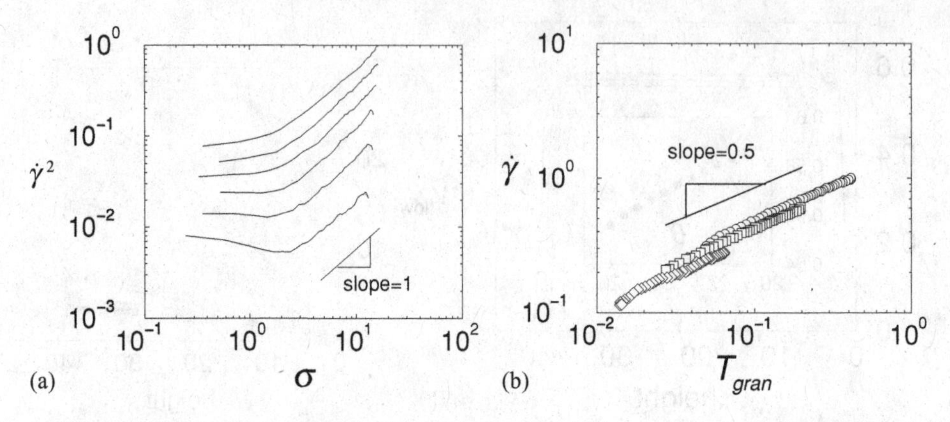

Figure 5.7 Rheology data obtained over a range of steady state inclined plane flows. (a) Shear stress, σ, vs. square of the shear rate, $\dot{\gamma}^2$ (on log-log axes). (b) Square-root dependence of the shear rate, $\dot{\gamma}$, on the granular temperature, T_{gran}.

Bagnold form from which we obtain A_{Bag} as a fitting parameter, which we discuss further below.

From the velocity profiles, we also obtain the shear rates, which we plot parametrically against the shear stress as shown in Figure 5.7(a). Even for these dense flows, over a range $\dot{\gamma}$ we find that the shear rate scales quadratically with shear stress, consistent with Bagnold scaling. Interestingly, at smaller stresses, corresponding to the upper region of the pile, Bagnold scaling breaks down and the shear stress appears to approach zero at a finite shear rate. The reason for this is unclear, but suggests that free surface effects extend well into the constant density region of the flowing pile [41]. We can also determine the relationship between the granular temperature and the shear rate. The results of Figure 5.7(b) demonstrate that $\dot{\gamma} \sim T^{1/2}$, as expected for Bagnold rheology.

5.4.6 Analogy to dense fluids

These data present an intriguing scenario. The flows are sufficiently dense to allow the formation of system-spanning force networks of particles experiencing multiple contacts. However, there is minimal evidence for the existence of actual force chains that carry a significant fraction of the stress. Yet, the rheology is dominated by features often ascribed to dilute flows, whereby Bagnold scaling persists and the relationship between the granular temperature and shear rate satisfies the Bagnold requirement. There are two additional aspects of dense granular flows that we explore in this and the following sections. The first is the analogy between steady state granular flows and dense equilibrium fluids. The other is the relation between

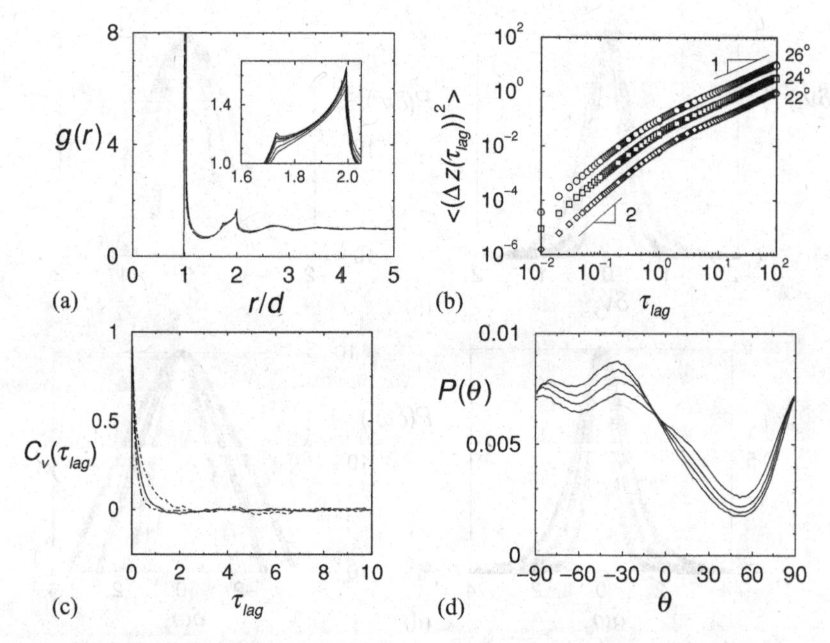

Figure 5.8 (a) Pair correlation function $g(r)$ for different inclination angles. (b) Mean squared displacement in the z-direction computed over time windows τ_{lag}. (c) Velocity autocorrelation function, $C_v(\tau_{lag})$ computed over time windows τ_{lag} (d) Distribution, $P(\theta)$, of two-particle contact angles θ, projected onto the flow-gradient plane.

the approach toward the angle of repose and jamming phenomena as described in Chapter 2.

In this first section, we discuss the liquid-like properties of granular flows. In liquid state theories of fluids [78], there are several variables and statistical analyses that are useful to describe the properties of the system. We present some of these measures in Figure 5.8: the radial distribution function $g(r)$, mean squared displacement in the z-direction, velocity autocorrelation function for the z-component of the velocity, and the two-particle contact angle. The radial distribution functions for these granular flows are similar to that for dense fluids. In particular, we note a sharp first nearest neighbor peak located at a separation of a particle diameter, $r = d$, and the emergence of a split second peak at separations of $\sqrt{3} \leq r/d \leq 2$, which is indicative of an amorphous structure.[1] The split second peak grows as the inclination angle decreases. Such features are also observed in jammed static packings, see Chapter 2, as the jamming transition is approached [13]. The mean squared displacement is calculated in the z-direction to avoid the influence of the mean flow directed along the x-direction. Similar results to those shown here are

[1] The second peak is skewed compared to an isotropic system due to the flow geometry.

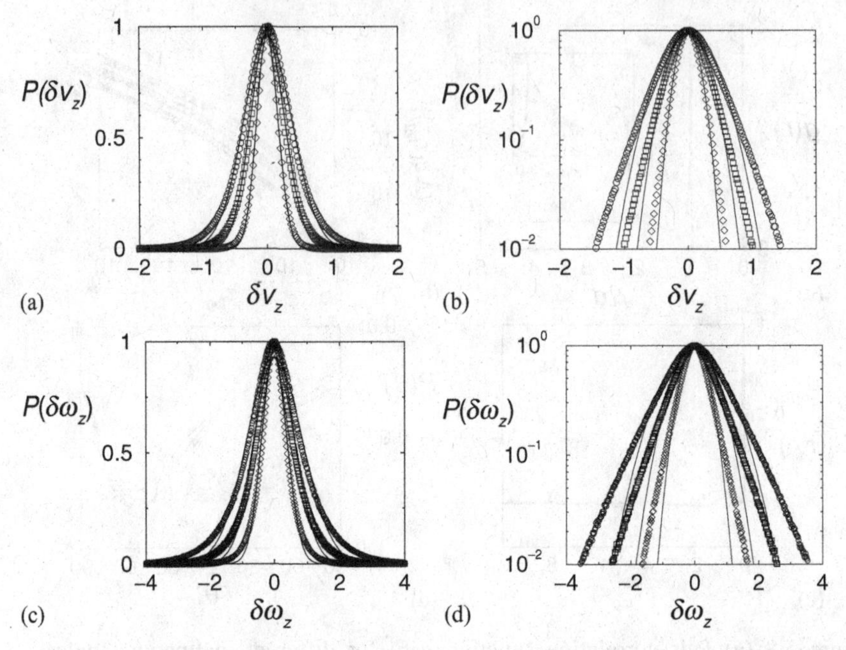

Figure 5.9 Distributions of velocity fluctuations for three different angles; 26° (○), 24° (□), and 22° (◇). Distribution of velocity fluctuations, $P(\delta v_z)$, for the component of the translational velocity in the z-direction using (a) linear–linear and (b) linear–log scales. Distribution of velocity fluctuations, $P(\delta \omega_z)$, for the component of the rotational velocity vector in the z-direction using (c) linear–linear and (d) linear–log scales. The lines are fits to Gaussians: $P(x) = \frac{1}{\sqrt{2\pi\sigma^2}} e^{-x^2/2\sigma^2}$. Here, σ is the variance of the distribution. In both cases exponential tails develop.

obtained in the y-direction. The mean squared displacement provides a measure of the diffusion of particles, and again the behavior shown in Figure 5.7 is analogous to dense fluids. At short times, the particles undergo unimpeded ballistic motions characterized by $< (\Delta z)^2 >\sim \tau_{lag}^2$, which crosses over to the diffusive regime at later times, $< (\Delta z)^2 >\sim \tau_{lag}$. The velocity autocorrelation function, $C_v(\tau_{lag})$:

$$C_v(\tau_{lag}) = \left\langle \frac{1}{N} \sum_{i=1}^{N} v_{i_z}(\tau_{lag}) v_{i_z}(0) \right\rangle \tag{5.24}$$

is computed over an ensemble average over time windows τ_{lag}. This quantity measures correlations between the velocities (z-component) at some origin of time $v_{i_z}(0)$, and their velocities, $v_{i_z}(\tau_{lag})$, at some time τ_{lag} later. Again, the qualitative features are consistent with those seen in dense fluids. As the inclination angle decreases, the correlations die off more slowly. Finally, the two-particle contact angle is a projection of the angle between a pair of contacting particles projected onto the flow-gradient plane. These results clearly show how the imposition of flow in the inclined plane geometry strongly influences the geometry of the contact network. The effect of the shearing throughout the bulk induces a relative increase in contacts along the compressive direction at $\approx -45°$, and a relative depletion of contacts along the extensional direction at $\approx 45°$.

We can push this analogy between granular flow and fluids further by measuring quantities that have well-known properties in equilibrium systems, such as distributions of velocity fluctuations. In Figure 5.9, we show these distributions over a range of angles for the z-component of the translational and rotational velocities. For a thermal system in thermodynamic equilibrium, these distributions follow the Maxwell–Boltzmann distribution for velocities [78], and therefore should be Gaussian. The lines in Figure 5.9 are Gaussian fits to the data. Although the peak region is accurately described, the tails of the distributions deviate strongly from Gaussian and decay exponentially. This breakdown of Gaussian behavior is usually attributed to the presence of correlations in the system [79]. Interestingly, the rotational velocity fluctuations are significant in this flow geometry and should not be ignored in future theoretical developments.

5.4.7 Approach to jamming

In this second section, we probe the onset of jamming as the inclination angle is decreased to the point where flow ceases. The parameters studied in the previous section provide further evidence of this slowing down, such as the increased splitting in the second peak of $g(r)$ and the decrease in the mean squared displacement. In this study, we do not decrease the angle below the intermittency angle θ_i, where the rheology becomes spatially and temporally heterogeneous, so we do not get infinitely close to the angle of repose. However, we can extrapolate several parameters down to the point where the flow stops.

In Figure 5.10(a), we present data for the Bagnold coefficient A_{Bag} defined in Equations (5.16) and (5.21), which we obtain from fitting the velocity profile results to the expected Bagnold expression. We plot A_{Bag} versus ϕ, the bulk value of the constant packing fraction for a given inclination angle. We find that A_{Bag} decreases linearly with increasing ϕ. Extrapolating these data down to $A_{Bag} = 0$, we obtain $\phi_0 = 0.59$. Notice that this value of the packing fraction is well below the random close packed value for frictionless spheres ($\phi_{rcp} = 0.64$) but larger than the lowest reported value for random loose packing ($\phi_{rlp} = 0.55$). In fact, the extrapolated packing fraction obtained here, ϕ_0, coincides closely with the value of a static packing, with $\mu = 0.5$, at the jamming transition, $\phi_J(\mu = 0.5) \approx 0.57$ [19,23].

We also see a similar trend in the bulk averaged coordination number, n_c, which was computed in a large slab of the flowing system away from the bottom and

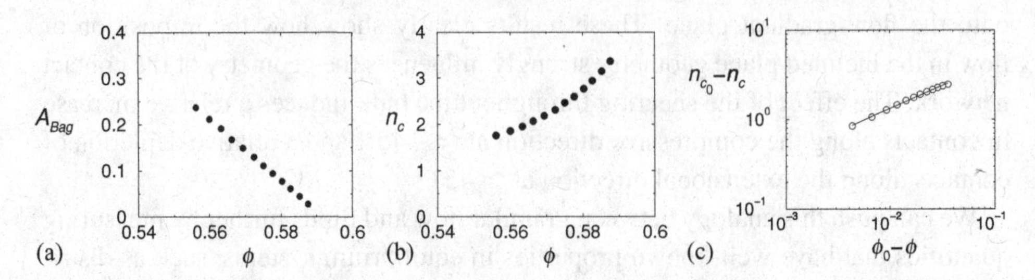

Figure 5.10 Onset of jamming in granular flows. (a) Dependence of the Bagnold coefficient, A_{Bag}, on the packing fraction, ϕ. (b) Dependence of coordination number on packing fraction. (c) Coordination number and packing fraction relative to the jammed values. Line is a square root power law fit to the data.

top surfaces. As the inclination angle decreases, and correspondingly ϕ increases, the coordination number also increases. Motivated by recent studies of jammed systems [11, 13, 31], we fit the data to the following power law:

$$(n_{c_0} - n_c) \propto (\phi_0 - \phi)^{1/2}, \tag{5.25}$$

Here ϕ_0 is the value of the packing fraction obtained from extrapolating $A_{Bag} \to 0$ in Figure 5.10(a), and n_{c_0} is a fitting parameter that provides the best fit to the square root power law behavior. We find that for the system studied here, $n_{c_0} \approx 4.2$. Therefore, from extrapolating our data to the jamming limit, we characterize the granular flow–jammed transition by the values $\phi_0 \approx 0.59$ and $n_{c_0} = 4.2$. These results are quite remarkable in that they are close in value to those for the jamming transition of an isotropic packing with $\mu = 0.5$ [19, 23] and poured frictional particles [20]. Thus, we can conclude that studying granular flows can also shed light on the jamming transition and the likely stability properties, for example the coordination number and packing fraction of the jammed state.

The distribution of forces, $P(f)$, is also known to be a good indicator of the onset of jamming in driven systems [80, 81]. In Figures 5.11(a) and (b), we plot the distributions of the normal and tangential contact forces between pairs of interacting particles. Although the tails of the distributions, corresponding to large forces, are somewhat insensitive to the inclination angles, the most noticeable changes occur at small forces. At higher inclination angles, the low-force region of the normal force distribution is almost flat. As the inclination angle is decreased towards the angle of repose, a peak develops between $0.5 \leq f \leq 1.0$, as more and more particles achieve mechanical equilibrium. This jamming notion is further emphasized in Figure 5.11(c), which shows the distributions of the ratios of the largest to next largest magnitudes of the total contact forces [82]. Again, as the inclination angle is reduced, an increasing fraction of the particles experience

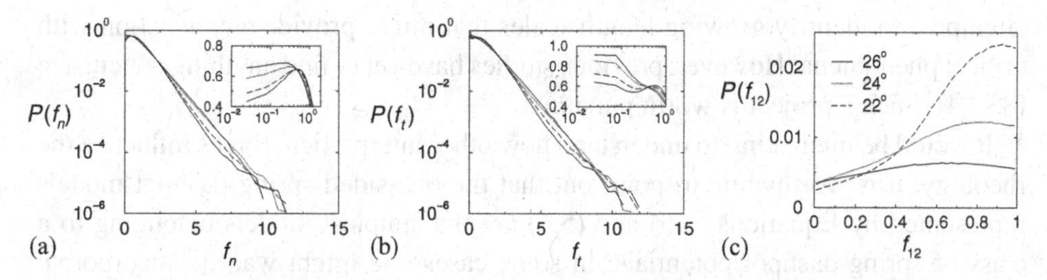

Figure 5.11 Contact force statistics at various inclination angles. Distributions $P(f)$, of (a) normal and (b) tangential contact forces, where $f = F/ < F >$, is the contact force normalized by the average contact force. Insets: low force region. (c) Distributions, $P(f_{12})$, of the ratio, f_{12}, of the largest to next largest total contact force experienced by a particle.

forces that balance, $f_{12} \approx 1$, which indicates that the system is approaching a state of mechanical equilibrium, or in other words it is starting to jam.

5.5 Summary

We have provided details on methods for granular dynamics simulations and applied these to a particular class of dense granular flows down an inclined plane. The simulations not only provide macroscopic information that can be compared directly with experiment, but also shed light on microscopic properties at the particle scale that are typically inaccessible experimentally in 3D systems, such as force networks and coordination numbers.

The particle and force model used here was the simplest type of granular particles: monodisperse spheres with the same mass density, interacting through a linear repulsive force law and static friction with one value of the friction coefficient. The force model appears to have little affect on the qualitative and quantitative behavior described here [42]. However, smaller values of the particle friction coefficient are likely to shift the region over which steady state flow is observed. One would expect the angle of repose to decrease with decreasing friction.

We also used the simulation results to connect with other theoretical avenues, in particular liquid state theory and jamming. The approach to the angle of repose shares strong similarities with the onset of jamming in static packings. The features reported here for the radial distribution function, mean squared displacement, and velocity autocorrelation functions are similar to the behavior of dense fluids. Moreover, we find similar power-law behavior in the relative coordination number and packing fraction with respect to the jammed values. One unanswered question not explored here is the notion of a characteristic length scale that controls the rheology. Numerical and theoretical studies on inclined plane flows have attempted to identify growing length scales that might provide a connection with critical phenomena. However, previous studies have yet to find anything conclusive [83, 84]. Such a project is worth pursuing.

It would be interesting to understand how other interparticle forces influence the rheology. It is worthwhile to point out that the one-sided spring-dashpot models represented by Equations (5.5) and (5.6) are the simplest models belonging to a class of spring-dashpot potentials. In some cases one might want to incorporate hysteretic springs, where the values of the spring constants depend on the history of the contact force and/or the signs of the relative velocities between two interacting particles [34, 39, 85]. Such models are guided by plastic deformations of the contact point.

The results presented here neglect particle cohesion. Damp or humid sand can be modeled through the introduction of a cohesive force, $\mathbf{F}^c$, acting between the grains. Cohesive forces that mimic the presence of liquid bridges have been developed for inclined plane flow simulations [71]. This uses a rather simple model that can be incorporated into the simulations. The interaction potential is modeled as an attractive Gaussian well:

$$\mathbf{F}_{ij}^{c} = 2A\frac{(r_c - |\mathbf{r}_{ij}|)}{\ell^2} e^{-(r_c - |\mathbf{r}_{ij}|)^2/\ell^2}\mathbf{n}_{ij}. \tag{5.26}$$

The short-range cohesive force is set to zero for distances greater than some predefined cutoff, $r_c = 1.02d$. The strength of the attraction is controlled by A and the width of the attractive well by ℓ.

Powders, composed of small particles $d \leq 100\mu$m, can also experience cohesion through van der Waals attraction. If the powders become charged, long-ranged electrostatic forces might be required, perhaps implemented through some Ewald summation procedure [35]. This is computationally expensive, so suitable parallelized algorithms might be required to simulate large systems.

Another area where computational studies have made little progress is in the study of slurries, where the granular particles are suspended in some background fluid, such as water. Accurately taking into account the coupling to hydrodynamic fields of the suspending medium also introduces long-ranged interaction coupled effects, which are also computationally problematic.

Finally, another area of research that has received little attention is the study of complex particle shapes and particle composites. Chapter 2 presents some of the first results on simulation studies of ellipsoidal particles. This needs to be extended to include friction forces in a physically suitable manner that allows the study of frictional complex particles that would provide further steps towards simulations of real granular materials.

Acknowledgments

I dedicate this chapter to Gary S. Grest with whom I continue to enjoy a fruitful collaboration on granular matter research. This work was supported by the National Science Foundation under Grant No. CBET-0828359.

References

[1] R. L. Brown and J. C. Richards, *Principles of Powder Mechanics* (Pergamon Press, Oxford, 1970).

[2] S. B. Savage, "Gravity flow of cohesionless granular materials in chutes and channels," *J. Fluid Mech.* **92**, 53 (1979).

[3] R. M. Iverson, "The physics of debris flows," *Rev. Geophys.* **35**, 245 (1997).

[4] J. Duran, *Sands, Powders, and Grains An Introduction to the Physics of Granular Materials* (Springer-Verlag, New York, 2000).

[5] R. M. Nedderman, *Statics and Kinematics of Granular Materials* (Cambridge University Press, Cambridge, 1992).

[6] *Challenges in Granular Physics*, T. Halsey and A. Mehta (eds.) (World Scientific, Singapore, 2002).

[7] H. M. Jaeger, S. R. Nagel, and R. P. Behringer, "Granular solids, liquids, and gases," *Rev. Mod. Phys.* **68**, 1259 (1996).

[8] I. S. Aranson and L. S. Tsimring, "Patterns and collective behavior in granular media: theoretical concepts," *Rev. Mod. Phys.* **78**, 641 (2006).

[9] G. D. Scott, A. M. Charlesworth, and M. K. Max, "On the random packing of spheres," *J. Chem. Phys.* **40**, 611 (1964).

[10] J. D. Bernal, "The Bakerian Lecture, 1962: the structure of liquids," *Proc. Roy. Soc. London. Series A* **280**, 299 (1964).

[11] C. S. O'Hern, L. E. Silbert, A. J. Liu, and S. R. Nagel, "Jamming at zero temperature and zero applied stress: the epitome of disorder," *Phys. Rev. E* **68**, 011306 (2003).

[12] L. E. Silbert, A. J. Liu, and S. R. Nagel, "Vibrations and diverging length scales near the unjamming transition," *Phys. Rev. Lett.* **95**, 098301 (2005).

[13] L. E. Silbert, A. J. Liu, and S. R. Nagel, "Structural signatures of the unjamming transition," *Phys. Rev. E* **73**, 041304 (2006).

[14] J. D. Bernal and J. Mason, "Co-ordination of randomly packed spheres," *Nature* **188**, 910 (1960).

[15] G. D. Scott, "Packing of equal spheres," *Nature* **188**, 908 (1960).

[16] J. G. Berryman, "Random close packing of hard spheres and disks," *Phys. Rev. A* **27**, 1053 (1983).

[17] G. Y. Onoda and E. G. Liniger, "Random loose packings of uniform spheres and the dilatancy onset," *Phys. Rev. Lett.* **64**, 2727 (1990).

[18] S. Torquato, T. M. Truskett, and P. G. Debenedetti, "Is random close packing of spheres well defined?," *Phys. Rev. Lett.* **84**, 2064 (2000).

[19] L. E. Silbert, "Jamming of frictional spheres and random close packing," (to appear in *Soft Matter*, 2010).

[20] L. E. Silbert, D. Ertaş, G. S. Grest, T. C. Halsey, and D. Levine, "Geometry of frictional and frictionless sphere packings," *Phys. Rev. E* **65**, 031304 (2002).

[21] H. A. Makse, D. L. Johnson, and L. M. Schwartz, "Packing of compressible granular materials," *Phys. Rev. Lett.* **84**, 4160 (2000).

[22] H. P. Zhang and H. A. Makse, "Jamming Transition in emulsions and granular materials," *Phys. Rev. E* **72**, 011301 (2005).
[23] C. Song, P. Wang, and H. A. Makse, "A phase diagram for jammed matter," *Nature* **453**, 629 (2008).
[24] I. Goldhirsch, "Rapid granular flows," *Ann. Rev. Fluid Mech.* **35**, 267 (2003).
[25] K. To, P.-Y. Lai, and H. K. Pak, "Jamming of granular flow in a two-dimensional hopper," *Phys. Rev. Lett.* **86**, 71 (2001).
[26] L. E. Silbert, "Temporally heterogeneous dynamics in granular flows," *Phys. Rev. Lett.* **94**, 098002 (2005).
[27] A. Ferguson and B. Chakraborty, "Spatially heterogeneous dynamics in dense, driven granular flows," *Europhys. Lett.* **78**, 28003 (2007).
[28] L. E. Silbert, G. S. Grest, S. J. Plimpton, and D. Levine, "Boundary effects and self-organization in dense granular flows," *Phys. Fluids* **14**, 2637 (2002).
[29] J. Gollub, G. Voth, and J. Tsai, "Internal granular dynamics, shear induced crystallization, and compaction steps," *Phys. Rev. Lett.* **91**, 064301 (2003).
[30] K. E. Daniels and R. P. Behringer, "Hysteresis and competition between disorder and crystallization in sheared and vibrated granular flow," *Phys. Rev. Lett.* **94**, 168001 (2005).
[31] P.-A. Lemieux and D. J. Durian, "From avalanches to fluid flow: a continuous picture of grain dynamics down a heap," *Phys. Rev. Lett.* **85**, 4273 (2000).
[32] L. Bocquet, W. Losert, D. Schalk, T. C. Lubensky, and J. P. Gollub, "Granular shear flow dyanmics and forces: experiment and continuum theory," *Phys. Rev. E* **65**, 011307 (2001).
[33] P. A. Cundall and O. D. L. Strack, "A discrete numerical model for granular assemblies," *Géotechnique* **29**, 47 (1979).
[34] O. R. Walton, "Numerical simulation of inclined chute flows of monodisperse, inelastic, frictional spheres," *Mech. Mat.* **16**, 239 (1993).
[35] M. P. Allen and D. J. Tilldesley, *Computer Simulations of Liquids* (Oxford University Press, Oxford, 1987).
[36] D. C. Rapaport, *The Art of Molecular Dynamics Simulation* (Cambridge University Press, Cambridge, 1995).
[37] K. L. Johnson, *Contact Mechanics* (Cambridge University Press, Cambridge, 1987).
[38] L. D. Landau and E. M. Lifshitz, *Theory of Elasticity*, Vol. 7 of *Course of Theoretical Physics*, 3rd edn (Elsevier, Oxford, 1986).
[39] J. Schafer, S. Dippel, and D. E. Wolf, "Force schemes in simulations of granular materials," *J. Phys. I France* **6**, 5 (1996).
[40] T. Pöschel and T. Schwager, *Computational Granular Dynamics* (Springer, Berlin, 2005).
[41] D. Ertaş, G. S. Grest, T. C. Halsey, D. Levine, and L. E. Silbert, "Gravity-driven dense granular flows," *Europhys. Lett.* **56**, 214 (2001).
[42] L. E. Silbert, D. Ertaş, G. S. Grest, T. C. Halsey, D. Levine, and S. J. Plimpton, "Granular flow down an inclined plane: Bagnold scaling and rheology," *Phys. Rev. E* **64**, 051302 (2001).
[43] R. Brewster, G. S. Grest, and A. J. Levine, "Effects of cohesion on the surface angle and velocity profiles of granular material in a rotating drum," *Phys. Rev. E* **79**, 011305 (2009).
[44] S. J. Plimpton, "Fast parallel algorithms for short-range molecular dynamics," *J. Comp. Phys.* **117**, 1 (1995).
[45] http://lammps.sandia.gov/.

[46] O. Pouliquen, "Scaling laws in granular flows down rough inclined planes," *Phys. Fluids* **11**, 542 (1999).
[47] C. Ancey, "Dry granular flows down an inclined channel: experimental investigations on the frictional-collisional regime," *Phys. Rev. E* **65**, 011304 (2001).
[48] E. Azanza, F. Chevoir, and P. Moucheront, "Experimental study of collisional granular flows down an inclined plane," *J. Fluid Mech.* **400**, 199 (1999).
[49] T. Borzsonyi, T. C. Halsey, and R. E. Ecke, "Two scenarios for avalanche dynamics in inclined granular layers," *Phys. Rev. Lett.* **94**, 208001 (2005).
[50] L. E. Silbert, J. W. Landry, and G. S. Grest, "Granular flow down a rough inclined plane: transition between thick and thin piles.," *Phys. Fluids* **15**, 1 (2003).
[51] W. T. Bi, R. Delannay, P. Richard, N. Taberlet, and A. Valance, "Two- and three-dimensional confined granular chute flows: experimental and numerical results," *J. Phys.: Condens. Matter* **17**, S2457 (2005).
[52] P. Mills, D. Loggia, and M. Tixier, "Model for a stationary dense granular flow along an inclined wall," *Europhys. Lett.* **45**, 733 (1999).
[53] C. Ancey and P. Evesque, "Frictional-collisional regime for granular suspension flows down an inclined channel," *Phys. Rev. E* **62**, 8349 (2000).
[54] I. S. Aranson and L. S. Tsimring, "Continuum description of avalanches in granular media," *Phys. Rev. E* **64**, 020301 (2001).
[55] L. Bocquet, J. Errami, and T. C. Lubensky, "Hydrodynamic model for a dynamical jammed-to-flowing transition in gravity driven granular media," *Phys. Rev. Lett.* **89**, 184301 (2002).
[56] A. Lemaître, "Origin of a repose angle: kinetics of rearrangement for granular media," *Phys. Rev. Lett.* **89**, 064303 (2002).
[57] D. Ertas and T. C. Halsey, "Granular gravitational collapse and chute flow," *Europhys. Lett.* **60**, 931 (2002).
[58] M. Y. Louge, "Model for dense granular flows down bumpy inclines," *Phys. Rev. E* **67**, 061303 (2003).
[59] V. Kumaran, "Kinetic theory for the density plateau in the granular flow down an inclined plane," *Europhys. Lett.* **73**, 232 (2006).
[60] J. T. Jenkins, "Dense shearing flows of inelastic disks," *Phys. Fluids* **18**, 103307 (2006).
[61] N. Menon and D. J. Durian, "Diffusing-wave spectroscopy of dyanmics in a three-dimensional granular flow," *Science* **275**, 1920 (1997).
[62] N. Mitarai and H. Nakanishi, "Hard-sphere limit of soft-sphere model for granular materials: stiffness dependence of steady granular flow," *Phys. Rev. E* **67**, 021301 (2003).
[63] D. Volfson, L. S. Tsimring, and I. S. Aranson, "Partially fluidized shear granular flows: continuum theory and molecular dynamics simulations," *Phys. Rev. E* **68**, 021301 (2003).
[64] G. Midi, "On dense granular flows," *Eur. Phys. J. E* **14**, 341 (2004).
[65] F. da Cruz, S. Emam, M. Prochnow, J.-N. Roux, and F. Chevoir, "Rheophysics of dense granular materials: discrete simulation of plane shear flows," *Phys. Rev. E* **72**, 021309 (2005).
[66] R. A. Bagnold, "Experiments on a gravity-free dispersion of large solid spheres in a newtonian fluid under shear," *Proc. R. Soc. London. Series A* **225**, 49 (1954).
[67] G. Lois, A. Lemaître, and J. M. Carlson, "Numerical tests of constitutive laws for dense granular flows," *Phys. Rev. E* **72**, 051303 (2005).
[68] C. S. Campbell, "Stress-controlled elastic granular shear flows," *J. Fluid Mech.* **539**, 273 (2005).

[69] N. Mitarai and H. Nakanishi, "Bagnold scaling, density plateau, and kinetic theory analysis of dense granular flow," *Phys. Rev. Lett.* **94**, 128001 (2005).
[70] L. E. Silbert, G. S. Grest, R. Brewster, and A. J. Levine, "Rheology and contact lifetimes in dense granular flows," *Phys. Rev. Lett.* **99**, 068002 (2007).
[71] R. Brewster, G. S. Grest, J. W. Landry, and A. J. Levine, "Plug flow and the breakdown of Bagnold scaling in cohesive granular flows," *Phys. Rev. E* **72**, 061301 (2005).
[72] R. Brewster, L. E. Silbert, G. S. Grest, and A. J. Levine, "Relationship between interparticle contact lifetimes and rheology in gravity-driven granular flows," *Phys. Rev. E* **77**, 061302 (2008).
[73] G. Lois, A. Lemaître, and J. M. Carlson, "Emergence of multi-contact interactions in contact dynamics simulations of granular shear flows," *Europhys. Lett.* **76**, 318 (2006).
[74] G. Lois, A. Lemaître, and J. M. Carlson, "Spatial force correlations in granular shear flow, I: numerical evidence," *Phys. Rev. E* **76**, 021302 (2007).
[75] G. Lois, A. Lemaître, and J. M. Carlson, "Spatial force correlations in granular shear flow, II: theoretical implications," *Phys. Rev. E* **76**, 021303 (2007).
[76] C. S. Campbell, "Granular shear flows at the elastic limit," *J. Fluid Mech.* **465**, 261 (2002).
[77] G. Lois and J. M. Carlson, "Force networks and the dynamic approach to jamming in sheared granular media," *Europhys. Lett.* **80**, 58001 (2007).
[78] D. A. McQuarrie, *Statistical Mechanics* (University Science Books, Sausalito, 2000).
[79] D. L. Blair and A. Kudrolli, "Velocity correlations in dense granular gases," *Phys. Rev. E* **64**, 050301 (2001).
[80] C. S. O'Hern, S. A. Langer, A. J. Liu, and S. R. Nagel, "Force Distributions near jamming and glass transitions," *Phys. Rev. Lett.* **86**, 111 (2001).
[81] E. I. Corwin, H. M. Jaeger, and S. R. Nagel, "Structural signature of jamming in granular media," *Nature* **435**, 1075 (2005).
[82] K. A. Reddy and V. Kumaran, "Applicability of constitutive relations from kinetic theory for dense granular flows," *Phys. Rev. E* **76**, 061305 (2007).
[83] O. Pouliquen, "Velocity correlations in dense granular flows," *Phys. Rev. Lett.* **93**, 248001 (2004).
[84] O. Baran, D. Ertas, T. C. Halsey, G. S. Grest, and J. B. Lechman, "Velocity correlations in dense gravity-driven granular chute flow," *Phys. Rev. E* **74**, 051302 (2006).
[85] L. Pournin, T. M. Liebling, and A. Mocellin, "Molecular-dynamics force models for better control of energy dissipation in numerical simulations of dense granular media," *Phys. Rev. E* **65**, 011302 (2001).

6

Rheological and microrheological measurements of soft condensed matter

JOHN R. DE BRUYN AND FELIX K. OPPONG

6.1 Introduction

Fluids deform irreversibly under shear; in other words, they flow. In contrast, solids deform elastically when subjected to a small shearing force and recover their original shape when the force is removed. The behavior of what is termed soft matter is somewhere in between. Soft matter systems are typically viscoelastic, that is they display a combination of viscous (fluid-like) and elastic (solid-like) behavior. Measuring the flow behavior and the mechanical response to deformation of viscoelastic materials provides us with information that can be interpreted in terms of their small-scale structure and dynamics.

The mechanical properties of soft materials depend on the length scale probed by the measurements due to the fact that the materials are structured on length scales intermediate between the atomic and bulk scales [1]. For example, a colloidal suspension has structure on the scale of the spacing between the colloidal particles; a concentrated polymer system, on the scale of the entanglements between large molecules. As a result, their bulk properties can be quite different from properties on length scales smaller than or comparable to the structural scale. Making measurements on both macroscopic and microscopic length scales can help us to develop a better understanding of the relationship between microstructure and bulk properties in soft materials.

Following a brief introduction to viscoelasticity, this chapter will focus on two methods of measuring the viscoelastic properties of soft matter. On the macroscopic scale, rotational shear rheometry provides a well-established set of techniques for determining the mechanical properties of complex fluids. We will summarize the principles of rheology and describe the types of measurements one can perform [2–4]. Typically shear rheometry will be performed using a commercial

Experimental and Computational Techniques in Soft Condensed Matter Physics, ed. Jeffrey Olafsen.

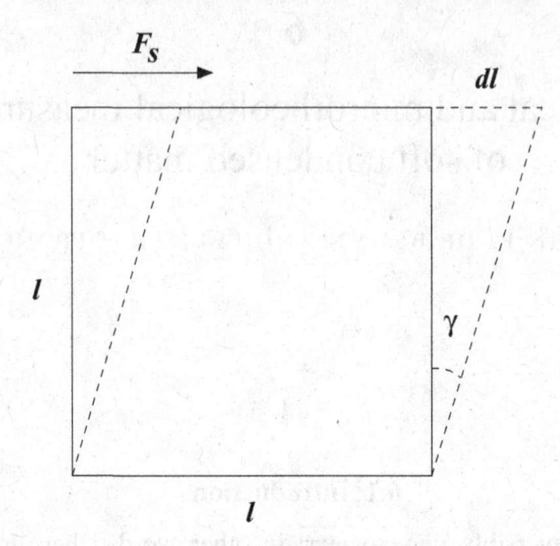

Figure 6.1 A small element of material subject to a shear force F_s. The bottom edge of the material is held fixed. The material experiences a shear strain $\gamma = d\ell/\ell$.

rheometer, and we will discuss the types of instrument commonly available. We will not discuss extensional rheology or other rheometrical methods; the interested reader is referred to a number of excellent textbooks [2–4]. Techniques for making rheological measurements on the microscopic scale have been developed more recently, and the field of microrheology is still evolving. Several reviews of the field have been published recently [5–9]. We will summarize the theoretical background and describe a frequently used experimental method based on tracking the motion of micron-sized tracer particles as they diffuse through the material of interest [6, 7, 10, 11]. Such experiments are typically home-built, and we will describe one implementation. In both cases, we will provide examples that illustrate how rheological measurements can provide information about the structure and dynamics of soft materials.

Consider a small solid cube having sides of length ℓ and area $A = \ell^2$, as shown in Figure 6.1. Suppose that the bottom surface is held in place, and that we apply a shear force F_s parallel to plane of the top surface. The shear stress σ is the shear force divided by the surface area:

$$\sigma = F_s/A. \tag{6.1}$$

In response to the applied shear force, the solid will deform by an amount dl. The shear strain γ is the relative deformation, given by

$$\gamma = d\ell/\ell. \tag{6.2}$$

For small strains, this is the same as the angle γ shown in Figure 6.1. For a perfectly elastic solid, Hooke's Law tells us that stress and strain are linearly proportional, that is,

$$\sigma = G'\gamma, \tag{6.3}$$

where the elastic modulus G' is a measure of the elastic energy stored in the material when it is deformed. Hooke's Law implies reversibility – the deformation immediately returns to zero when the shear stress is removed.

Now imagine applying the same shear stress to a cubic element of a Newtonian fluid. In this case, the material flows continuously – the strain continues to increase for as long as the stress is applied. If the stress is increased, the fluid flows faster. Newton's law of viscosity states that

$$\sigma = \eta\dot{\gamma}, \tag{6.4}$$

where $\dot{\gamma} = d\gamma/dt$ is the strain rate. Here the strain rate, not the strain, is proportional to the stress. Equation (6.4) can be considered to define the viscosity η, which is related to the dissipation of energy in the flow. η can be thought of as the tendency of a material to resist flow.

The discussion above (and throughout this chapter) is very much simplified by our neglect of the fact that stress, strain, and strain rate are all, in general, tensor quantities. For example, the shear stress σ referred to in the above equations is really just one component of a symmetric second-rank stress tensor. Since the focus of this book is on experimental techniques, we will ignore this rather important detail and use a simple scalar description. This implies a neglect of any material anisotropy and, more seriously, of several interesting and important physical phenomena, some of which will be mentioned very briefly below. More mathematically rigorous discussions can be found in textbooks such as references [2, 3, 12].

In general, the response of soft matter systems or complex fluids to an applied deformation or stress is time-dependent – they typically behave as solids on short time scales or at high frequencies and as viscous liquids over long times or at low frequencies. For example, when a viscoelastic material is subjected to a step increase in strain by an amount γ_0, the stress σ in the material undergoes a step change followed by, in the simplest case, an exponential relaxation. For small strains, one can define a relaxation modulus $G(t) = \sigma(t)/\gamma_0$, which is independent of γ_0, so

$$d\sigma = G(t)d\gamma. \tag{6.5}$$

Analogously, in a creep experiment, the stress is suddenly increased from zero to some value σ_0. In this case, the strain will initially change rapidly in response, then continue to evolve more slowly as a function of time. The creep compliance $J(t)$

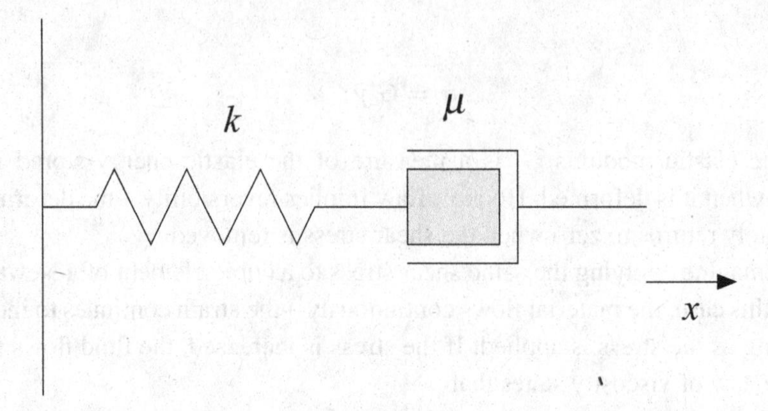

Figure 6.2 The spring-dashpot representation of the Maxwell model for a viscoelastic medium. The viscous and elastic elements are combined in series and stretched by a distance x.

is defined as

$$J(t) = \gamma(t)/\sigma_0, \tag{6.6}$$

and is conceptually equivalent to the reciprocal of the relaxation modulus $G(t)$. The range of strains in which stress and strain are linearly related is called the linear viscoelastic regime. In this regime, the strain is small enough that it does not seriously disrupt any structure in the material, and one thus probes the response of the material to small perturbations of its equilibrium configuration. At higher strains, the structure is disrupted and the response of the material changes accordingly. In particular, the material properties become functions of strain or strain rate, and often functions of time as well. For example, outside of the linear regime, the viscosity of polymer solutions typically decreases as the strain rate is increased (shear thinning), because the shear stretches out the polymer molecules and allows them to partially disentangle. On the other hand, a concentrated clay suspension, which is a gel at rest, will become liquid when subjected to a large enough strain as the small-scale arrangement of the interacting clay particles is changed. It will then slowly gel again over time [13] as the clay particles rearrange and reorient. The interpretation of rheological measurements in the nonlinear regime is complex, and a subject of substantial current research [14, 15].

The simplest model of a viscoelastic material is originally due to Maxwell [16], and is mechanically equivalent to a spring in series with a viscous damper (a dashpot), as in Figure 6.2. If we stretch this combined system by a displacement x, a force F is developed that obeys the differential equation

$$F + \frac{\mu}{k}\frac{dF}{dt} = \mu\frac{dx}{dt}, \tag{6.7}$$

where k is the spring constant and μ the viscosity of the dashpot. In terms of the viscoelastic properties of a complex fluid, the Maxwell model takes the form

$$\sigma + \varphi \frac{d\sigma}{dt} = \eta \frac{d\gamma}{dt}, \tag{6.8}$$

where the viscosity η is assumed to be constant and φ is a time scale for the relaxation of the material.

We can derive the integral form of the Maxwell model by imagining that the stress in a material depends on the strain to which it has been subjected over all time, from the distant past to the present [2]. The stress due to an arbitrary deformation can be calculated by integrating over contributions due to a series of small deformations as long as the relaxation modulus $G(t)$ depends only on time and not on γ. In this case, we can rewrite Equation (6.5) in the form

$$d\sigma = G\frac{d\gamma}{dt}dt = G\dot{\gamma}dt. \tag{6.9}$$

Integrating, we have

$$\sigma = \int_{-\infty}^{t} G\left(t - t'\right)\dot{\gamma}\left(t'\right)dt', \tag{6.10}$$

where t' runs from $-\infty$ to the present time t, and in general the strain rate $\dot{\gamma}$ is a function of time. In this context, the relaxation modulus $G(t - t')$ is referred to as a memory function. If $G(t)$ can be represented by an exponential decay with a decay time φ, that is, $G(t) = G_0 e^{-t/\varphi}$, then Equation (6.10) becomes

$$\sigma = \int_{-\infty}^{t} G_0 e^{-(t-t')/\varphi}\dot{\gamma}(t')dt', \tag{6.11}$$

which gives the stress as a sum of infinitesimal contributions weighted by the memory function. Differentiating Equation (6.11) with respect to time recovers Equation (6.8), showing that the two formulations are equivalent. At steady state, $d\sigma/dt = 0$ and Equation (6.8) becomes Newton's law of viscosity, while for rapid changes, $d\sigma/dt \gg \sigma/\varphi$ and Hooke's law is recovered.

This model captures the basic physics of linear viscoelasticity and provides a reasonable description of some of the viscoelastic behavior of some "simple" materials such as wormlike micelles, some emulsions, and linear polymers at frequencies low compared to the slowest relaxation time of the system [2, 3]. Most soft materials are more complex, however. In a typical soft matter system, several different processes contribute to the relaxation of stress, each operating on a different time scale, and a model with a single relaxation time is inadequate. This more complicated behavior is often described empirically by fitting data to a "generalized Maxwell model" involving a sum of several exponential relaxation terms with different time constants [2, 3]. Much more sophisticated rheological models based on microscopic physics have been developed to describe the behavior of a range of complex fluids, and are discussed, for example, in References [2,3,12]. Briefly, a physically reasonable constitutive model must treat stress as a second-rank tensor and must thus be coordinate-invariant; beyond that it must be "materially objective," meaning that the predicted material response must be independent of the reference frame of the observer. These very reasonable physical constraints result in significant mathematical complexity that would quickly take us beyond the scope of this chapter.

6.2 Shear rheometry

The principle of shear rheometry is straightforward: the material of interest is subjected to shear in a controlled way, and the mechanical response measured. Typically one either applies a controlled shear strain and measures the resulting shear stress, or applies a known shear stress and measures the strain. Commercial rheometers are available that can do either, and many modern instruments do a good job of both.

A schematic representation of a layer of fluid subjected to simple shear is shown in Figure 6.3. The fluid to be studied is sandwiched between two parallel plates separated by a gap d. In a strain-controlled experiment (Figure 6.3(a)), one plate – for concreteness, say the upper plate – is moved to the right at a constant speed v_0, while the other plate is fixed. The no-slip boundary condition ensures that the flow velocity is $v = 0$ at the bottom plate, while $v = v_0$ at the top plate. As a result, a uniform shear is set up in the fluid, and the strain rate within the fluid layer is $\dot{\gamma} = \partial v/\partial y = v_0/d$. The horizontal (shear) force required to hold the lower plate fixed in position is measured by a transducer and the corresponding shear stress calculated from the geometry of the experiment. In a stress-controlled measurement, shown in Figure 6.3(b), the lower plate is again fixed. A known force applied to the upper plate causes it to move, and its displacement and velocity are measured.

Most commercial shear rheometers are rotational rheometers, meaning that the rheometer tool has cylindrical symmetry and shear is applied by rotating the tool about its axis. In a strain-controlled shear rheometer, a motor rotates the lower element of the rheometer tool in a precisely controlled manner. Depending on the type of measurements being performed, this can mean rotation at a constant angular velocity, a sinusoidally varying angular velocity, or a rapid jump to a

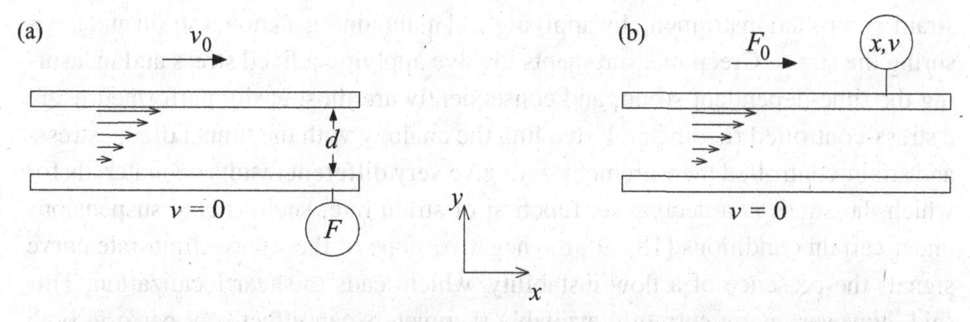

Figure 6.3 The distinction between strain- and stress-controlled deformation. (a) In a strain-controlled measurement, a fluid sample sandwiched between two plates is deformed by moving the upper plate at a controlled velocity v_0, while the bottom plate is kept stationary. The force required to hold the lower plate fixed is measured by a transducer, indicated by the circle. (b) In a stress-controlled measurement, a controlled shear force is applied to the top plate, and the resulting velocity is measured.

specified angular displacement. The upper element of the tool is connected to a sensitive torque transducer. The rotation of the lower element induces deformation in the fluid, which in turn causes a torque to be exerted on the upper element. This torque is used to calculate the shear stress, as outlined below for three common tool geometries. In the case of a stress-controlled rheometer, the lower element is rigidly fixed. A motor applies a controlled torque to the upper tool, causing it to rotate, and the resulting angular displacement and angular velocity are measured.

Many rheometers can also measure the normal (upward or downward) force produced in response to the shear. Normal forces are a tensor phenomenon and will not be discussed in detail here, but they are significant in many complex fluids. The normal force – or, more correctly, the normal stress difference – is the cause of the well-known rod-climbing (or Weissenberg) effect seen when a rotating rod is immersed in an elastic polymer solution [17].

The difference between stress- and strain-controlled measurements is analogous to the difference between voltage- and current-controlled measurements in electronics. In many cases, the same information can be obtained with either technique – a resistor has the same resistance no matter whether one measures it using a voltage supply or a current supply. On the other hand, the mode of measurement can make a difference for complex systems. The current–voltage curve for a tunnel diode, which has a negative differential resistance over a range of bias voltages, would look quite different in the two cases. It is thus important to be aware of the mode of operation and limitations of any instrument being used. There is really no "better" or "worse" mode, but some things are more easily measured in one mode than the other. For example, stress relaxation by definition involves measuring a time-dependent stress. This is most straightforwardly measured with a strain-controlled instrument, by applying and maintaining a known strain and measuring the stress. Creep measurements involve applying a fixed stress and measuring the time-dependent strain, and consequently are most easily performed using a stress-controlled rheometer. Extending the analogy with the tunnel diode, stress- and strain-controlled measurements will give very different results on materials for which the stress is a decreasing function of strain rate, such as clay suspensions under certain conditions [18, 19]; the negative slope of the stress–strain-rate curve signals the presence of a flow instability, which leads to shear localization. This said, however, many currently available rheometers can effectively perform both stress- and strain-controlled measurements by making use of fast computer control. For example, an instrument might actually function by applying a known torque to the tool, but its control system can continuously adjust the torque so that the net result is a constant strain rate. Given this, the distinction is perhaps not so much whether the instrument is stress- or strain-controlled, but whether the measurement transducer is on the part of the tool that is moved by the motor, or that which is fixed.

There are several high-quality rotational rheometers available commercially. Published specifications for high-end instruments quote minimum torques of a few times 10^{-9} Nm with a resolution of 10^{-11} Nm, minimum angular velocities on the order of 10^{-9} rad/s, and 10^{-8} rad resolution in angular position. As with any instrument, there is a trade-off between cost and performance, and less expensive instruments will have less impressive specifications. For the most accurate torque measurements, the body or frame of the rheometer should be very stiff, so that the torques applied to or generated by the sample act only on the rheometer tool and do not twist the instrument itself. High sensitivity and low frictional losses are extremely important for accurate measurements, particularly at low stresses or strain rates. Typically an air bearing or, in some commercial instruments, a magnetic thrust bearing, is used to reduce friction and inertial effects to a very low level. Other important features include the available range of applied shear rates or torques, the range of tool geometries, temperature control capabilities, and ease of use. Some rheometers have the capability for *in-situ* optical visualization of the sample while it is under shear, or even for small angle light scattering, which is useful for studies of structure under shear. All commercial instruments come with control software, which allows for several standard pre-programmed measurement sequences and has some data analysis capability; rheological software packages that perform more detailed data analysis can also be purchased.

There are three primary types of measurements that one can perform with a rotational rheometer – oscillatory shear, steady shear, and transient measurements. These are directly analogous to ac, dc, and transient electronic measurements.

The first two will be described in the context of controlled strain experiments, but analogous controlled stress measurements can also be performed.

Small-amplitude oscillatory shear measurements are used primarily to determine the linear viscous and elastic moduli. When a material is subjected to a small sinusoidal strain – small enough that the material remains in the linear viscoelastic regime – the resulting stress oscillates sinusoidally at the same frequency, but in general with a phase shift ϕ with respect to the strain. Thus, if γ is given by

$$\gamma = \gamma_0 \sin \omega t, \tag{6.12}$$

the resulting stress is

$$\sigma = \sigma_0 \sin (\omega t + \phi). \tag{6.13}$$

(If the applied strain takes the material beyond its linear regime, the response will include higher harmonics, as discussed in References [14, 15].) The stress in Equation (6.13) can be decomposed into a component in phase with the strain and another 90° out of phase:

$$\sigma = \sigma' + \sigma'' = \sigma_0' \sin \omega t + \sigma_0'' \cos \omega t. \tag{6.14}$$

The in-phase component is the elastic part of the response, with the elastic modulus defined as

$$G' = \frac{\sigma_0'}{\gamma_0}, \tag{6.15}$$

while the out-of-phase component is the viscous part, with the viscous modulus given by

$$G'' = \frac{\sigma_0''}{\gamma_0}. \tag{6.16}$$

Note that the viscous response is out of phase with the strain, but in phase with the strain rate $\dot{\gamma}$. It is common to define a complex modulus G^* as

$$G^* = G' + iG''. \tag{6.17}$$

G' is a measure of the elastic energy stored per unit volume per cycle, while G'' is a measure of the energy dissipated per unit volume per cycle [2]. Note that $G'(\omega)$ and $G''(\omega)$ are not independent quantities, but satisfy the Kramers–Kronig relations [20]; similarly, both are related to the relaxation modulus $G(t)$ discussed above by a Fourier transform [21].

For a Maxwell model with a single relaxation time φ, it is straightforward to show that G' and G'' depend on the frequency of oscillation as

$$G'(\omega) = G_0 \frac{\omega^2 \varphi^2}{1 + \omega^2 \varphi^2} \tag{6.18}$$

and

$$G''(\omega) = G_0 \frac{\omega \varphi}{1 + \omega^2 \varphi^2}. \tag{6.19}$$

Thus at low frequencies, $G' \propto \omega^2$ and $G'' \propto \omega$. This is in agreement with observations for linear polymer melts and micelle solutions [2]. More generally, the frequency dependence of the elastic and viscous moduli reflects the time dependence of the small-scale dynamics within the material.

In steady shear experiments, shear is applied at a series of constant strain rates and, once the system has reached a steady state, the stress is measured at each rate. Such experiments are inherently nonlinear, since the strain increases steadily with time for the duration of the experiment. The material typically exhibits a transient stress response as its microscopic structure is driven from its equilibrium state, then settles to a constant stress once the microstructure has reorganized into a new nonequilibrium configuration, which depends on the strain rate. The result of such a measurement is a plot of stress versus strain rate, commonly called a flow curve. Since the viscosity $\eta = \sigma/\dot{\gamma}$, $\eta(\dot{\gamma})$ is also determined from steady shear experiments.

In a transient measurement, a step strain is applied, and the resulting stress recorded as it relaxes. Similarly, a step stress can be applied and the strain measured as a function of time. The former is a stress relaxation experiment, and the latter a creep experiment. One can also apply shear at a fixed strain rate and record the stress transient discussed above. A Newtonian fluid dissipates – but does not store – energy, and so responds instantaneously to an applied stress or strain. The presence of a transient response is thus an indication of elastic effects, and consequently measurements of stress relaxation or other transient effects are useful as probes of the elastic behavior of a material. In particular, stress relaxation is a fairly direct method of determining the relaxation times of a material – φ in the simple Maxwell model described by Equation (6.8) above.

Several different geometries can be used for the rheometer tool, which contains and deforms the sample being studied. Three commonly used tool geometries are illustrated in Figure 6.4. Figure 6.4(a) shows a cone-and-plate tool. One part of the tool – here the lower part – is a horizontal plane, while the other part is a truncated cone. The gap between the cone and plate is set such that the extrapolated tip of the cone would just touch the surface of the plate. The angle between the surface

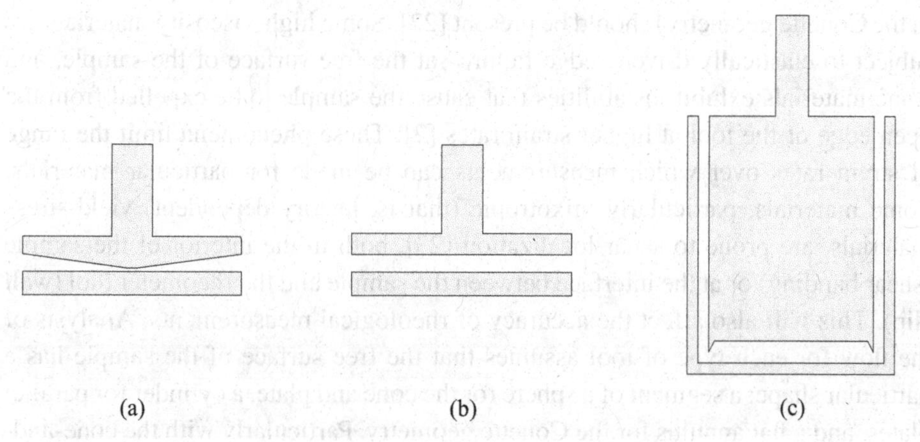

Figure 6.4 Three commonly used rheometer tool geometries. (a) Cone-and-plate; (b) parallel plates; (c) the Couette, or concentric cylinder, geometry. In all cases, one part of the tool is fixed and the other rotates about its axis.

of the cone and the plate is small, typically a few degrees (that is, so that the small-angle approximation is valid). The sample fills the gap between the two parts of the tool, remaining confined by surface tension. A parallel-plate tool is shown in Figure 6.4(b). As the name implies, both parts of the tool are planar, and the sample again fills the gap between them. In this case, the separation between the two parts of the tool can be varied. The Couette geometry, shown in Figure 6.4(c), consists of an outer cup of radius R_o and a coaxial inner cylinder of radius R_i and length L. The sample fluid fills the annular gap between the two cylinders, as well as the volume below the end of the inner cylinder. The shape and depth of this lower volume should be designed to minimize its effects on the measurements.

In practice, loading the sample into the rheometer tool cannot be done without causing some disruption to its structure. As an example, to load the cone-and-plate tool, the upper cone is raised well above the lower plate, an appropriate amount of the sample is placed on the plate, and the cone is then lowered to give the correct gap. As the cone is lowered, the sample is squeezed between the two parts of the tool. This clearly shears the sample in a somewhat uncontrolled way. This is addressed either by giving the sample adequate time to recover its equilibrium structure before measurements are started, or by subjecting it to a controlled "pre-shear" to prepare it in a reproducible initial state before starting to take measurements. It is also important to load the tool in such a way that the shape of the free surface at the perimeter of the cone-and-plate or parallel plate tools, or at the top of the annulus in the Couette geometry has the appropriate symmetry and is reproducible.

In all cases, the flow within the material under study must be steady and laminar, and no secondary flows due to hydrodynamic instabilities (such as Taylor vortices in the Couette geometry) should be present [22]. Some high-viscosity materials are subject to elastically driven "edge failure" at the free surface of the sample, and some materials exhibit instabilities that cause the sample to be expelled from the open edge of the tool at higher strain rates [2]. These phenomena limit the range of strain rates over which measurements can be made for particular materials. Some materials, particularly thixotropic (that is, history dependent) yield-stress materials, are prone to shear localization [23], both in the interior of the sample (shear banding) or at the interface between the sample and the rheometer tool (wall slip). This will also affect the accuracy of rheological measurements. Analysis of the flow for each type of tool assumes that the free surface of the sample has a particular shape: a segment of a sphere for the cone and plate, a cylinder for parallel plates, and a flat annulus for the Couette geometry. Particularly with the cone-and-plate and parallel-plate geometries, care must be taken to avoid evaporation at the free surface, which will change not only its shape, but also the fluid properties. In some cases, viscous heating of the sample, and the resulting change in fluid properties with temperature, must also be considered.

The relationship between the shear stress, which is the rheological quantity of interest, and the torque, which is the quantity actually measured (or applied), depends on the geometry of the tool. In the cone-and-plate geometry, Figure 6.4(a), simple shear flow is produced in the azimuthal direction when the plate is rotated at a constant angular velocity Ω, and, as long as the cone angle θ is small, the strain rate is the same everywhere in the sample. It is given by

$$\dot{\gamma} = -\frac{\Omega}{\theta}. \tag{6.20}$$

If the plate is rotating, the torque M on the cone is

$$M = \int_A (\text{stress})(\text{lever arm})\, d\mathrm{A} = \int_0^{2\pi} \int_0^R \sigma r^2 \, d\phi dr, \tag{6.21}$$

where σ here is the shear stress on the surface of the cone, and the integral is over that surface. Since the strain rate is constant, the shear stress is also constant and can be removed from the integral, in which case it is straightforward to show that

$$\sigma = \frac{3M}{2\pi R^3}. \tag{6.22}$$

Because the deformation of the sample is homogeneous and the strain rate and stress are independent of position, the cone-and-plate is the most commonly used geometry for studying soft materials.

In the parallel-plate geometry of Figure 6.4(b), the strain rate is simply given by

$$\dot{\gamma} = \frac{\Omega r}{h}, \tag{6.23}$$

where h is the spacing between the plates and r the radial coordinate. While h can be varied, it should be small compared to the radius R of the plates and large compared to the scale of any structure in the material (for example, the size of any suspended particles). In this geometry, the deformation is not uniform: Equation (6.23) shows that the strain rate depends on r. The shear stress can be calculated in terms of the torque to be

$$\sigma = \frac{M}{2\pi R^3}\left(3 + \frac{d \ln M}{d \ln \dot{\gamma}_R}\right), \tag{6.24}$$

where $\dot{\gamma}_R$ is the strain rate the the edge of the plate [2]. Note that in principle the derivative in Equation (6.24) has to be measured to permit an accurate determination of the stress (see, for example, Reference [24]), while rheometer software typically calculates an effective stress based on an approximation to Equation (6.24) which assumes a particular value for $d \ln M / d \ln \dot{\gamma}_R$ [2]. One should, in general, be aware how the numbers produced by the software are actually determined! Despite this, parallel-plate tools are simple to use, and by making h small can be used for measurements at high strain rates. Measurements at different values of h can also be used to check and correct for the presence of slip at the surfaces of the tool [25].

In the Couette geometry with the inner cylinder fixed and the outer cylinder rotating at a constant angular velocity Ω, the velocity field in the fluid is purely azimuthal. In this case, the shear stress σ on the inner cylinder at $r = R_i$ can straightforwardly be shown to be [2]

$$\sigma = \frac{M}{2\pi R_i^2 L}, \tag{6.25}$$

where R_i is the radius of the inner cylinder and L its length. If the gap between the cylinders is small, that is, if $R_o - R_i \ll R_i$, then the strain rate in the gap is uniform and is given by

$$\dot{\gamma} = \frac{\bar{R}\Omega}{R_o - R_i}, \tag{6.26}$$

where R_o is the radius of the outer cylinder, and $\bar{R} = (R_o + R_i)/2$. If the gap is not small (as is usually the case in practice), the strain rate varies across the gap and the analysis becomes more complex. As with the parallel-plate geometry, the strain rate then involves a derivative, in this case $d \ln M_i / d \ln \Omega_i$. In principle, this must be measured, but in practice an approximation is often used. As above, one should be aware of exactly how the rheometer software handles this situation. The Couette geometry is well-suited for measurements on low viscosity fluids because of the large surface area of the tool, and for measurements at high shear rate, as the fluid cannot easily be expelled from the gap between the cylinders.

We end this section with a few examples of rheological measurements on soft materials of interest in our research group. Figure 6.5 shows rheological data for Carbopol 2050, a commercial polymer microgel [26]. These data were obtained using a cone-and-plate tool on a strain-controlled rheometer. Carbopol consists of micron-sized particles of cross-linked polymer molecules [27]. The sample used here was prepared by dispersing Carbopol powder in deionized water at a concentration of 0.5 wt%. When initially mixed, the pH is about 3 and the microgel particles are about 2 μm in diameter. As the pH is increased, the polymer molecules become charged, causing the microgel particles to expand dramatically. The data shown in Figure 6.5 were taken at a pH of 6. At this pH and concentration, the Carbopol forms a fairly stiff gel that keeps its shape; it has a consistency similar to hair gel (of which it is often a major ingredient). The flow curve for this material is shown in Figure 6.5(a). An important feature of these data is that the stress approaches a nonzero value at low strain rates. This finite stress is called the yield stress. Since the viscosity $\eta = \sigma/\dot{\gamma}$, the finite yield stress implies that η diverges as $\dot{\gamma}$ approaches zero. This is illustrated in Figure 6.5(b). (As an aside, we note that there has been some controversy for several years as to whether the yield stress is "real," that is whether the viscosity of so-called yield-stress fluids at low stress is actually infinite, or just extremely high [28]. Certainly, for all practical purposes, the yield stress is real [29], and very recent experimental results strongly support the existence of a real yield stress in materials (like Carbopol), which do not show significant rheological aging [30].) The viscous and elastic moduli of Carbopol determined from small-amplitude (that is, linear) oscillatory shear experiments are plotted as a function of the angular frequency ω in Figure 6.5(c). The response of the material is solid-like, in the sense that $G' \gg G''$ over the entire frequency range accessible to the rheometer. Both G' and G'' are more-or-less independent of frequency, although the viscous modulus shows a weak minimum. This behavior is characteristic of a large class of so-called soft glassy materials, including foams, slurries, gels, and emulsions, and has been shown to result from small-scale structural disorder [31]. The lines plotted in Figure 6.5(c) were obtained from microrheological measurements and will be discussed in Section 3.

Figure 6.6 shows the relaxation modulus $G(\gamma, t) = \sigma(t)/\gamma$ of 2-hydroxyethylcellulose, a linear-chain polymer with a molecular weight of 720 000 g/mol, for several different values of the strain γ [33]. The sample was prepared in a reproducible initial state, strain was applied at time zero, and the stress $\sigma(t)$ was

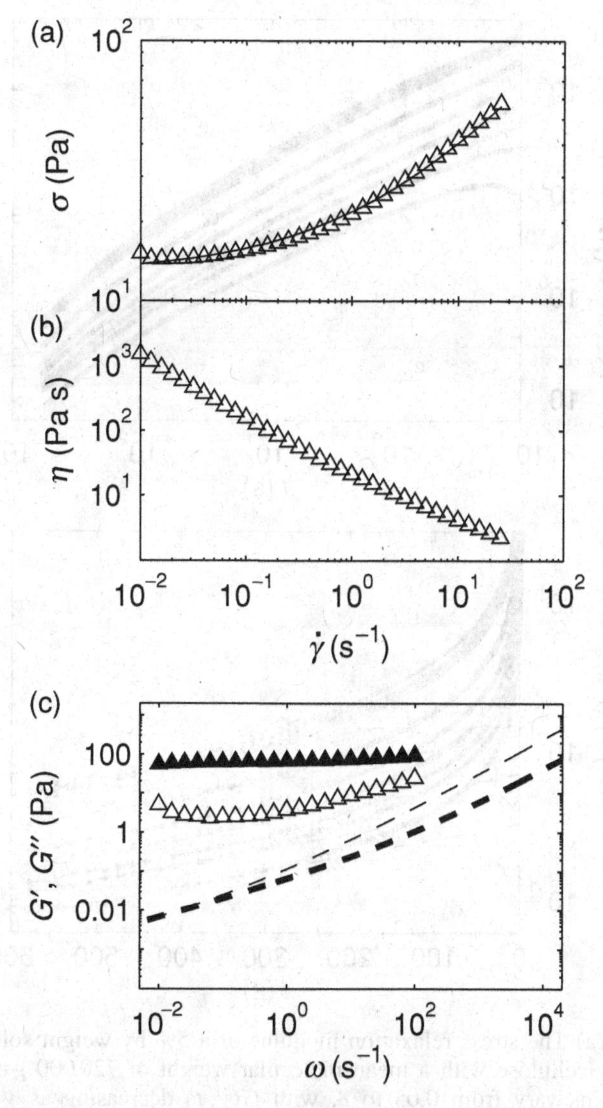

Figure 6.5 The rheological behavior of a pH-neutralized 0.5% by weight aqueous dispersion of Carbopol 2050, a polymer microgel. (a) The stress σ as a function of strain rate (flow curve), measured under steady shear with a controlled-strain rheometer using a cone-and-plate tool. The data approach a finite yield stress at zero strain rate. The curve is fit to a particular model for yield-stress behavior (the Herschel–Bulkley model), which describes the data reasonably well. (b) The viscosity $\eta = \sigma/\dot{\gamma}$ as a function of strain rate. (c) The solid and open triangles are the elastic and viscous moduli, G' and G'' respectively, plotted as a function of angular frequency. They were measured using small-amplitude oscillatory shear with a controlled-strain rheometer. The heavy and light dashed lines are the elastic and viscous moduli, respectively, determined from microrheological measurements [32]. They differ from the bulk measurements because the Carbopol is inhomogeneous on the micron scale, as discussed in the text.

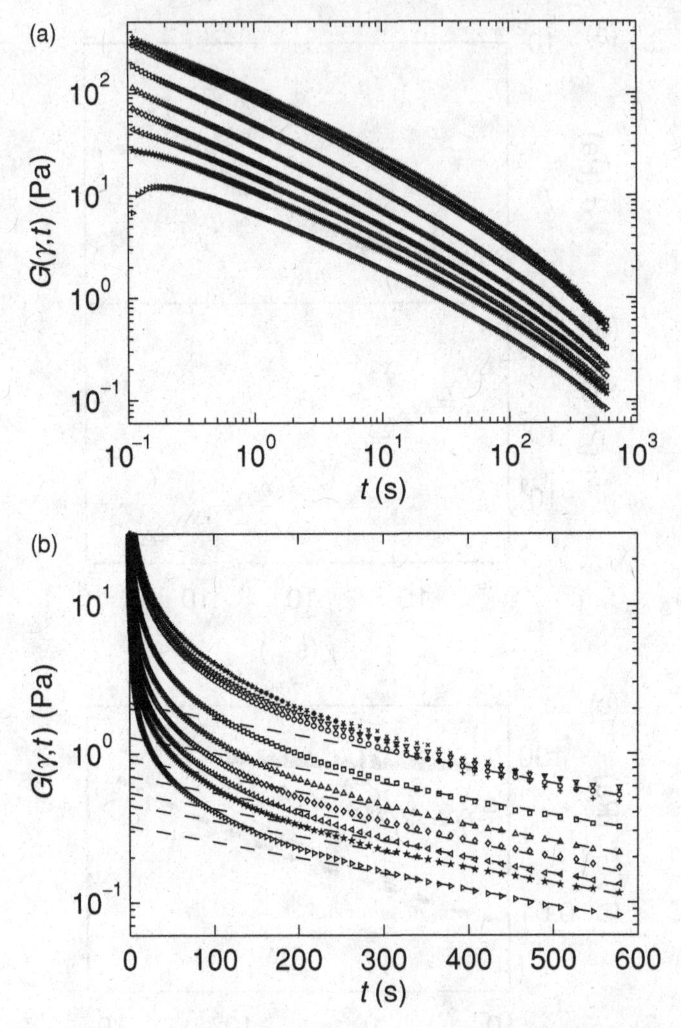

Figure 6.6 (a) The stress relaxation modulus of a 5% by weight solution of 2-hydroxyethylcellulose with a mean molecular weight of 720 000 g mol^{-1}. The applied strains vary from 0.05 to 8, with $G(\gamma, t)$ decreasing as γ increases. (b) The same data are shown on a semi-logarithmic plot. The dashed lines in (b) are fits of the late-time data to an exponential decay, which give a long-time relaxation time of 400 ± 10 s, independent of γ. The measurements were made with a controlled-strain rheometer using a cone-and-plate tool [33].

then recorded as a function of time. For small strains, the material remains in the linear regime, so $\sigma(0)$ is proportional to γ and $G(\gamma, t)$ is thus independent of γ. For larger strains, however, the structure of the material is changed – in this case, some of the entanglements between polymer molecules are lost – the material properties change, and so $G(\gamma, t)$ becomes dependent on γ. The behavior of $G(\gamma, t)$ shown in

Figure 6.6(b) is clearly not a simple exponential decay, reflecting the fact that the stress in this system relaxes through a variety of different processes, which have a large range of time scales. Fitting the data to a sum of five exponential decays (a generalized Maxwell model) gives relaxation times ranging from less than one second up to 400 s. The longest relaxation time is independent of γ, and is characteristic of the time required for the polymer molecules to completely re-establish their original state of entanglement [33, 34].

6.3 Microrheology

Shear rheometry provides information about the viscoelastic behavior of soft matter systems using samples that are typically a few milliliters in volume, and the gap between the parts of the rheometer tool is on the order of a millimeter. The rheological data that one obtains are thus some sort of average over that millimeter length scale. Since soft matter systems are structured on a much smaller length scale, however, their properties will be inhomogeneous on the same scale. Experimental techniques which probe the material properties and dynamics on that structural length scale, or, even better, as a function of length scale, can therefore provide a great deal of information that is not available from bulk measurements.

Microrheology involves the determination of the viscoelastic properties of materials on the micrometer scale. Here we will focus on multiple particle-tracking microrheology, which is arguably the most straightforward microrheological technique, both conceptually and experimentally. Particle-tracking microrheology involves directly following the thermally-driven motion of neutrally-buoyant microscopic tracer particles suspended in a viscoelastic fluid. This is a passive technique – no external forces are applied, and the particle motion is driven only by equilibrium thermal fluctuations. As a result, their motion automatically reflects the linear response of the system [35]. Microrheological measurements can often be performed on very small quantities of material – on the order of tens of microliters is easy – and so are well-suited for the study of expensive or hard-to-obtain materials such as some biopolymers. Direct imaging of the thermally-driven motion of the tracer particles provides information on the rheological properties of the material on the length scale of the particle size and in the region in which each individual particle moves [10, 11, 35, 36]. If the particles are large compared to the structural length scale of the soft material, bulk rheological parameters are recovered, while if they are small, the material's microstructure and spatial heterogeneity [11,32,37–40] can be explored. Correlations in the motion of pairs of tracer particles can be analyzed to give information about the viscoelastic properties as a function of length scale [41]. The diffusion of tracer particles has also been studied using interference microscopy [42, 43].

A variety of other microrheological techniques have also been developed. Dynamic light scattering from suspended tracer particles can measure viscoelastic properties to much higher frequencies than can be attained by mechanical rheometry [44,45]. Active techniques involving, for example, manipulation of suspended particles using laser tweezers [46,47] or magnetic forces [48] allow the study of both linear and nonlinear microrheology. A completely different class of methods for measuring micron-scale viscoelastic properties involves probing the material surface with the tip of an atomic force microscope [49,50]. Rheological techniques drawing on microfluidics have also been developed [9], while small-angle neutron and x-ray scattering are well-established and powerful techniques for studying the small-scale structure of soft materials [51]. Several recent reviews of microrheology provide an exposition of the development of the field as well as more detailed discussions of other microrheological techniques [5–9,52].

It is well-known that a microscopic particle suspended in a Newtonian fluid such as water undergoes Brownian motion, that is, random motion due to thermal fluctuations in the forces acting on the particle due to its collisions with liquid molecules. In 1905, Einstein argued that Brownian motion was direct evidence for the atomic nature of matter [53], and showed that the mean squared displacement of a particle of radius a undergoing Brownian motion grows linearly with time

$$\langle r^2(\tau)\rangle = 2dD\tau, \tag{6.27}$$

where r is the displacement, τ is the time, d the dimensionality of the motion, and D the diffusion constant. The angle brackets indicate an average over an ensemble of particles. This equation can be derived from a Langevin equation for the motion of a particle acted on by a randomly fluctuating force and a viscous drag [54]. D is related to the viscosity of the fluid by the Stokes–Einstein equation:

$$D = \frac{k_B T}{6\pi\eta a}, \tag{6.28}$$

which is itself a result of the fluctuation-dissipation theorem [54]. Here k_B is the Boltzmann constant and T the temperature.

In a viscoelastic fluid, however, the motion of a suspended particle is subdiffusive, that is, $\langle r^2(\tau)\rangle$ increases more slowly than linearly. The subdiffusive behavior is due to elasticity of the fluid; the fact that some energy is stored by elastic deformations of the medium leads to correlations in the particle's velocity fluctuations and consequently a slower increase in the mean squared displacement [10]. Alternatively, the motion of the suspended particles can be thought of as being restricted by the small-scale structure of the material. Note that from a practical point of view, the medium must be soft enough that there is enough energy in the thermal motion of the suspended particles to deform it by a measurable amount. In practice, this requires an elastic modulus less than of order 10 Pa for multiple particle-tracking experiments [6].

If $\langle r^2(\tau)\rangle$ does not vary too rapidly with τ, it is convenient to write

$$\langle r^2(\tau)\rangle \propto \tau^{\alpha}, \tag{6.29}$$

where the power law exponent α can itself be a (slowly varying) function of τ and can also depend on the length scale being probed. For soft-matter systems, $0 \leq \alpha \leq 1$. If $\alpha = 1$, the particle motion is purely diffusive, while if $\alpha = 0$, the mean squared displacement does not grow at all with time and the particle is trapped, either temporarily "caged" within transient pores formed by the microstructural elements of the material [55, 56] or constrained by the material's elasticity [11]. Which mechanism is at play can be determined from the way in which the plateau in $\langle r^2(\tau)\rangle$ depends on the particle size a [11]. A power law exponent $\alpha > 1$ would indicate the presence of a mean flow, which is possible in nonequilibrium or biological systems [57], but in the present context would typically be caused by a leak in the experimental cell.

One can account for the elastic forces exerted on the particle as it moves in response to the fluctuating collision force by adding a term involving a memory function into the Langevin equation, in analogy to the approach taken in the Maxwell model of viscoelasticity in Equation (6.10) above. This leads to a generalized Stokes–Einstein relation involving a frequency-dependent complex modulus, whose magnitude is related to the mean squared displacement [10]. The complex modulus is most rigorously calculated by taking a Laplace transform of the Langevin equation to obtain a modulus $\tilde{G}(s)$ in the Laplace domain, then using analytic continuation to obtain the complex modulus $G^*(\omega)$ in the Fourier domain [6, 7, 10, 35]. In practice, performing numerical Laplace or Fourier transforms on limited data sets introduces errors at either end of the frequency range. It is more accurate (and numerically simpler) to use Equation (6.29) to approximate $\langle r^2(\tau)\rangle$ by a power law locally, in which case the magnitude of the complex modulus can be calculated as

$$|G^*(\omega)| = \frac{k_B T}{\pi a \langle r^2(1/\omega)\rangle \Gamma(1+\alpha(\omega))}. \tag{6.30}$$

Here Γ is the gamma function and $\alpha(\omega)$ is the local power-law exponent of $\langle r^2(\tau)\rangle$, that is; the logarithmic derivative $d\ln\langle x^2(\tau)\rangle/d\ln\tau$, evaluated at $\tau = 1/\omega$. More details of the reasoning leading to Equation (6.30) can be found in References [6, 10, 35, 58, 59]. The real part of G^* is the elastic modulus of the material on the scale of the particle motion, and the imaginary part the viscous modulus, that is

$$G'(\omega) = |G^*| \cos(\pi\alpha(\omega)/2) \tag{6.31}$$

and

$$G''(\omega) = |G^*| \sin(\pi\alpha(\omega)/2). \tag{6.32}$$

It is possible to rewrite Equation (6.30) as

$$\langle r^2(\tau)\rangle = \frac{k_B T}{\pi a J(\tau)} \tag{6.33}$$

to give the mean squared displacement in terms of the creep compliance $J(\tau)$. Larsen and Furst [60] have recently used microrheology to study the gel transition in physical and chemical polymer gels using $J(\tau)$ directly rather than converting $\langle r^2(\tau)\rangle$ into frequency-dependent elastic and viscous moduli, thereby avoiding the numerical issues mentioned above.

The moduli calculated using Equations (6.31) and (6.32) describe the material of interest on the length scale of the particle size a. Suppose the length scale characterizing the microscopic structure of the material is r. If $a \gg r$, the diffusing particles effectively average over the structural heterogeneities of the material, and the microrheological values of the viscous and elastic moduli will be the same as the bulk values. However, if $a \ll r$, different particles will sample a range of microrheological environments, and the properties of these micro-environments may not be the same as those of the bulk material. In this case the microrheological moduli will in general differ from those measured in bulk experiments, and variations in the microscopic moduli among a group of tracer particles can be used to study local inhomogeneities in the structure and properties of the material [11, 13, 37, 38, 40]. Similarly, measurements covering a range of values of a/r provide information about the micron-scale structure [39, 41].

It has been shown theoretically that the response function of a particle diffusing in a viscoelastic fluid is well approximated by the generalized Stokes–Einstein relation as long as inertial effects can be neglected and the size of the particle is large compared to both the amplitude of its oscillations and the structural length scale of the material, so that the material can be treated as a continuum [58, 59]. These conditions are met over a range of frequencies, limited at the high end by the viscous dissipation rate, which has been estimated to be at least 10^5 Hz, and at the low end by the decay rate of the longitudinal compression mode of the medium, which is on the order of 10 Hz [58, 59]. Under these conditions, microrheology will measure bulk properties. It is worth noting that the case $a \ll r$ discussed above explicitly violates the assumption of homogeneity, and also that the frequency range covered by most particle-tracking experiments extends below 10 Hz. Some caution is thus required in quantitatively interpreting data obtained under these conditions, but such data nonetheless contain useful information about the small-scale structure and properties of the material.

The apparatus required to perform multiple particle-tracking microrheology measurements is straightforward to assemble and relatively economical, the most expensive component being a research-quality optical microscope. Tracer particles are suspended in the fluid of interest, which is placed in a sample holder that is typically fabricated from microscope slides as described below. The sample is placed on the stage of the microscope and the particles imaged using a video camera mounted on the microscope and interfaced to a computer. While particle-tracking measurements can be done using a standard upright microscope, an inverted microscope is more convenient simply because its design provides more space on the microscope stage. This allows for flexibility in the design of the holder and can simplify sample handling. All modern research microscopes can accommodate a video camera. It is convenient to mount the camera on one microscope port, while using a second port for direct viewing of the sample by eye.

We have typically worked with sample volumes of the order of 100 μl in cells 1 cm by 1 cm by 1 mm thick, but it would be straightforward to reduce the horizontal dimensions of the cell substantially when working with scarce or expensive materials. Since this technique is based on actually imaging the tracer particles, the fluid being studied must be sufficiently transparent that the particles can be visualized through a sample on the order of a millimeter thick. We have fabricated sample chambers from microscope slides as illustrated in Figure 6.7. Three small glass bars are cut from a microscope slide and glued with optical cement onto a second slide to form a U-shaped wall. A cover slip is then glued on top of these bars to form a covered chamber with one end open. Once the glue has set, the experimental fluid is injected into the open end of the chamber using a dropper or other appropriate tool, and the open side is sealed with an inert silicone-based grease. Other chamber geometries can easily be fabricated. Pre-made microscope well slides and cover slips could also be used. Note that evaporation or a small leak in the chamber will likely lead to a mean flow in the sample fluid, which will in turn result in a nonzero mean displacement of the particles from one image to the next. This will affect the measurement of $\langle r^2 \rangle$. If the mean flow is small, it can be compensated for by subtracting the mean displacement from all individual particle displacements when the images are analyzed, but it is far better to build a chamber that does not leak.

The tracer particles should be neutrally buoyant in the fluid of interest, at least to the extent that they do not settle significantly over the duration of the experiment. Polystyrene latex spheres with a density of 1.05 g/cm^3 are suitable for aqueous

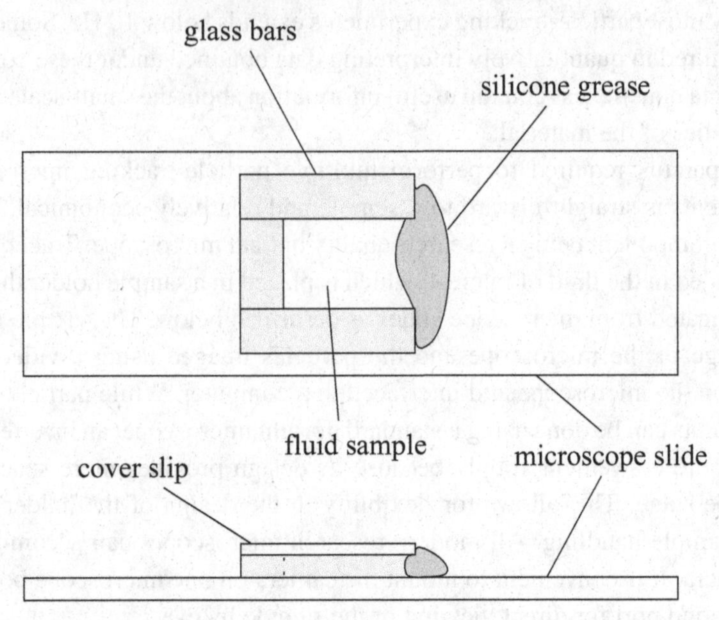

Figure 6.7 A sketch of a sample cell used in a particle-tracking microrheology experiment. See the text for more details.

solutions and are available in a range of sizes from several sources. These can be obtained with a variety of chemical coatings, and care should be taken to ensure that the tracer particles are truly passive, that is, that they do not interact either chemically or electrostatically with the material being probed [45, 61]. This can usually only be confirmed by experiment. Smaller particles display larger and more easily measurable displacements, and performing measurements as a function of particle size allows characterization of the microstructural length scale in the complex fluid [39], as noted above. If direct bright-field microscopy is used, the minimum usable particle diameter determined by the diffraction limit of the microscope optics is typically a bit less than 0.5 μm. An alternative is to image fluorescently dyed particles with a fluorescence microscope, in which case the motion of particles smaller than 0.1 μm in diameter can be followed. The operation of a fluorescence microscope is shown schematically in Figure 6.8. The fluorescent dye is excited by light in a particular wavelength band and fluoresces at longer wavelengths. Light in the excitation band is reflected by a dichroic mirror through the microscope objective and onto the sample. The longer-wavelength emitted light is collected by the objective and passes directly through the dichroic mirror to the detector. Filter sets corresponding to the excitation and emission wavelengths of a range of fluorescent dyes are available as accessories for microscopes designed for fluorescent imaging.

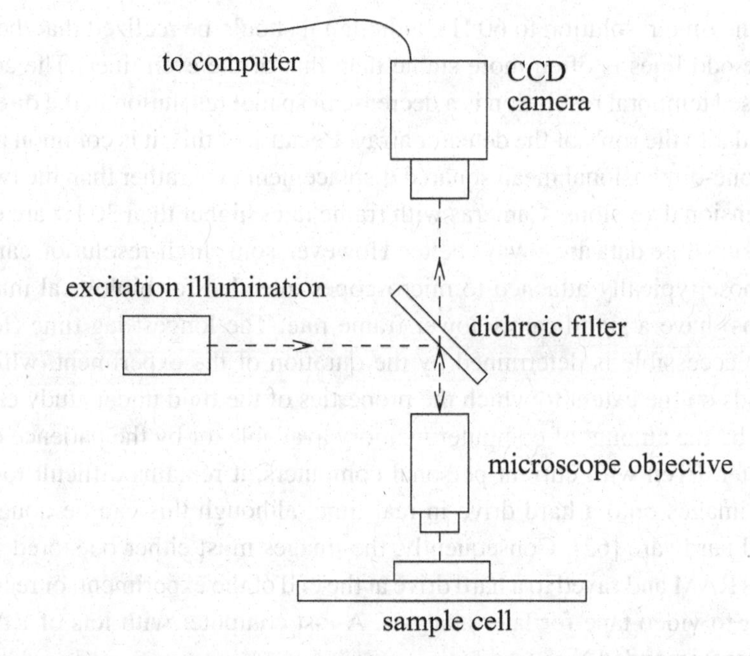

Figure 6.8 A schematic illustration of the setup for a particle-tracking microrheology experiment using a fluorescence microscope.

To get good statistics in the calculation of the mean squared displacement, the trajectories of a large number of particles must be averaged, so there should be many tracer particles in the microscope's field of view. At the same time, if the particles are too close together, it is difficult to unambiguously determine their trajectories, as discussed below. We have used particle concentrations of the order of several times 10^{-5} by volume, giving roughly 50 particles in the field of view with a 40× objective. With this objective, the spatial resolution in our system is about three pixels per micron. The field of view should be located away from the walls of the sample cell so that they do not affect the motion of the particles. Similarly, the microscope should be focused at approximately mid-height within the sample cell.

The minimum lag time that can be studied, or equivalently the maximum frequency, is set by the frame rate of the video camera being used. The standard (North American) video frame rate is 30 frames per second. CCD cameras normally record an image as two interlaced fields, one consisting of rows 1, 3, 5, . . . in the CCD array, and the other of rows 2, 4, 6, ... A tracer particle can easily move a distance corresponding to one pixel in 1/60 s, so it is important to separate the two fields and analyze them individually to avoid smearing of the particle image. (Similarly, the shutter speed of the camera must be fast enough that particles do not move significantly while the image is being captured.) Treating fields separately increases the time resolution to 60 Hz, although it should be realized that the scan time of the odd lines is often more stable than that of the even lines. The cost of this increased temporal resolution is a decrease in spatial resolution in the direction perpendicular to the rows of the detector array. Because of this, it is common to calculate the one-dimensional mean squared displacement $\langle x^2 \rangle$ rather than the two- or three-dimensional versions. Cameras with frame rates higher than 30 Hz are easily available, and more data are always better. However, some high-resolution cameras (such as those typically attached to microscopes intended for biological imaging applications) have a significantly lower frame rate. The longest lag time (lowest frequency) accessible is determined by the duration of the experiment, which in turn depends on the extent to which the properties of the fluid under study change with time, by the amount of computer memory available, or by the patience of the experimenter. Even with current personal computers, it remains difficult to store full video images onto a hard drive in real time, although this can be done with specialized hardware [62]. Consequently, the images must either be stored in the computer's RAM and saved to a hard drive at the end of the experiment, or recorded digitally or to video tape for later analysis. A fast computer with lots of RAM is therefore recommended.

Once the video images of the diffusing particles have been recorded, they must be analyzed to extract the trajectory of each particle, and ultimately the mean squared displacement of an ensemble of particles. Several freely available and commercial image processing software packages exist that can be used to do this analysis. Particle-tracking software written in the IDL programming language by Crocker *et al.* [63] is freely available [64] and has been used extensively within the soft matter community. MATLAB versions of the same package are also available [65]. We will briefly describe the method used by this software [63, 64] (see also Chapters 7 and 10 in this volume).

The steps involved in the image analysis process are illustrated in Figure 6.9. Digitized images may contain imperfections such as nonuniform contrast and background intensity variations. The first step in the image processing filters the image and subtracts a background image (recorded with the same optical system and illumination, but without tracer particles) to reduce these variations. The brightness of the images of the individual tracer particles will vary depending on how well focused they are and, for fluorescent spheres, can be affected by photobleaching. Pixels brighter than a threshold level are considered as candidate particles if no other pixel within a specified distance, chosen to be larger than the apparent radius of a single particle but smaller than the average separation between them, is brighter. The software allows the use of parameters such as eccentricity and image size to further discriminate true particles from doubtful ones, and can produce an image with circles drawn around the identified particles to reassure the user. Once the

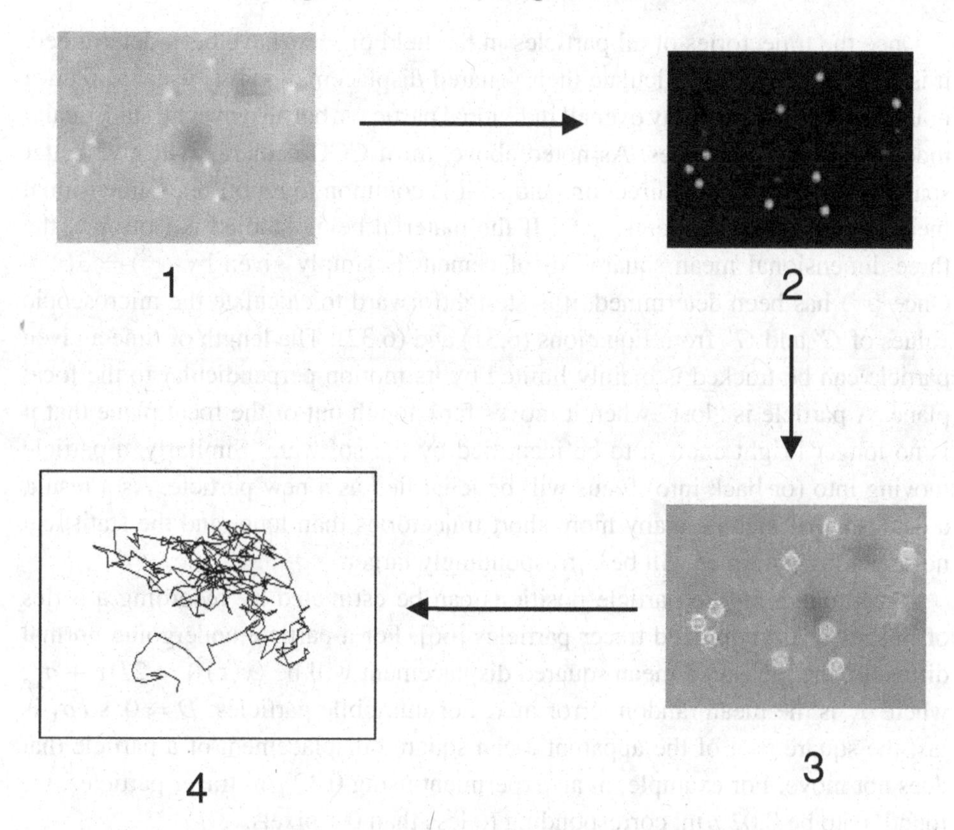

Figure 6.9 Steps in the image processing. 1 The raw image captured by the CCD camera. The white spots are possible particle images. 2 The image after filtering and subtracting a background image. 3 The original image with circles drawn around the features identified as particle images. In 4, the positions of one particle obtained from a series of successive images are linked to give its trajectory.

particles have been satisfactorily identified, the software refines their positions to sub-pixel accuracy by calculating the centroid of the brightness distribution of each particle's image.

The final stage in the procedure is to match particles in each image with their most likely counterparts in later images. For indistinguishable particles, this is only possible by proximity, and unambiguous identification of particles from one frame to the next is possible only if the typical particle displacement at each time step is sufficiently small compared to the typical interparticle spacing b. The most likely assignment of particle labels from one frame to the next is the one which minimizes $\sum_{i=1}^{N} dr_i^2$, where dr_i is the displacement of particle i between frames and N is the number of particles. The dr_i are constrained to be shorter than a length L, chosen based on a plot of the distribution of displacements to be in the range $dr_i < L < b/2$.

Once the trajectories of all particles in the field of view have been determined, it is straightforward to calculate their squared displacements. It is usual to reduce noise by averaging not only over all individual particles, but also over all statistically independent starting times. As noted above, most CCD cameras will give better spatial resolution in one direction, and so it is common to report one-dimensional mean squared displacements, $\langle x^2 \rangle$. If the material being studied is isotropic, the three-dimensional mean squared displacement is simply given by $\langle r^2 \rangle = 3\langle x^2 \rangle$. Once $\langle r^2 \rangle$ has been determined, it is straightforward to calculate the microscopic values of G' and G'' from Equations (6.31) and (6.32). The length of time a given particle can be tracked is mainly limited by its motion perpendicular to the focal plane. A particle is "lost" when it moves far enough out of the focal plane that it is no longer bright enough to be identified by the software. Similarly, a particle moving into (or back into) focus will be identified as a new particle. As a result, a data set will include many more short trajectories than long, and the statistical noise at long lag times will be correspondingly larger.

Uncertainties in the particle positions can be estimated by recording a series of images of immobilized tracer particles [66]. For a particle undergoing normal diffusion, the measured mean squared displacement will be $\langle x(\tau)^2 \rangle = 2D\tau + \sigma_x^2$, where σ_x is the mean random error in x. For immobile particles, $D = 0$, so σ_x is just the square root of the apparent mean squared displacement of a particle that does not move. For example, in an experiment using 0.49 μm tracer particles, we found σ_x to be 0.02 μm, corresponding to less than 0.1 pixels.

Figure 6.10 shows data obtained in a particle-tracking experiment on a 0.4% by volume suspension of Laponite clay particles in water [13]. Laponite is a synthetic clay consisting of disk-shaped particles about 30 nm in diameter by 1 nm thick [67]. When initially mixed, the suspension is very fluid, but it forms a stiff gel over a period of about an hour. Figure 6.10 shows the one-dimensional mean squared displacement $\langle x(\tau)^2 \rangle$ for a series of times t after the start of the experiment. The sample was prepared at $t = 0$ by sonicating it to destroy any pre-existing structure. The video images from which $\langle x(\tau)^2 \rangle$ was calculated were recorded over time intervals short enough that the material properties are constant for a given set of data, but Figure 6.10 indicates that they change significantly over longer times. $\langle x(\tau)^2 \rangle$ is well described by a power law in τ for all ages t, but the power law exponent α decreases from close to one initially to zero as the material gels [13]. Figure 6.11 shows selected tracer particle trajectories from the same experiment. At early times, the particles move quite freely and their trajectories are close to Brownian random walks, although, as can be seen in Figure 6.11, they tend to spend substantial periods of time stuck in one region before wandering over a relatively short time from one "cage" to another. This behavior, which has also been observed in other soft matter systems [39, 55, 56], is the cause of the sublinear growth

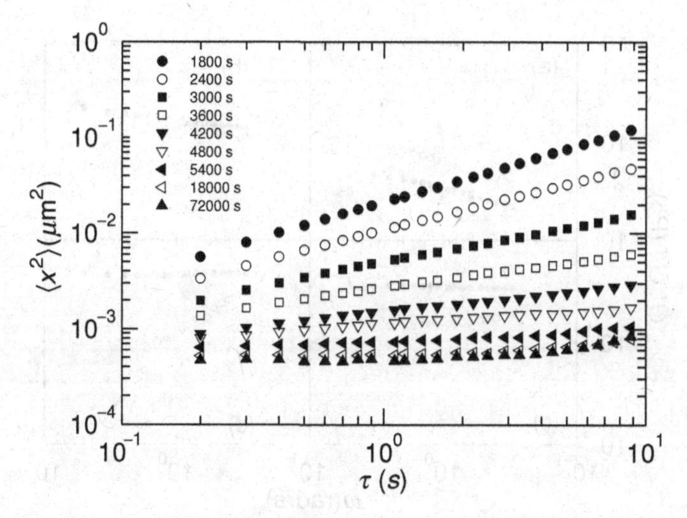

Figure 6.10 The mean squared displacement of 0.5 μm radius fluorescent spheres in an aqueous suspension of Laponite clay particles, measured at a series of times after the sample was prepared at $t = 0$. The logarithmic slope of $\langle x(\tau)^2 \rangle$ decreases over time as the material forms a gel [13].

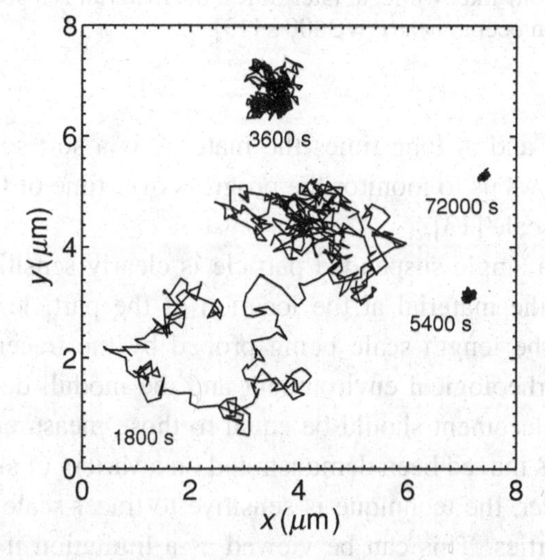

Figure 6.11 The trajectories of individual particles in the Laponite suspension recorded at different times as the material gels. At early times, the particles move quite freely, while, at later times, they are completely immobilized by the small-scale structure of the material [13].

of the mean squared displacement. At later times, the particles become completely immobilized. The microrheological moduli calculated from $\langle x(\tau)^2 \rangle$ are plotted in Figure 6.12 for four different times from the start of the experiment. Initially $G'' > G'$ and the tracer particles a viscous fluid. As time passes, however, the

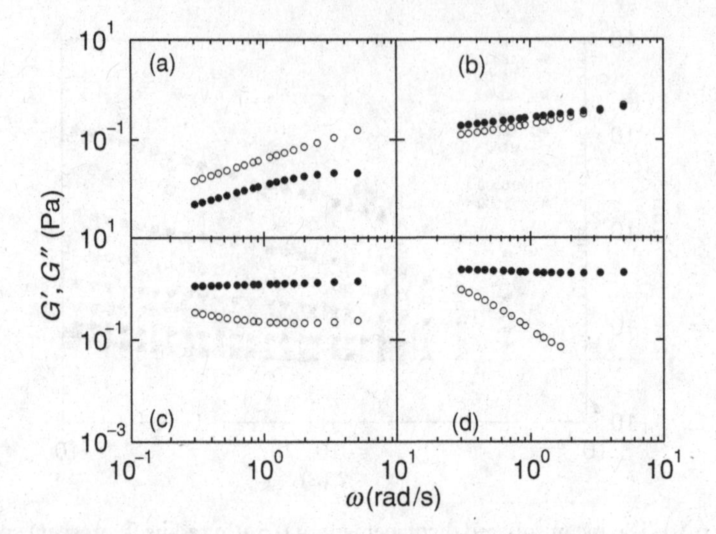

Figure 6.12 The microscopic elastic and viscous moduli of Laponite, determined via Equations (6.31) and (6.32) from the mean squared displacements plotted in Figure 6.10. The data were obtained (a) 1800 s, (b) 3600 s, (c) 5400 s, and (d) 72000 s after the start of the experiment. At early times, $G'' > G'$ and the suspension is liquid-like, while, at later times, the material is a gel with $G' > G''$. The gel transition occurs near $t = 3600$ s [13].

moduli cross over, and at long times the material is a soft solid with $G' > G''$. Microrheology allows us to monitor the progress over time of the gel transition at the micron length scale [13].

The motion of a single suspended particle is clearly sensitive to the rheological properties of the material at the location of the particle. If the material is homogeneous on the length scale being probed by the tracers, then all tracers will see a similar rheological environment and the moduli determined from the mean squared displacement should be equal to those measured using bulk shear rheometry. This has indeed been demonstrated on a variety of simple and complex fluids [11]. However, the technique is sensitive to tracer-scale heterogeneities in the material properties. This can be viewed as a limitation if the goal is to use microrheology to determine bulk rheological properties [41], but one can also take advantage of this sensitivity to probe the heterogeneity itself, and to learn about the structure of the material on the micron length scale [11, 13, 26, 37, 38, 40]. For example, in a homogeneous material, the squared displacements of a collection of individual tracer particles at a given lag time should be normally distributed, and averaging them would give a mean squared displacement that is characteristic of the material as a whole. In an inhomogeneous material, in contrast, the distribution of squared displacements will not be normal, and statistical techniques can be used to investigate the range of rheological microenvironments that are present [11, 38].

The lines plotted in Figure 6.5(c) above show the microscopic elastic and viscous moduli of a 0.5% dispersion of Carbopol [32]. These data were obtained from the mean squared displacement of an ensemble of tracer particles suspended in the Carbopol using Equations (6.31) and (6.32). While at lower Carbopol concentrations the microscopic and bulk viscoelastic moduli are in agreement (data not shown), at this concentration they differ substantially. This is because the dispersion is inhomogeneous on the micron scale, and, as discussed above, the tracer particles see a rheological microenvironment that has very different properties from the bulk material. In particular, the material is predominantly viscous over most of the frequency range plotted in Figure 6.5(c), while on the bulk scale it is a soft elastic solid.

This result is examined in more detail in Figure 6.13, which shows histograms of the distribution $N(\alpha)$ of the logarithmic slopes of the squared displacements of each individual tracer particle moving in dispersions of Carbopol [26]. The Carbopol concentration increases as one goes from panel (a) to panel (d) in Figure 6.13. At the lowest concentration, $N(\alpha)$ has a single peak close to $\alpha = 1$, corresponding to diffusive tracer motion. This indicates that the dispersion is more-or-less homogeneous and essentially purely viscous at this concentration. As the microgel concentration is increased, however, a second peak at a lower value of α appears in the distribution. In this range of concentrations, some of the particles still move diffusively, while others move subdiffusively. In other words, the material has become inhomogeneous, with some particles seeing a purely viscous environment and others seeing a viscoelastic environment. This inhomogeneity is reflected in the moduli calculated from Equations (6.31) and (6.32). If one neglects the heterogeneity and average over all tracer particles, one obtains the microscopic elastic and viscous moduli shown by the dashed lines in Figure 6.5(c), which are substantially lower than those of the bulk material. At the highest concentration shown in Figure 6.13, $N(\alpha)$ is sharply peaked at $\alpha = 0$, indicating that the material is again homogeneous (at least on the length scale probed by the tracer particles) and completely elastic; the tracer particles are immobilized.

The motion of an individual tracer particle in a viscoelastic fluid sets up a fluctuating strain field, which in turn affects the motion of other tracer particles. Because of this, correlations in the motion of pairs of particles can provide information about the rheological properties on the scale of the separation between them. This is the basis of two-particle microrheology [41]. This technique has the advantage that it gives moduli independent of the way the tracer particles couple to the medium, meaning that interactions between the tracer particles and the material under study can be neglected [41].

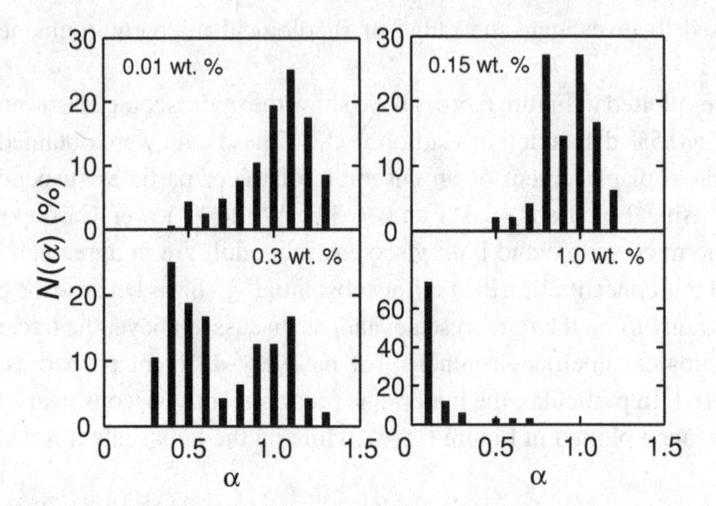

Figure 6.13 Histograms showing the distribution $N(\alpha)$ of the logarithmic slopes of the squared displacements of particles suspended in a Carbopol dispersion. At low Carbopol concentration, $N(\alpha)$ has a single peak near $\alpha = 1$, indicative of diffusive motion. As the concentration is increased, a second peak appears at a lower value of α, indicating that a fraction of the tracer particles experience a more elastic environment [33].

The apparatus required to do two-particle microrehology is identical to that described above for single-particle measurements; the only difference is in the data analysis. Since two-particle measurements involve calculating correlations between the motion of pairs of particles, they typically require substantially more data than in single-particle measurements for the same level of accuracy. Crocker *et al.* [41] confirmed that two-particle microrheology gave results that were consistent with bulk measurements in heterogeneous materials, while the moduli obtained from single-particle measurements were quite different from the bulk values. Two-particle microrheology has been used to study length-scale dependent rheology in inhomogeneous materials such as actin networks [41, 68].

6.4 Summary

Rheological measurements probe the bulk mechanical properties of soft materials directly, while microrheology provides a window on their local properties and structure on length scales one thousand times smaller. In combination, these experimental techniques provide one path towards addressing one of the key goals and challenges of research on soft materials, which is to understand the connection between their bulk material properties and their microscopic structure.

Acknowledgments

Portions of the research presented here were carried out with support from the Natural Sciences and Engineering Research Council of Canada and the Canadian Space Agency.

References

[1] R. G. Larson, *The Structure and Rheology of Complex Fluids* (Oxford University Press, New York, 1999).
[2] C. W. Macosko, *Rheology: Principles, Measurments, and Applications* (Wiley-VCH, New York, 1994).
[3] F. A. Morrison, *Understanding Rheology* (Oxford University Press, New York, 2001).
[4] R. W. Whorlow, *Rheological Techniques* (Ellis Horwood, New York, 1992).
[5] F. C. MacKintosh and C. F. Schmidt, "Microrheology," *Curr. Opin. Coll. Interf. Sci.* **4**, 300 (1999).
[6] M. L. Gardel, M. T. Valentine, and D. A. Weitz, "Microrheology," in K. Breuer (ed.), *Microscale Diagnostic Techniques* (Springer, New York, 2005).
[7] T. A. Waigh, "Microrheology of complex fluids," *Rep. Prog. Phys.* **68**, 685 (2005).
[8] N. Willenbacher and C. Oelschlaeger, "Dynamics and structure of complex fluids from high frequency mechanical and optical rheometry," *Curr. Opin. Colloid Interface Sci.* **12**, 43 (2007).
[9] C. Pipe and G. H. McKinley, "Microfluidic rheometry," *Mech. Res. Comm.* **36**, 110 (2009).
[10] T. G. Mason and D. A. Weitz, "Optical measurements of frequency-dependent linear viscoelastic moduli of complex fluids," *Phys. Rev. Lett.* **74**, 1250 (1995).
[11] M. T. Valentine, P. D. Kaplan, D. Thota, J. C. Crocker, T. Gisler, R. K. Prud'homme, M. Beck, and D. A. Weitz, "Investigating the microenvironments of inhomogeneous soft materials with multiple particle tracking," *Phys. Rev. E* **64**, 061506 (2001).
[12] R. Bird, R. Armstrong, and O. Hassager, *Dynamics of Polymeric Liquids, Volume 1: Fluid Mechanics* (Wiley, New York, 1987).
[13] F. K. Oppong, P. Coussot, and J. R. de Bruyn, "Gelation on the microscopic scale," *Phys. Rev. E* **78**, 021405 (2008).
[14] M. Wilhelm, "Fourier transform rheology," *Macromol. Mater. Eng.* **287**, 83 (2002).
[15] R. H. Ewoldt, A. E. Hosoi, and G. H. McKinley, "New measures for characterizing nonlinear viscoelasticity in large amplitude oscillatory shear," *J. Rheol.* **52**, 1427 (2008).
[16] J. C. Maxwell, "On the dynamical theory of gases," *Phil. Trans.* **157**, 49 (1867).
[17] D. V. Boger and K. Walters, *Rheological Phenomena in Focus* (Elsevier, Amsterdam, 1993).
[18] P. Coussot, A. I. Leonov, and J.-M. Piau, "Rheology of concentrated dispersed systems in a low molecular weight matrix," *J. Non-Newtonian Fuid Mech.* **46**, 179 (1993).
[19] F. Pignon, A. Magnin, and J.-M. Piau, "Thixotropic colloidal suspensions and flow curves with minimum: identification of flow regimes and rheometric consequences," *J. Rheol.* **40**, 573 (1996).
[20] P. M. Chaikin and T. C. Lubensky, *Principles of Condensed Matter Physics* (Cambridge University Press, Cambridge, 1995).

[21] D. J. Ferry, *Viscoelastic Properties of Polymers*, 3rd edn (Wiley, New York, 1980).
[22] R. G. Larson, "Instabilities in viscoelastic flows," *Rheol. Acta* **31**, 213 (1992).
[23] P. C. F. Møller, S. Rodts, M. A. J. Michels, and D. Bonn, "Shear banding and yield stress in soft glassy materials," *Phys. Rev. E* **77**, 041507 (2008).
[24] H. Tabuteau, J. R. de Bruyn, and P. Coussot, "Drag force on a sphere in steady motion through a yield-stress fluid," *J. Rheol.* **51**, 125 (2007).
[25] A. Yoshimura and R. K. Prud'homme, "Wall slip corrections for Couette and parallel disk viscometers," *J. Rheol.* **32**, 53 (1988).
[26] F. K. Oppong and J. R. de Bruyn, unpublished.
[27] Noveon technical data sheet 216 (2002); www.pharma.noveon.com/literature/tds/tds216.pdf
[28] H. A. Barnes and K. Walters, "The yield stress myth?" *Rheol. Acta* **24**, 323 (1985).
[29] G. Astarita, "The engineering reality of the yield stress," *J. Rheol.* **34**, 275 (1990).
[30] P. C. F. Moller, A. Fall, and D. Bonn, "No steady state flows below the yield stress: a true yield stress at last?," arXiv:0904.1467v1 (2009); to be published.
[31] P. Sollich, F. Lequeux, P. Hébraud, and M. E. Cates, "Rheology of soft glassy materials," *Phys. Rev. Lett.* **78**, 2020 (1997).
[32] F. K. Oppong, L. Rubatat, B. J. Frisken, A. E. Bailey, and J. R. de Bruyn, "Microrheology and structure of a yield-stress polymer gel," *Phys. Rev. E* **73**, 041405 (2006).
[33] F. K. Oppong, Ph.D. thesis, University of Western Ontario (2009).
[34] M. Doi and S. F. Edwards, *The Theory of Polymer Dynamics* (Oxford University Press, New York, 1986).
[35] T. G. Mason, "Estimating the viscoelastic moduli of complex fluids using the generalized Stokes-Einstein equation," *Rheol. Acta*, **39**, 371 (2000).
[36] T. G. Mason, K. Ganesan, J. H. van Zanten, D. Wirtz, and S. C. Kuo, "Particle tracking microrheology of complex fluids," *Phys. Rev. Lett.* **79**, 3282 (1997).
[37] J. Apgar, Y. Tseng, E. Federov, M. B. Herwig, S. C. Almo, and D. Wirtz, "Multiple-particle tracking measurements of heterogeneities in solutions of actin filaments and actin bundles," *Biophys. J.* **79**, 1095 (2000).
[38] Y. Tseng and D. Wirtz, "Mechanics and multiple-particle tracking microheterogeneity of α-actinin-cross-linked actin filament networks," *Biophys. J.* **81**, 1643 (2001).
[39] I. Y. Wong, M. L. Gardel, D. R. Reichman, E. R. Weeks, M. T. Valentine, A. R. Bausch, and D. A. Weitz, "Anomalous diffusion probes microstructure dynamics of entangled f-actin networks," *Phys. Rev. Lett.* **92**, 178101 (2004).
[40] F. K. Oppong and J. R. de Bruyn, "Diffusion of microscopic tracer particles in a yield-stress fluid," *J. Non-Newtonian Fluid Mech.* **142**, 104 (2007).
[41] J. C. Crocker, M. T. Valentine, E. R. Weeks, T. Gisler, P. D. Kaplan, A. G. Yodh, and D. A. Weitz, "Two-point microrheology of inhomogenous soft materials," *Phys. Rev. Lett.* **85**, 888 (2000).
[42] B. Schnurr, F. Gittes, F. C. MacKintosh, and C. F. Schmidt, "Determining microscopic viscoelasticity in flexible and semiflexible polymer networks from thermal fluctuations," *Macromol.* **30**, 7781 (1997).
[43] F. Gittes, B. Schnurr, P. D. Olmstead, F. C. MacKintosh, and C. F. Schmidt, "Microscopic viscoelasticity: shear moduli of soft materials determined from thermal fluctuations," *Phys. Rev. Lett.* **79**, 3286 (1997).
[44] J. L. Harden and V. Viasnoff, "Recent advances in DWS-based micro-rheology," *Curr. Opin. Colloid Interface Sci.* **6**, 438 (2001).

[45] B. R. Dasgupta, S.-Y. Tee, J. C. Crocker, B. J. Frisken, and D. A. Weitz, "Microrheology of polyethylene oxide using diffusing wave spectroscopy and single scattering," *Phys. Rev. E*, **65**, 051505 (2002).
[46] M. T. Valentine, L. E. Dewalt, and H.D. Ou-Yang, "Forces on a colloidal particle in a polymer solution: a study using optical tweezers," *J. Phys. Condens. Matter* **8**, 9477 (1996).
[47] E. M. Furst, "Applications of laser tweezers in complex fluid rheology," *Curr. Opin. Colloid Interface Sci.* **6**, 438 (2005).
[48] T. R. Strick, M. N. Dessinges, G. Charvin, N. H. Dekker, J. F. Allemand, D. Bensimon, and V. Croquette, "Stretching of macromolecules and proteins," *Rep. Prog. Phys.* **66**, 1 (2003).
[49] R. E. Mahaffy, C. K. Shih, F. C. MacKintosh, and J. Käs, Scanning probe-based frequency-dependent microrheology of polymer gels and biological cells," *Phys. Rev. Lett.* **85**, 880 (2000).
[50] N. Yang, K. K.-H. Wong, J. R. de Bruyn, and J. L. Hutter, "Frequency-dependent viscoelasticity measurement by atomic force microscopy," *Meas. Sci. Technol.* **20**, 025703 (2009).
[51] T. Zemb and P. Lidner (eds.), *Neutrons, X-rays and Light: Scattering Methods Applied to Soft Condensed Matter* (Elsevier, Amsterdam, 2002).
[52] A. Mukhopadhyay and S. Granick, *Curr. Opin. Colloid Interface Sci.* **6** 423 (2001).
[53] A. Einstein, *Annalen der Physik* **17**, 549 (1905).
[54] G. D. J. Phillies, *Elementary Lectures in Statistical Mechanics* (Springer, New York, 2000).
[55] E. R. Weeks and D. A. Weitz, "Subdiffusion and the cage effect studied near the colloidal glass transition," *Chem. Phys.*, **284**, 361 (2002).
[56] E. R. Weeks and D. A. Weitz, "Properties of cage rearrangements observed near the colloidal glass transition," *Phys. Rev. Lett.* **89**, 095704 (2002).
[57] A. M. Reynolds, *Phys. Lett. A* **342**, 439 (2005).
[58] A. J. Levine and T. C. Lubensky, "One- and two-particle microrheology," *Phys. Rev. Lett.* **85**, 1774 (2000).
[59] A. J. Levine and T. C. Lubensky, "Response function of a sphere in a viscoelastic two-fluid medium," *Phys. Rev. E* **63**, 041510 (2001).
[60] T. H. Larsen and E. M. Furst, "Microrheology of the liquid-solid transition during gelation," *Phys. Rev. Lett.* **100**, 146001 (2008).
[61] M. T. Valentine, Z. E. Perlman, M. L. Gardel, J. H. Shin, P. Matsudaira, T. J. Mitchison, and D. A. Weitz, "Colloid surface chemistry critically affects multiple particle tracking measurements of biomaterials," *Biophys. J.* **84**, 4004 (2004).
[62] M. Keller, J. Schilling, and E. Sackmann, *Rev. Sci. Instrum.* **72**, 3626 (2001).
[63] J. C. Crocker and D. G. Grier, *J. Coll. Interf. Sci.* **179**, 298 (1996).
[64] E. Weeks and J. C. Crocker, "Particle tracking using IDL," http://www.physics.emory.edu/~weeks/idl/
[65] M. Kilfoil, http://www.physics.mcgill.ca/~kilfoil/downloads.html; D. Blair and E. Dufresne, http://physics.georgetown.edu/matlab/.
[66] D. S. Martin, M. B. Forstner, and J. A. Käs, "Apparent subdiffusion inherent to single particle tracking," *Biophys. J.* **83**, 2109 (2002).
[67] Southern Clay Products, Inc., Gonzales, Texas; http://www.laponite.com
[68] J. Liu, M. L. Gardel, K. Kroy, E. Frey, B. D. Hoffman, J. C. Crocker, A. R. Bausch, and D. A. Weitz, "Microrheology probes length scale dependent rheology," *Phys. Rev. Lett.* **96**, 118104 (2006).

7

Particle-based measurement techniques for soft matter

NICHOLAS T. OUELLETTE

7.1 Introduction

Optical measurement techniques are ubiquitous across many disciplines of science. Since so much of our everyday interaction with the world around us is based on vision, optical investigations feel very natural and, when possible, are often the preferred interrogation technique.

While the sophistication of optical measurement has certainly increased with time, the basic idea remains the same: the physical system of interest is illuminated with visible light (which is assumed to interact with the system only passively) and is imaged by a photosensitive detector. In the early days of optical measurement, film was the preferred medium, and optical studies were by and large static. The few dynamic measurements that were made involved long or multiple exposures and tedious reconstructions done by hand [1]. With the advent of digital imaging and large-scale digital storage, however, optical measurements can now easily address dynamic questions, with enough temporal and spatial resolution to measure a wide range of quantities and sufficient storage to gather enough samples for well-converged statistics. In physics, we are often interested in velocities and accelerations, which give us access to momentum, energy, and force; such dynamic quantities are now accessible with imaging techniques.

Before discussing particular imaging techniques, let us draw the distinction between qualitative and quantitative optical measurement. Qualitative measurements have been made for a very long time, particularly in fluid mechanics, where they date back to Prandtl [2]. Also known as *flow visualization*, qualitative techniques seek to visualize otherwise invisible information, such as, say, the streamline pattern of a flow, much as iron filings can be used to "see" the field of a permanent magnet. Pure visualization, however, does not give *quantitative* information about

Experimental and Computational Techniques in Soft Condensed Matter Physics, ed. Jeffrey Olafsen. Published by Cambridge University Press.

the flow. To make an optical measurement quantitative, it must be well calibrated, robust against noise, and we need to know what exactly is being measured.

In this chapter, we discuss some of the ideas behind quantitative optical measurement, as well as some specific techniques that have become standard tools. The first technique we describe, in Section 3, is Particle Image Velocimetry (PIV), which has become ubiquitous in fluid mechanics and is rapidly being adopted in other areas of soft condensed matter physics. PIV relies on the assumption that nearby elements of a continuum system move similarly, and uses this information to generate velocity fields.

Sometimes, however, we want to study the time-resolved motion of individual points rather than the statistical properties of regions; in these cases, the tracking techniques described in Section 4 can be applied. True particle tracking gives us not only the instantaneous positions, velocities, and accelerations of points in the system, but also their full histories. We describe some ways in which this extra information is currently being used to deepen our understanding of dynamically complex systems.

Most real physical systems are three-dimensional; cameras, however, record two-dimensional projections of the three-dimensional world. In Section 5, we describe several methods that have been developed for reconstructing the full three-dimensional information we often desire.

Finally, in Section 6, we discuss several future directions for optical measurement science. We focus on the difficult problem of measuring three-dimensional, highly resolved field information in rapidly evolving systems, on extensions of particle-based measurements to nonkinematic quantities, and on potential technological improvements.

First, however, in Section 2 we discuss some generic considerations of digital imaging, as they are common to all the measurement techniques presented here.

As digital imaging and computing technology continues to advance at a rapid pace, the techniques described here (as well as their future iterations) will become more and more widely adopted standard tools, and will help us to better characterize the rich dynamics of soft condensed matter systems.

7.2 Digital imaging

7.2.1 Film versus pixels

Traditionally, images were recorded on film. When the mechanical shutter of a camera was opened, its lens system focused light onto a photosensitive emulsion, which would later be chemically developed. The spatial resolution of such a system (meaning, essentially, the smallest feature that could be resolved accurately) can be very high, determined by the size of the crystals in the emulsion.

Analog imaging, however, has many serious drawbacks as a scientific tool. Chief among these is the requirement that the film images must either be processed by hand, which is time consuming and often subjective, or must later be digitized to be put on the computer, a process that is lossy and that reduces the spatial resolution of the image. In addition, since the imaging process is mechanical, the imaging speed (and therefore the temporal resolution of the measurement) is severely limited by the physical components of the camera.

In a digital imaging system, the physical film is replaced by an electronic photosensitive array that converts the incident light to electric current, which can then be amplified and measured. The quantum efficiencies of digital image sensors tend to be much higher than those for film: digital sensors perform much better in low-light conditions since they register more of the incoming photons. Each element of the array (corresponding to a pixel in the output image), however, integrates the light that it collects, coarse-graining the image at the pixel scale. The spatial resolution of the resulting image is therefore limited by the size and spacing of the pixels. Because there are (in principle) no moving parts, however, the imaging rate and therefore temporal resolution is limited only by the sensitivity of the photosensitive elements and the rate at which information can be read from them, and commercial cameras that can operate in the kilohertz range are now commonplace. Since digital images never exist in hard-copy form, the number of images that can be collected and stored is limited only by hard-drive capacity, leading to an enormous advantage over film imaging. The images can also be processed easily by computational means, as we discuss below. In addition, electronic detectors can be constructed to image not only visible light, but also other ranges of the electromagnetic spectrum.

Modern digital cameras fall into two categories: charge-coupled devices (CCDs) and complimentary metal-oxide semiconductor (CMOS) sensors. As a rule of thumb, CCD cameras tend to have smaller pixels and somewhat greater sensitivity, while CMOS sensors feature on-pixel electronics (so that they are sometimes called "active pixel sensors") and faster readout times; there are, of course, exceptions in each category. The fastest cameras on the market today tend to be based on CMOS sensors; cameras designed for increased sensitivity tend to be CCD based. From the standpoint of the techniques presented here, the underlying technology on which the image sensor is based is relatively unimportant: either CCD or CMOS may be used with equal success.

7.2.2 Noise and imaging errors

As with any measurement device, digital cameras are noisy. We can roughly break the sources of noise into three classes: photon counting, intrinsic noise from the detector, and pixelation effects.

Digital detectors work by collecting photons on each pixel and converting them to electrons, which can then be read out. This photon collection is a random process, and is governed by Poisson statistics. There is therefore an intrinsic noise associated with photon counting, which is independent of the quality of the detector. Photon-counting noise is a small effect for well-lit situations, but can be nonnegligible for low light conditions.

Since both CCDs and CMOS sensors rely on semiconductor physics, thermal effects can also play a role and produce noise. As the temperature rises, thermally excited electrons and phonons can interact with the electrons that carry the photocurrent, leading to noisy images. Additionally, even when no light falls on the detector, a digital sensor at finite temperature will produce a signal known as *dark current*. The solution to both of these problems is cooling: as the camera temperature decreases, both the dark current and thermal noise also decrease. Typically, cameras designed for high sensitivity are therefore actively cooled.

Finally, the physical arrangement of the photosensitive elements into finite-size pixels can lead to undesired effects. In theory, all the incident light falling onto a pixel is integrated to produce the signal. In practice, however, pixels are not uniformly sensitive; their sensitivity tends to fall off near the edges. There are also unavoidable dead regions between pixels. Because of these effects, position measurements from digital images (made via the various methods discussed below) are biased to lie nearer to the centers of the pixels than to their edges. This effect is known as *peak locking* and is unavoidable, as it derives from the physical construction of the detector. With properly chosen imaging parameters, however, it can be minimized.

7.3 Particle image velocimetry

In many situations of interest, the dynamics of a physical system can be well characterized by its velocity. Velocity measurements allow access to the momentum, and tell us directly about the evolution of the relative positions of various points. For a continuum system like a fluid, we desire the velocity field, that is, a mapping of a velocity vector to (ideally) each point in space. Once we know the velocity field, its spatial gradients give us information about strain rate and therefore stress, and we can in principle completely determine the system dynamics.

In this section, we describe a technique for measuring the full velocity field that has become a powerful and widely used tool in fluid mechanics [3], and is beginning to find application more generally in soft matter physics. Based on optical imaging of the motion of many closely spaced flow tracers, this technique of PIV is noninvasive and under proper conditions can be highly accurate.

In this section, we discuss the ideas that underlie PIV, as well as its basic algorithmic implementation. As PIV has become more widely used, several extensions to the basic algorithm have been proposed to increase its accuracy; we therefore also describe some of these extensions. Finally, we discuss the advantages and disadvantages associated with the closely related technique of particle tracking velocimetry (PTV).

7.3.1 Tracers

It is almost tautological to say that optical measurement depends on having something to see. If our system of interest has spatial inhomogeneities in, say, material composition (such as granular media), optical tags (such as specific fluorescent labels in a biological system), or the index of refraction (as in a thermal convection system), imaging of these inhomogeneities can give us the dynamical information we desire. But very frequently we wish to study a homogeneous material, such as a single-phase fluid. In that case, all points look the same; so what can we see?

For homogeneous media, we are forced to make approximations. Instead of imaging the fluid itself, we study the motion of small, solid *tracer particles* that are assumed to follow the flow of the continuum system faithfully. While the full dynamics of a small solid particle in a fluid flow are quite complex [4], the tracer assumption is valid when the particles are both neutrally buoyant and very small compared to the relevant dynamical length scales. For small particles, the degree to which the particles follow the flow is captured by the Stokes number:

$$\mathrm{St} = \frac{2}{9}\frac{\rho_p}{\rho_f}\frac{a^2}{\nu\tau}, \tag{7.1}$$

where ρ_p and ρ_f are the densities of the particle and the suspending fluid, respectively, a is the particle radius, ν is the kinematic viscosity, and τ is the smallest dynamical time scale of the flow. Typically, particles with Stokes numbers of order 10^{-3} or smaller are valid tracers. Note, however, that recent work has suggested that the effects of finite size and finite density can be quite different [5], and that the minimum Stokes number required to make a particle a good flow tracer may depend on the quantity being measured [6].

Common materials for tracer particles include plastics (primarily polystyrene or polymethylmethacrylate) or hollow glass spheres, all available commercially in a wide range of sizes. We note that it is very difficult to find good tracers for gas flows, since the density of most solids is so much larger than typical gas densities. Under some conditions, however, either hollow glass spheres [7] or helium-filled soap bubbles [8] can be used.

7.3.2 Basic PIV algorithm

Unlike some of techniques we will discuss below, PIV does not attempt to measure the motion of individual tracers. Rather, it is a statistical measurement that reports the average motion of groups of nearby tracers that are within the same subregion, or *interrogation region*, of the image. Implicit in the method is the assumption that particles in an interrogation region move similarly, and that apparent differences in their behavior can be ascribed to noise rather than to real dynamical effects. As we note below, under some conditions this assumption is inappropriate, and PIV will provide untrustworthy results; but, when used correctly, PIV is very effective at providing reliable, quantitative velocity fields. For robust PIV, then, the size of the interrogation regions must be determined by the properties of the velocity field. Subsequently, the loading density of tracer particles is adjusted so that each interrogation region contains many particles, leading to converged statistics.

Single-exposure optical systems can only measure position information; without time-resolved measurements, dynamics cannot be measured. The canonical PIV method therefore records two images separated by a short time lag, δt. Measurement of the displacement field between the images then yields the velocity field when scaled by δt, in a finite-difference approximation. For such a measurement, δt must be kept small for two reasons. First, the quality of the finite-difference derivatives increases as the time lag between the images decreases. Additionally, PIV systems typically illuminate thin slices of the flow of interest rather than thick volumes; the high seeding density of tracer particles required by the method makes volumetric illumination impossible. Keeping δt small therefore minimizes uncertainties due to out-of-plane motion of the tracers.

Early PIV measurements often used multiple exposures on photographic film in order to maximize the spatial resolution of the image and to minimize δt [9]. In this type of experiment, δt is determined solely by the pulse rate of the light source. More modern systems based on digital cameras[1] typically record two images in rapid succession. Commercial PIV systems provide image pairs with very small values of δt by using specialized "PIV cameras" and dual-head lasers sent through cylindrical lenses to compress the beams into laser sheets. Most such systems are not, however, capable of *sustained* high-speed acquisition, and so the time lag between individual velocity fields is often much longer than δt; note, though, that as camera technology develops, systems are becoming available that are capable of time-resolved or "cinema" PIV.

The problem now becomes one of computing the displacement field between the two images. We desire a single displacement vector for each interrogation region of the image. To find this vector, we compute the cross-correlation function:

$$R_{12}(\mathbf{x}) = \iint W_1(\mathbf{x}) I_1(\mathbf{x}) W_2(\mathbf{x}+\mathbf{s}) I_2(\mathbf{x}+\mathbf{s}) d\mathbf{s}, \tag{7.2}$$

[1] Such techniques are sometimes known as "DPIV," for digital particle image velocimetry.

where I_1 and I_2 are the spatially resolved intensities of the two images and W_1 and W_2 are window functions that specify the interrogation domain in each image [10]. If the PIV conditions are met, R_{12} will show a pronounced peak corresponding to the mean displacement of the particles in the interrogation region. The displacement vector for the interrogation region is then given by the center of this peak, typically estimated either by taking its centroid or by fitting it to a Gaussian profile. In either case, the displacement vector can generally be found to sub-pixel accuracy, usually to roughly 0.1 pixels or better [3, 10, 11].

7.3.3 Error control

The uncertainty of the velocity field returned by PIV is primarily determined by the accuracy with which we can determine the displacement field, which in turn comes from the accuracy with which we can determine the center of the mean-displacement peak in the cross-correlation function $R_{12}(\mathbf{x})$ defined in Equation (7.2). In general, the image intensities I_1 and I_2 can be broken into the sum of their mean values $\langle I_1 \rangle$ and $\langle I_2 \rangle$ and the fluctuations about these means I_1' and I_2'. R_{12} can then be broken into the sum of three contributions [3, 10]:

$$R_{12}(\mathbf{x}) = R_D(\mathbf{x}) + R_C(\mathbf{x}) + R_F(\mathbf{x}). \tag{7.3}$$

R_D is the cross-correlation of I_1' and I_2'; it is the part we are interested in, since it tells us how the particles are moving. R_C is the cross-correlation of $\langle I_1 \rangle$ and $\langle I_2 \rangle$, which tells us only about characteristics of the global image illumination. It contains no dynamical information, and yet may obscure the R_D signal that we want to measure. R_F contains the two cross terms, namely the cross-correlations of one mean image with the other fluctuation. As with R_C, R_F contains only information that we do not want.

In practice, we can often eliminate R_C and R_F by subtracting the mean illumination from each image. If the illumination is uniform and if the particle seeding is homogeneous, then removing the background intensity will force $\langle I_1 \rangle = \langle I_2 \rangle = 0$, and R_C and R_F will vanish. Note, however, that if either of these conditions is not met, these unwanted terms will remain. If the illumination is not uniform, removing the spatially averaged mean image intensity will not remove the mean intensity of any individual interrogation region; instead $\langle I_1 \rangle$ and $\langle I_2 \rangle$ become functions of space, and cannot be made to vanish everywhere by subtracting a constant value. If the particle seeding is not homogeneous, then $\langle I_1 \rangle$ and $\langle I_2 \rangle$ cannot be removed by subtraction of a single value even if the illumination is uniform, since the scattered light will still be nonuniform; a signal is only recorded when particles are physically present.

Now let us consider R_D. Just as we did with the image intensities, we can break R_D into a mean $\langle R_D \rangle$, known as the displacement-correlation peak, and a fluctuation R'_D, known as the random correlation. The displacement-correlation peak contains the average displacement of the interrogation region; the random correlation contains the fluctuations about this average, due both to noise and to nonuniform movement of the particles in the region. PIV will only provide a reasonable result if the peak in $\langle R_D \rangle$ is significantly higher than the maximum value of R'_D; otherwise, the algorithm will report a spurious value. The ratio of these two peak heights gives a lower bound on the signal-to-noise ratio of PIV.

We note finally that it is common practice to locate such spurious vectors in a postprocessing step by comparing the results found for nearby interrogation regions. If the velocity vector found for one interrogation region is significantly different from its nearest neighbors, it is most likely incorrect (assuming, as always, that nearby points in the flow move similarly). The correct vector can then be approximated by interpolation [3].

7.3.4 Applicability and limitations

When the assumptions underlying PIV are met, it can be a very useful tool. It is not, however, an appropriate technique for all situations, including several that are very common in soft matter systems.

As we have seen, PIV assumes locally correlated tracer motion. If the nearby motion of particles is, however, uncorrelated, the algorithm will produce spurious results. In particular, PIV should be applied to Brownian particles only with great caution. If the Brownian particles are embedded in a shear flow, for example, PIV can often successfully find the shear velocity field, though with a lower signal-to-noise ratio than the corresponding non-Brownian case [12]: the random motion of the individual tracers enhances R'_D relative to $\langle R_D \rangle$ in the image cross-correlation. If PIV is applied to a pure Brownian system without an imposed flow, however, it will return very poor results: the random correlation term will dominate, and any velocity vectors found will be spurious.

PIV also assumes that the tracer seeding is homogeneous. In bulk flow, this is a reasonable assumption: nothing breaks the spatial symmetries, and the particles will remain (approximately) homogeneously distributed. Near an interface, however, the homogeneity assumption may fail. The interface may be one between two fluids or may be a solid wall; in either case, the tracer seeding density will fall off close to the interface, leading to biased correlation functions (due to the cross terms between the mean image intensities and the intensity fluctuations). PIV is therefore only fully valid away from such interfaces.

7.3.5 Extensions

When PIV is implemented exactly as described above, using a simple single-pass image cross-correlation analysis, its quantitative accuracy can be compromised by many different effects. Since it has become so ubiquitous in fluid mechanics, many extensions have been proposed in order to compensate for the shortcomings of the basic algorithm. We describe some of these methods briefly here; a more complete discussion can be found in [13] or [3].

In the basic algorithm, the ring of pixels around the edge of each interrogation region contributes very little to the measurement: in the interframe time δt, some particles will leave the region, while others will enter it. In both cases, they will not contribute to the displacement-correlation peak. If the assumption of locally correlated motion holds, however, we should be able to recover the information contained in these edge particles by appropriately shifting the interrogation region in the second image relative to its position in the first image. This idea is at the heart of iterative window-shifting methods. A first cross-correlation pass determines the rough displacement vector associated with the interrogation region. The region in the second image is then shifted based on this displacement vector, and the cross-correlation analysis is performed a second time. In principle, this window-shifting can be performed multiple times until the change in the determined displacement vector drops below some threshold. As the interrogation region is shifted, we may also choose to refine the windowing, changing the number of pixels in the interrogation region to capture the local motion most effectively.

In most situations, the assumption that all particles in the interrogation region move identically will be false; instead, there is likely to be some velocity gradient across the interrogation region. We can attempt to correct for such a gradient by *deforming* the interrogation region between the two images in addition to shifting it. Typically, one applies a linear deformation to the region, amounting to keeping the first term in the Taylor expansion for the velocity gradient.

Most commercial PIV packages have support for these and other analysis techniques beyond the standard PIV algorithm, in order to increase the robustness of the method. We also note that new PIV extensions are rapidly being proposed in order to increase the accuracy of the technique.

7.3.6 Particle Tracking Velocimetry

Particle Tracking Velocimetry (PTV) is the low-seeding-density limit of PIV, and relies on a fundamentally different technique to generate similar data. The end result is the same as in traditional PIV: a velocity field defined on a regular grid. Instead of measuring the average displacement vectors of interrogation regions that contain many particles, however, PTV measures the displacements of individual tracer particles by explicitly locating them in the image pairs. Since we seek to locate the individual particles, the seeding density must be lower: particles must be well separated in space so that their images are nonoverlapping. To find the displacements, either a nearest-neighbor approach is used (described more fully below), or an iterative method based on prior cross-correlation analysis. This latter technique is sometimes known as "super-resolution" PIV, since it returns multiple velocity vectors for each PIV interrogation region [14].

7.4 Particle tracking and Lagrangian measurement

PIV, as described above, is a powerful technique for measuring *field* variables; it gives us direct access to the velocity field, which under suitable conditions can be integrated or differentiated to give us strain or acceleration fields. But since PIV measures only statistical averages of the motion of tracers, it cannot give us information about individual tracers. PTV, the low-density limit of PIV, does measure the displacement of individual particles, but does not maintain long-time information, and still leaves us with a velocity field as the output information.

In many cases, however, the individual motion of each particle is the real quantity we want, rather than the mean velocity field. Consider, for example, a granular material or a colloidal suspension. The most natural description of such particulate systems is in terms of the particles themselves, rather than a coarse-grained field description. And even for true continuum systems, such as homogeneous fluids, problems of mixing and transport are most elegantly described in a "Lagrangian" framework based on the dynamics of hypothetical "material points" or "fluid elements." What we desire for Lagrangian measurement, therefore, is not explicitly the field variables but rather the long-time trajectories of individual tracers. If the density of tracers is sufficient, we can still construct fields from them (as in classical PTV), but we retain more information that can be used to extract more dynamical information, as we describe below. Additionally, since we have more history for each tracer, we can apply higher-order time differentiation schemes than are possible with PIV, including polynomial fitting [15,16] or convolution with a smoothing and differentiating kernel [17,18].

A Lagrangian tracking system must accomplish two main tasks: we must locate each tracer spatially at every instant of time, and we must then solve the correspondence problem to match successive positions of the same tracer together in time. If the number of tracers in view is very small and if they are well separated in space, tracking is not a difficult problem. If, however, the tracer particles are densely seeded or if they move quickly or randomly, tracking can become very challenging. In this section, we describe strategies for solving the correspondence

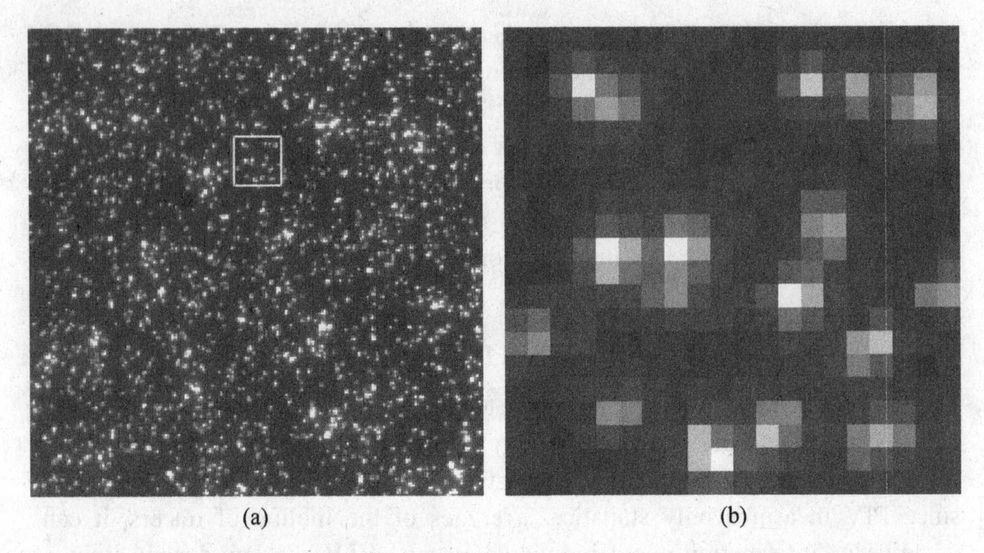

Figure 7.1 (a) Sample image of tracer particles. (b) Detailed view of several particle images from the white box in part (a).

problem for different types of systems. First, however, we briefly review methods for locating individual tracers with sub-pixel resolution.

7.4.1 Tracer identification

In principle, any well-defined feature in an image may be tracked. Suppose, for instance, that we want to track a square object as it moves through space, we could then follow the motion of each of its corners as time evolves, thereby capturing both the translational and rotational degrees of freedom. Most commonly, though, tracking algorithms are applied to the case of spherical, featureless tracer particles, such as those used for PIV; that is the case on which we will focus here.

A sample image of 80 micron-diameter spherical tracer particles in water is shown in Figure 7.1. Several features are immediately apparent. First, the particles appear as bright spots on a dark background; this is the typical case for imaging either by scattering or fluorescence. Second, as shown in more detail in Figure 7.1(b), the brightness of the particle image is largest at its center and falls off towards its edges. Because of this feature, we associate each local intensity maximum with a particle.

Several complicating factors arise in this kind of imaging that make accurately locating the particles a difficult problem. The discretization of the particle image by the camera pixels can obscure the expected smooth intensity falloff of the particle image, and can lead to large fluctuations as the particle moves over pixel boundaries due to peak locking. The experimental illumination of the imaging may be nonuniform, and particles passing through dimly lit regions will change their appearance on the camera sensor; similar effects are seen for particles that move into or out of the focal region of the camera. By far the largest source of difficulty in particle identification, however, arises when two particle images overlap. In general, particles move in some three-dimensional volume. The camera image is a perspective projection of this volume onto a plane; therefore, even particles that are well separated in the depth direction may appear to be extremely close to one another in the camera image. If the particles overlap too much, they appear to be a single, albeit distorted, particle. This situation can lead to significant uncertainty and error. PIV solves this problem by using laser sheets rather than volumetric imaging. Laser sheets do not work well for Lagrangian particle tracking, however, since particles pass out of the sheet rapidly, leading to short trajectories. We are therefore forced to use volumetric imaging and the face the problem of overlap.

Any particle-finding method must be robust against the unavoidable problems described above. Additionally, an algorithm should ideally also perform well in situations when there is a significant amount of background noise in the image, as is the case in low-light situations, such as biological imaging or in field measurements of fluid flows where background signals cannot be eliminated. We also desire an algorithm that is computationally efficient, so that we can process hundreds of thousands of images containing thousands or tens of thousands of particles in a reasonable amount of time. Many different particle-finding algorithms have been proposed that fulfill some of these criteria. We here review several, including two methods that are widely used.

In general, the more we know about the image and the cleaner it is, the better we can design a particle-finding algorithm. For example, if we knew the particle-image intensity profile, we could fit that profile to the image and extract the coordinates of the center. Even though we do not formally know the full profile, we do know that it is peaked in the center; it is therefore reasonable to approximate it as a two-dimensional Gaussian [19]. Fitting the measured intensity profile can then give us the position of the particle center that can be as accurate as 0.01 pixels. Two-dimensional nonlinear least-squares fits (using, for example, the Levenberg–Marquardt algorithm [20]) are, however, computationally expensive, and the computational cost can become prohibitive as the particle seeding increases. Additionally, since a two-dimensional Gaussian is defined by five independent parameters (the two coordinates of the peak, the width in each direction, and the height of the peak), a large number of pixels are required to converge the fit satisfactorily. In principle, this fitting procedure can handle weakly overlapping particle images by fitting N Gaussians to a pixel cluster with N intensity maxima; we then, however, need a number of pixels that increases rapidly with N in order to get a reasonable fit.

Alternatively, if we empirically know what most particle images look like, we can cross-correlate the particle image with a template image, with which we can account for the discretization of the particle intensity profile by the pixilated camera [21]. Template matching can work well if the particle images are all the same size and shape, but loses accuracy rapidly due to defocusing, which changes the apparent size of particle images, and overlap, which distorts their apparent shape. Template matching is also somewhat computationally expensive, requiring discrete Fourier transforms for each particle in the camera image.

To avoid both the computational cost problem and the sensitivity to the exact particle size and shape, several simpler methods have been proposed and have found widespread use. If the particle image is large and covers many pixels, a simple intensity-weighted average (equivalent to locating the center of mass of the pixel group) may be used. The horizontal position of the particle center is computed as:

$$x_c = \frac{\sum_p x_p I(x_p, y_p)}{\sum_p I(x_p, y_p)}, \tag{7.4}$$

where x_c is the horizontal coordinate of the particle center, x_p and y_p are the positions of the centers of the pixels making up the particle image, and $I(x_p, y_p)$ is the image intensity of the pixel at (x_p, y_p). The vertical coordinate of the particle center is determined analogously. While center-of-mass averaging can work well for large particle images, its accuracy drops as the particle image becomes smaller. Additionally, center-of-mass averaging does not handle overlap well: while overlapping particle images can be segmented based on lines of minimum intensity in a pixel cluster [22], the overlapping images tend to shift the particle center away from the true value, since the weighting factors are modified. Segmenting the pixel cluster also introduces additional computational work. If the particle images are large and nonoverlapping, however, center-of-mass averaging can be a good choice for a particle identification scheme [23].

For smaller particle images, an approximate method that combines some of the accuracy of function fitting with the computational simplicity of center-of-mass averaging works well. As mentioned above, peaked functions can be well approximated as Gaussian near their centers. Full 2D fitting is computationally difficult; fitting a 1D Gaussian is much simpler. If we choose only a local intensity maximum and its two directly adjacent neighbors, we can solve the system of equations:

$$I_i = I_0 \exp\left[-\frac{1}{2}\left(\frac{x_i - x_c}{\sigma_x}\right)^2\right] \tag{7.5}$$

and determine the position of the particle center x_c as:

$$x_c = \frac{1}{2}\frac{(x_1 - x_2)^2 \ln(I_2/I_3) - (x_2 - x_3)^2 \ln(I_1/I_2)}{(x_1 - x_2)\ln(I_2/I_3) - (x_2 - x_3)\ln(I_1/I_2)}; \tag{7.6}$$

here x_i are the horizontal coordinates of the three pixels, where $i = 1, 2, 3$ (x_2 is the coordinate of the local maximum), I_i are the corresponding pixel intensities, I_0 is the peak value of the Gaussian fit, and σ_x is the Gaussian width. The vertical coordinate y_c is found analogously. Since the image is digital with a known bit-depth, the required logarithms can all be precomputed; finding the particle center then only requires a few multiplications, and is computationally efficient. To handle overlap, we simply fit one Gaussian to each local intensity maximum; the error increases for overlapping images, but remains acceptable. This method has proved to be both computationally efficient and reasonably accurate, to at least 0.1 pixels and often much better [11,23].

7.4.2 Basic algorithm: nearest-neighbor tracking

Once the positions of the tracer particles (or other features) have been determined, we must match the instantaneous positions together in time. This *correspondence problem* is an instance of the more general multidimensional assignment problem. For a pair of images, this problem is computationally tractable; for more than two images, however, multidimensional assignment is computationally intractable [24], and approximations must be made. Typically, we must limit the temporal scope of our matching, in what is termed a *greedy matching* approximation. By considering only a few frames at a time, we make tracking computationally feasible and hope to find the correct solution to the correspondence problem, despite ignoring large amounts of our raw data. In addition, the tracking problem can be simplified by restricting the distance we allow a correctly matched particle to move between frames, effectively limiting the spatial scope of the problem as well.

At its heart, tracking is an optimization problem: we seek to minimize the number of incorrect frame-to-frame links made. Suppose that $\mathbf{x}_i^n$ is the position of the ith particle in the nth frame. Generating a correct link to frame $n+1$ means choosing the $\mathbf{x}_j^{n+1}$ such that the particle identity is conserved. To solve this problem, we define a cost function ϕ_{ij}^n that is defined for each pair $\mathbf{x}_i^n$ and $\mathbf{x}_j^{n+1}$; ϕ_{ij}^n encodes the likelihood that $\mathbf{x}_i^n$ and $\mathbf{x}_j^{n+1}$ are images of the same particle in the two frames. We then seek to minimize the total cost $\sum_{n,i,j} \phi_{ij}^n$. The total cost is in principle summed over all frames and all particles; the greedy-matching approximation, however, limits the number of frames for which we perform the optimization.

A successful tracking algorithm depends crucially on choosing a proper cost function ϕ. In general, any slowly changing (or fixed) property can be included in ϕ^n_{ij}. For example, suppose that we wish to track two particles, and that one of the particles is red and one is blue. By choosing ϕ^n_{ij} to be the difference in color between particle images i and j, we can track these two particles perfectly, regardless of how quickly or irregularly they move: since the color does not change, it makes an excellent tracking variable. In many cases, however, our tracers do not have any distinguishing features, and we must instead use characteristics of their motion that change slowly from frame to frame. For featureless particles, it is convenient to define a nondimensional measure of the difficulty of the tracking problem as:

$$\xi = \frac{\Delta r}{\Delta_0}, \tag{7.7}$$

where Δr is the mean distance traveled by a particle between a pair of images and Δ_0 is the mean minimum distance between particles in a single image [23]. As ξ increases to order unity, tracking becomes nearly impossible for featureless particles: the imaging is slow enough and the particle seeding density high enough that the particle identities are irretrievably lost from frame to frame. If ξ is small, however, we can track accurately, and in the limit $\xi \to 0$, corresponding to a single particle per frame, tracking becomes trivial.

The simplest choice of cost function for featureless particles is the particle position: as long as we are imaging quickly compared to the velocity of the particles, their position will not change much from frame to frame. We can then define:

$$\phi^n_{ij} = |\mathbf{x}^{n+1}_j - \mathbf{x}^n_i|, \tag{7.8}$$

which we refer to as a *nearest neighbor* approach: minimizing the frame-to-frame distance traveled by each particle amounts to picking the particle in frame $n + 1$ that is closest to each measured position in frame n. This method is sketched in Figure 7.2(a).

The nearest-neighbor method works well if the imaging is very fast in comparison to the particle motion; as we will see below, however, more complex algorithms can perform better. But if the motion of a particle is unpredictable, as in the case of Brownian motion, the nearest-neighbor algorithm is often the only viable choice [25]. Nearest-neighbor algorithms also have the advantage of computational simplicity, both in terms of memory requirements and processing time. They tend to fail frequently, however, as ξ grows: in tests of tracking algorithms on simulated data in weakly turbulent flow fields, Ouellette, *et al.* found that the nearest-neighbor algorithm failed nearly 90% of the time for $\xi = 0.5$ [23]. For difficult tracking problems, we therefore need more complex and robust algorithms.

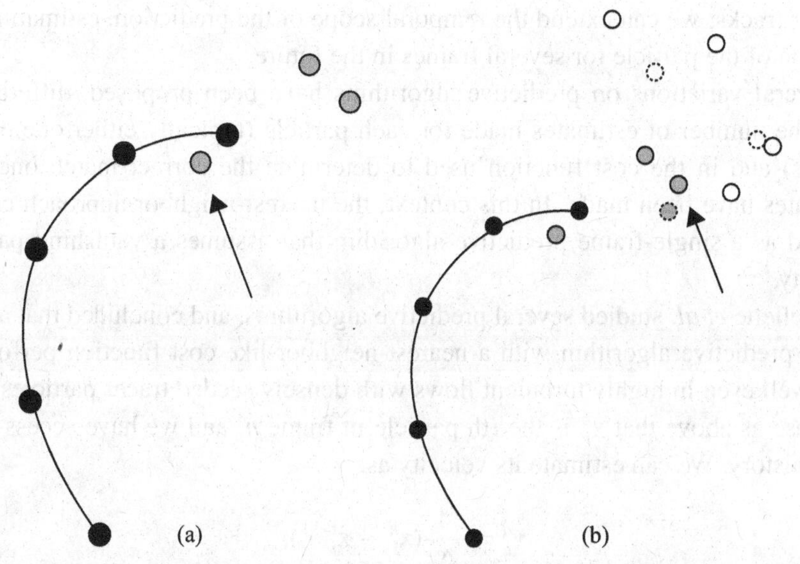

Figure 7.2 (a) Schematic of the nearest-neighbor tracking algorithm. Black circles are points already connected to a trajectory (shown by the solid black line); grey circles are points in the next frame. The arrow shows the point that will be chosen to continue the trajectory: in this case, it is the closest neighbor. (b) Schematic of the four-frame predictive "best estimate" algorithm. Here, circles with dashed edges show estimated positions (so that the grey, dashed circle is the estimated position one frame into the future and open circles are points found two frames into the future). Note that the point chosen to continue the trajectory is a better choice than the nearest-neighbor algorithm picked.

7.4.3 Predictive algorithms

If the motion of the tracer particles is not random and if we are imaging rapidly compared with their motion, then in principle the particle velocity and acceleration are also slowly changing variables and can be integrated into the cost function. In addition to making the tracking algorithm more accurate (in the sense of increasing the likelihood that the algorithm chooses the correct links), cost functions that use this type of additional information are capable of handling images with higher particle seeding densities; that is, ξ can be increased while still maintaining acceptable accuracy.

With the added velocity and acceleration information, however, we can make another change to the algorithm that has a larger effect. We can now predict where we expect a particle in frame n to be in frame $n+1$, which allows us to increase ξ yet higher and still maintain acceptable accuracy. Using both the particle acceleration and velocity makes this prediction more accurate, but it can be done with velocity alone with reasonably good results [23]. For added accuracy, at the cost of typically shorter tracks, we can extend the temporal scope of the prediction, estimating the position of the particle for several frames in the future.

Several variations on predictive algorithms have been proposed, differing in both the number of estimates made for each particle (typically either one or two frames) and in the cost function used to determine the correct match once the estimates have been made. In this context, the nearest-neighbor approach can be viewed as a single-frame predictive algorithm that assumes a vanishing particle velocity.

Ouellette *et al.* studied several predictive algorithms, and concluded that a two-frame predictive algorithm with a nearest-neighbor-like cost function performed very well even in highly turbulent flows with densely seeded tracer particles [23]. Suppose as above that $\mathbf{x}_i^n$ is the ith particle in frame n, and we have access to its prior history. We can estimate its velocity as:

$$\tilde{\mathbf{v}}_i^n = \frac{1}{\Delta t}(\mathbf{x}_i^n - \mathbf{x}_i^{n-1}), \tag{7.9}$$

where Δt is the time between frames and tildes denote estimated values. We then linearly extrapolate the particle position into frame $n + 1$, obtaining:

$$\tilde{\mathbf{x}}_i^{n+1} = \mathbf{x}_i^n + \tilde{\mathbf{v}}_i^n \Delta t. \tag{7.10}$$

Now, since we are restricting the spatial scope of the tracking algorithm, equivalent to setting a maximum speed for the particles, we consider only real particles $\mathbf{x}_j^{n+1}$ that fall within a distance R_s of $\tilde{\mathbf{x}}_i^{n+1}$. This search radius R_s is often easiest to define empirically: ideally, there should be a few particles (on average) less than R_s away from the estimate. For each of these $\mathbf{x}_j^{n+1}$, we project another frame into the future, using now both a velocity and an acceleration. Defining:

$$\tilde{\mathbf{v}}_i^{n+1} = \frac{1}{\Delta t}(\mathbf{x}_j^{n+1} - \mathbf{x}_i^n) \tag{7.11}$$

and:

$$\tilde{\mathbf{a}}_i^{n+1} = \frac{1}{\Delta t^2}(\mathbf{x}_j^{n+1} - 2\mathbf{x}_i^n + \mathbf{x}_i^{n-1}), \tag{7.12}$$

we generate an estimated position $\tilde{\mathbf{x}}_i^{n+2}$ for each $\mathbf{x}_j^{n+1}$ given by:

$$\tilde{\mathbf{x}}_i^{n+2} = \mathbf{x}_i^n + \tilde{\mathbf{v}}_i^{n+1}(2\Delta t) + \frac{1}{2}\tilde{\mathbf{a}}_i^{n+1}(2\Delta t)^2. \tag{7.13}$$

We then define our cost function to be:

$$\phi_{ij}^n = \min_k |\tilde{\mathbf{x}}_i^{n+2} - \mathbf{x}_k^{n+2}|, \tag{7.14}$$

that is, the smallest distance between our estimated position in frame $n+2$ and a real particle in frame $n+2$. We therefore will create a link between $\mathbf{x}_i^n$ and the $\mathbf{x}_j^{n+1}$ that leads to the best estimate of a particle position in frame $n+2$. This "best-estimate" algorithm is shown schematically in Figure 7.2, along with the nearest-neighbor algorithm. We note that this method differs from algorithms that use the minimum acceleration or change in acceleration as the cost function [26]. Even though the particle velocity and acceleration may be defined well enough to make estimates of the new particle position, they remain very uncertain quantities and therefore make poor choices for the cost function. Ouellette *et al.* found that such algorithms failed rapidly as ξ increased, performing nearly as poorly as the nearest-neighbor algorithm for $\xi \to 0.5$ [23]. In contrast, the predictive "best-estimate" algorithm presented here had a failure rate of only about 10% for $\xi \approx 0.5$.

7.4.4 Error control

Errors in Lagrangian particle tracking come from two primary sources: uncertainty in locating the particle centers and incorrect frame-to-frame links. As with PIV or PTV, particle centers can be found to an accuracy of at least 0.1 pixels, and often better if the imaging is very good. We are therefore primarily concerned with mistakes in the tracking, since they will play the dominant role in determining the accuracy of the technique.

Analyzing the accuracy of a particle-tracking algorithm mathematically is very difficult: unlike PIV, it is not a statistical process, and traditional error analysis methods are not readily applicable. Instead, we can discuss the accuracy only qualitatively. As described above, tracking algorithms tend to fail when the tracking difficulty ξ becomes large; in the limit of vanishing ξ (corresponding to well-separated particles moving slowly), any of the algorithms presented above will perform essentially perfectly [23]. As ξ grows, the number of possible links for each particle in the current frame increases, and the algorithm begins to make choices between multiple possible frame-to-frame links. Algorithms that make these choices poorly will be inaccurate. As discussed above, well-chosen predictive algorithms make good choices by using an extended temporal scope: it is highly unlikely that a decision that leads to a consistent set of future particle locations will be a mistake. Still, as ξ approaches 0.5, incorrect choices do occur, due to the rise of *tracking ambiguities* and *tracking conflicts*.

When the particle density is high or when the particles are moving quickly relative to the frame rate of the camera, multiple links may lead to equivalent tracking costs: a single particle in frame n links equally well to multiple particles in frame $n+1$, and we have a tracking ambiguity. We may also have the case where the best match for multiple particles in frame n is a single particle in frame

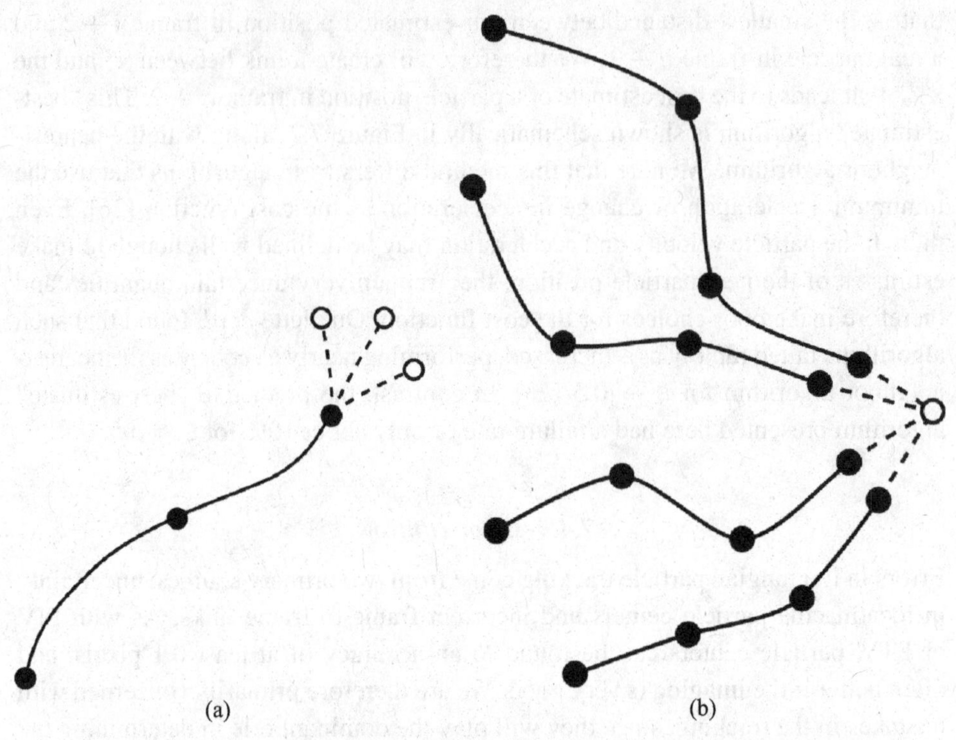

Figure 7.3 (a) A tracking ambiguity: one partially determined trajectory matches equally well to multiple particles in the next frame. Solid circles denote matched positions, and the solid line is the partially completed trajectory. The open circles are potential next matches, which would add the dashed lines to the trajectory. (b) A tracking conflict: several trajectories match equally well to a single particle in the next frame.

$n + 1$; then, we have a tracking conflict. These two cases are shown schematically in Figure 7.3.

Tracking conflicts can be handled in several ways. The most conservative choice is simply to give up. Instead of trying to decide which of the multiple tracks to assign the new particle to, we simply end all the conflicting tracks and start a new track with the particle in frame $n + 1$. This choice tends to limit the overall length of trajectories (though we can potentially recover the proper links in a postprocessing step; see below), but is guaranteed not to make an incorrect decision. A second choice is to optimize the tracking cost globally over the entire frame, often by adding some fixed penalty for ending a trajectory [27]. Efficient algorithms exist for computing such a global optimization [28]; in simulated test cases, however, such global optimization was found to degrade the performance of the tracking algorithm, leading to more incorrect links being made [23]. A third choice is to allow *all* of the conflicting tracks to match to the same particle in frame $n + 1$, with the hope that the predictive algorithm will not lead to all these tracks becoming superimposed for all future times. The implicit assumption in this case is that due to image overlap, what the particle-finding algorithm has identified as a *single* particle is in fact the superimposed images of *several* particles. This strategy can work well if overlap is a significant problem (due, for instance, to a very high particle loading with a large depth of field).

Tracking ambiguities are much more pernicious, since they are harder to detect. In a floating-point calculation, the tracking costs for multiple particles in frame $n+1$ are highly unlikely to be *exactly* the same, and setting a threshold below which two values are considered to be equal is uncontrolled. Typically, therefore, all that can be done is to keep ξ small enough that tracking ambiguities are rare, and will therefore have a negligible impact on the statistical analysis of the ensemble of trajectories. We can also try to remove ambiguities in a postprocessing step by looking for anomalously large accelerations along the trajectories, which may correspond to incorrect links. This approach, however, can be dangerous, particularly for systems such as turbulent flow where we expect enormously large short-lived acceleration fluctuations.

A third effect must be considered when we compute statistical analyses of particle-tracking data. In general, we image particles in a finite subvolume of the full system of interest. We must therefore contend with the problem of *finite-volume bias*. The bias is twofold. On the one hand, particles that move faster are more likely to enter our measurement subvolume. On the other, particles that stay in view for long periods of time (and which lead to long trajectories) are more likely to be moving slowly. These effects suggest that our short-time statistics will be biased towards fast velocities, while our long-time statistics will be biased towards slow velocities. Since these are true biases, collecting larger data sets cannot solve the problem; we must simply be cognizant of it, and try to estimate how it skews our statistics. For measurements of velocities and accelerations, Voth *et al.* [15,16] and Mordant *et al.* [29] used schemes based on weighting the contribution of a particle track to ensemble averages by the amount of time they spent in view to attempt to correct for the finite-volume bias. For position-based statistics, however, corresponding schemes are not known.

7.4.5 Extensions

As we have seen, inhomogeneities in illumination cause serious problems for PIV, by the inclusion of extra terms in the image cross-correlation that obscure the signal we want to measure. Illumination inhomogeneities are much less of a problem for particle tracking, in that they only marginally affect the accuracy of the technique (primarily through small variations in the detection of the particle centers). They may, however, lead to holes in the trajectories: if a particle is sitting in a dim region of the measurement volume, it may not be detected. Predictive algorithms lend themselves to a simple solution to this *drop-out* problem. In the case where we do not find an acceptable match for a trajectory from frame n to frame $n+1$, we store instead the estimated position $\tilde{\mathbf{x}}^{n+1}$, and try to extend the track between frames $n+1$ and $n+2$ with a real particle. We can in principle continue this process for several frames. As we try and push this drop-out handling to larger time windows, however, the probability of finding a true continuation of the track drops rapidly. In practice, keeping the number of frames we extrapolate through to four or fewer seems to work well.

We can also try to resolve drop-out and tracking conflicts in a postprocessing step. In an approach recently developed by Xu [30], we can use both the position information along trajectory and its velocity information, once we have differentiated it. As long as we are imaging fast enough, both of these quantities are slowly changing variables, and we can re-track our data in this higher-dimensional position-velocity phase space. The advantage to this procedure is that the extra dimensions spread out the particles: effectively, we reduce ξ. We can then increase our temporal and spatial scope without loss of accuracy, leading to more reliable tracking and the connection of previously broken trajectories. Xu showed that this method was effective at producing long tracks, both in a simulated data set and in real experimental measurements.

7.4.6 Field measurements

As we have described them so far, PIV and Lagrangian particle tracking seem to produce fundamentally different information: in PIV, we extract velocity fields, while in particle tracking we generate tracer-particle trajectories. As long as the seeding density of the tracer particles is high enough, however, the instantaneous velocities of all the Lagrangian tracers may be interpolated together to create a velocity field, just as they are in PTV. In general, such interpolation is reasonable if Δ_0, the mean minimum interparticle distance, is much smaller than the scale on which the velocity field varies.

Generating a velocity field from Lagrangian particle-tracking data has two major advantages over PIV. Since we have longer time-histories of position information for each particle, we can apply a higher-order differentiation scheme than a simple two-point finite difference. Each individual velocity that goes into generating the field can therefore be more accurate, irrespective on any statistical smoothing of the field. Additionally, we still retain and have access to the Lagrangian information associated with each particle. While the field of Lagrangian data analysis is not yet fully mature, Lagrangian field information has been exploited to study the spatially localized stretching of fluid flows [31, 32], a quantity related to the Lyapunov exponents, as well as the dynamics of flow topology [33, 34]. We anticipate that new uses for Lagrangian information will be discovered in the future.

7.5 Three-dimensional measurement

Physical phenomena are not, in general, confined to two dimensions. The measurement techniques we have discussed, however, rely on camera imaging: our raw data are particle images projected onto a two-dimensional plane. In order to measure fully three-dimensional dynamics, we therefore have to extend our measurement system.

Three-dimensional measurements are in general more difficult than their two-dimensional counterparts. Indeed, for some techniques, including PIV, fully three-dimensional measurement, in which we measure the three-dimensional dynamics of a physical system in a nearly isotropic volume, are not yet possible.

In this section, we describe three ways in which three-dimensional information may be gained from two-dimensional imaging systems. Multiple cameras may be used to collect three-dimensional information with proper calibration, much as our eyes do. Or, we may choose to collect many rapid two-dimensional slices (or one-dimensional spots) of the volume. We may also use holographic techniques to take advantage of the three-dimensional information stored in the diffraction pattern of each tracer particle. We discuss these ideas in turn below, as well as generic limitations on three-dimensional imaging as it is currently understood.

7.5.1 *Stereoimaging*

Biology gives us a clue as to an efficient way to reconstruct three-dimensional information from two-dimensional images: take several images from different angles and combine them. With our two eyes, we are able to compute the missing depth information. Unfortunately, our brains are much better at this process than the computer is: to implement such a stereoimaging system on the computer is nontrivial.

In general, information from at least two viewpoints is required to reconstruct the three-dimensional information using only image-based techniques. Each image contains overlapping information; the information redundancy combined with a knowledge of the relative positions of the cameras can be used to resolve the third coordinate. In practice, at least three views are needed in order to perform a good reconstruction, and four is better [35]. Further cameras lead to diminishing returns.

Each camera performs a perspective projection of the three-dimensional world onto its two-dimensional image sensor. Even though perspective projection is a nonlinear, lossy operation, we can still assert that a line of sight drawn through the camera's projective center and a point on the image plane will pass through the actual three-dimensional location of the imaged object; this is sometimes referred to as the *photogrammetric condition*. Furthermore, the lines of sight from all the different cameras in the measurement system will intersect at the true three-dimensional position of the object. This observation is the basis of stereoimaging: lines of sight are drawn through the position of each feature in each image, and their intersections are found [23]. In practice, lines of sight never intersect at a point, but rather come closer together than some small threshold; in that case, the centroid of the region defined by the lines of sight is chosen as the three-dimensional coordinate.

7.5.2 Tomography

Instead of attempting to image an entire volume using a set of simultaneous images recorded from different viewpoints, as in stereoimaging, we can generate three-dimensional information by recording a rapid sequence of two-dimensional images scanned through the desired measurement volume. Each image gives us a single slice of the volume; reconstruction of the full volumetric data is achieved by reassembling the individual slices. If we image very quickly compared to the fastest dynamical timescale, we can consider the sequence of images to be an approximately instantaneous snapshot of the full volume.

This type of *tomographic* imaging is very common in medical applications, where the subject is often static. Computed tomography (CT) scans are now a standard diagnostic tool. Rapid scanning in order to recover volumetric information is also the principle behind three-dimensional confocal microscopy, which has found widespread adoption in soft matter physics. For rapidly evolving systems such as fluid flows, however, tomographic techniques are only in their infancy, since the imaging rates required are often tremendously high.

Recently, building on earlier work with tomographic laser-induced fluorescence and PIV, Hoyer *et al.* have developed a tomographic particle-tracking technique, which they call scanning PTV [36]. They used a laser light sheet that passed through a rotating prism in order to scan the beam through the measurement volume. Owing to the speed of their camera and the mechanical rotation speed of the prism, however, they were limited to a volume scan rate of 50 Hz; as implemented, then, their system can be applied only to low-speed flows. Building a tomographic particle-tracking system for turbulent or other high-speed flows remains an open challenge.

7.5.3 Holography

In most particle-based measurements, lasers are the light sources of choice. Lasers can deliver very high power densities, since they are directed light sources, and they can be pulsed at high repetition rates. The ease of collimation of a laser beam makes alignment of optical systems much simpler. Lasers are also coherent light sources, but this coherence is typically not used from a measurement standpoint. Instead, the techniques we have described so far simply use the light scattered from the particles in order to make measurements.

We can gain more information about the particles, however, if we also store the phase information of the scattered light. If, instead of recording simply the scattering, we image the interference pattern between the scattered light and a reference beam, the three-dimensional spatial characteristics of each particle are encoded in the resulting hologram. For images with relatively low particle seeding density, we can use in-line holography: in this case, the reference beam is simply the unscattered light of the original illumination beam. By subsequently inverting the hologram [37], we can obtain three-dimensional information with a single camera. Holographic systems are therefore some of the most compact three-dimensional imagers as well as being simple to align, and are excellent choices for fieldwork installations where space is at a premium and complex alignment is a serious drawback [38]. Additionally, since only a single camera is required, in-line holographic systems are relatively inexpensive compared with multi-camera stereoimaging systems.

Holographic imaging does, however, have drawbacks. The particle density must be kept low, both to minimize the distortion of the reference beam and to increase the reliability of the real-space reconstruction [39]. Inverting the holograms is also tremendously computationally intensive, limiting the total amount of data that can be processed in a reasonable amount of time.

7.5.4 Limitations

As we have seen above, techniques that give us *field* information in two dimensions require very high particle seeding densities. PIV performs better when there are more particles in an interrogation region; PTV and Lagrangian particle tracking require closely spaced particles if their characteristics are to be reasonably interpolated. In volumetric measurements, however, the particle seeding required to measure three-dimensional fields is prohibitive, and techniques such as stereoimaging will fail due to the high rates of overlap and shadowing. For the present, then, three-dimensional measurements are limited to low-speed flows where we can use tomography or to low seeding densities where we cannot measure field quantities.

In Section 7.6, we discuss several strategies that may be able to overcome these obstacles.

7.6 Frontiers of particle-based measurement

While optical particle-based measurement science can be considered to be a fairly mature field, new developments are continually being made to address many of the challenges we have discussed above. Here, we describe a sampling of topics at the forefront of the field, and conclude with a brief discussion of features that we anticipate will be present in the next generation of digital cameras.

7.6.1 Three-dimensional fields

As discussed above, volumetric field measurements are not currently possible in rapidly evolving systems, leading to severe experimental limitations. For turbulent flow, for example, the large-scale three-dimensional velocity is currently only accessible via numerical simulation. The next major leap forward in particle-based measurement will most likely be the development of robust tools to measure three-dimensional fields.

At present, tomography appears to be the most viable candidate for making such measurements feasible. In principle, the technology for rapid tomographic imaging of high-speed flows already exists: while no single camera or laser is fast enough, multiple cameras and lasers could be interlaced to provide very rapid scanning. The cost of such a system, as well as the complexity of the synchronization, is currently prohibitive. Camera technology, however, is always improving, and prices correspondingly drop. Within a few years, therefore, we are hopeful that tomographic systems will become practical.

7.6.2 Measurement of additional quantities

Above, we have primarily discussed methods for extracting velocities from measurements of the motion of particles. In particular, PIV and PTV are used to construct velocity fields; Lagrangian particle tracking can also be used to measure fields, but is more often used to analyze statistics along particle trajectories. As these techniques have matured, however, researchers have begun to use these types of imaging methods to measure additional dynamical quantities, including accelerations and forces.

One particularly exciting new technology is traction force microscopy (TFM), designed for measuring wall shear stresses. In this technique, tracer particles are embedded in a thin elastic medium of known elastic modulus. By measuring the displacements of the particles from their equilibrium positions (using PIV or some other suitable technique), the strain field can be determined; subsequent inversion of the relationship between stress and strain can then be used to measure the stress in the elastic substrate [40]. Currently, researchers are using TFM to measure the forces exerted by cells as they crawl. In the future, it is expected that this technique will also be applied to fluid systems in order to measure wall shear stresses, which are difficult to determine with contemporary techniques.

7.6.3 Advances in camera technology

Camera hardware remains the single greatest limitation in digital imaging science. As camera sensors are pushed to ever larger pixel arrays and faster speeds, data bandwidth has become a bottleneck. There are two aspects to the bandwidth problem. First, information simply cannot be read out from large pixel arrays at very fast speeds: in general, higher-speed cameras use smaller image sensors. Second, once the image is read into internal camera RAM buffers, it cannot be transferred to computer hard drives quickly: typically the transfer to disk is orders of magnitude slower than the image acquisition rate. The great frontier of camera design is therefore solving these bandwidth problems.

A potential solution is on-chip processing. Chan *et al.* recently developed a real-time image compression system that tremendously improved their data bandwidth by partially processing the images before they were stored to disk [41]. They modified an off-the-shelf camera system by placing an extra digital processing stage between the camera and the recording computer; in principle, however, such processing methods could be implemented in the camera itself. The camera would lose some its appeal as a general-purpose imaging device, but could perform significantly better for a small range of tasks. In essence, the camera would become a specialized scientific instrument, just as microscopes have become specialized for different tasks over time. Such advances would have a considerable impact on digital imaging science.

7.6.4 Conclusions

Particle-based optical measurements have advanced significantly from the early days of manual analysis of long-exposed film. Digital PIV and particle tracking are fast becoming standard tools in soft matter physics, after already dominating experimental fluid mechanics. As we have discussed, these techniques are not without their problems; when the conditions are correct, however, they are quantitative, reasonably simple to implement, and powerful. With the promise of new developments and the capability of three-dimensional field measurement on the horizon, we anticipate that optically based measurements of tracer particles will continue to be standard tools for many years to come.

Acknowledgments

The Lagrangian particle-tracking procedures described here were developed in collaboration with H. Xu and E. Bodenschatz, as part of work supported by the National Science Foundation and the Max Planck Society.

References

[1] W.-C. Chiu and L. N. Rib, "The rate of dissipation of energy and the energy spectrum in a low-speed turbulent jet," *T. Am. Geophys. Union* **37**, 13–26 (1956).

[2] J. Kompenhans, "The 12th international symposium on flow visualization," *J. Visualization* **10**, 123–128 (2007).

[3] C. Tropea, F. Scarano, J. Westerweel, A. A. Cavone, J. F. Meyers, J. W. Lee, and R. Schodl, "Particle-based techniques," in C. Tropea, J. Foss, and A. Yarin (eds.), *Springer Handbook of Experimental Fluid Mechanics*, pp. 287–362 (Springer-Verlag, Berlin, 2007).

[4] M. R. Maxey and J. J. Riley, "Equation of motion for a small rigid sphere in a nonuniform flow," *Phys. Fluids* **26**, 883–9 (1983).

[5] N. M. Qureshi, U. Arrieta, C. Baudet, A. Cartellier, Y. Gagne, and M. Bourgoin, "Acceleration statistics of inertial particles in turbulent flow," *Eur. Phys. J. B* **66**, 531–6 (2008).

[6] N. T. Ouellette, P. J. J. O'Malley, and J. P. Gollub, "Transport of finite-sized particles in chaotic flow," *Phys. Rev. Lett.* **101**, 174504 (2008).

[7] A. Melling, "Tracer particles and seeding for particle image velocimetry," *Meas. Sci. Technol.* **8**, 1406–16 (1997).

[8] N. M. Qureshi, M. Bourgoin, C. Baudet, A. Cartellier, and Y. Gagne, "Turbulent transport of material particles: An experimental study of finite size effects," *Phys. Rev. Lett.* **99**, 184502 (2007).

[9] R. J. Adrian, "Particle-imaging techniques for experimental fluid mechanics," *Annu. Rev. Fluid Mech.* **23**, 261–304 (1991).

[10] J. Westerweel, "Fundamentals of digital particle image velocimetry," *Meas. Sci. Technol.* **8**, 1379–92 (1997).

[11] E. A. Cowen and S. G. Monismith, "A hybrid digital particle tracking velocimetry technique," *Exp. Fluids* **22**, 199–211 (1997).

[12] J. S. Guasto, P. Huang, and K. S. Breuer, "Statistical particle tracking velocimetry using molecular and quantum dot tracer particles," *Exp. Fluids* **41**, 869–80 (2006).

[13] F. Scarano, "Interative image deformation methods in PIV," *Meas. Sci. Technol.* **13**, R1–R19 (2002).

[14] R. D. Keane, R. J. Adrian, and Y. Zhang, "Super-resolution particle imaging velocimetry," *Meas. Sci. Technol.* **6**, 754–68 (1995).

[15] G. A. Voth, K. Satyanarayan, and E. Bodenschatz, "Lagrangian acceleration measurements at large Reynolds numbers," *Phys. Fluids* **10**, 2268–80 (1998).

[16] G. A. Voth, A. La Porta, A. M. Crawford, J. Alexander, and E. Bodenschatz, "Measurement of particle accelerations in fully developed turbulence," *J. Fluid Mech.* **469**, 121–60 (2002).
[17] N. Mordant, A. M. Crawford, and E. Bodenschatz, "Experimental Lagrangian acceleration probability density function measurement," *Physica D* **193**, 245–51 (2004).
[18] N. T. Ouellette, H. Xu, and E. Bodenschatz, "Measuring Lagrangian statistics in intense turbulence," in C. Tropea, J. Foss, and A. Yarin (eds.), *Springer Handbook of Experimental Fluid Mechanics*, pp. 789–99 (Springer-Verlag, Berlin, 2007).
[19] S. Ott and J. Mann, "An experimental investigation of the relative diffusion of particle pairs in three-dimensional turbulent flow," *J. Fluid Mech.* **422**, 207–23 (2000).
[20] W. H. Press, S. A. Teukolsky, W. T. Vetterling, and B. P. Flannery, *Numerical Recipes in C: The Art of Scientific Computing* (Cambridge University Press, Cambridge, 1992).
[21] Y. G. Guezennec, R. S. Brodkey, N. Trigui, and J. C. Kent, "Algorithms for fully automated three-dimensional particle tracking velocimetry," *Exp. Fluids* **17**, 209–19 (1994).
[22] H. G. Maas, A. Gruen, and D. Papantoniou, "Particle tracking velocimetry in three-dimensional flows. Part 1. Photogrammetric determination of particle coordinates," *Exp. Fluids* **15**, 133–46 (1993).
[23] N. T. Ouellette, H. Xu, and E. Bodenschatz, "A quantitative study of three-dimensional Lagrangian particle tracking algorithms," *Exp. Fluids* **40**, 301–13 (2006).
[24] C. J. Veenman, M. J. T. Reinders, and E. Backer, "Establishing motion correspondence using extended temporal scope," *Artif. Intell.* **145**, 227–43 (2003).
[25] J. C. Crocker and D. G. Grier, "Methods of digital video microscopy for colloidal studies," *J. Colloid Interf. Sci.* **179**, 298–310 (1996).
[26] N. A. Malik, T. Dracos, and D. A. Papantoniou, "Particle tracking velocimetry in three-dimensional flows. Part II. Particle tracking," *Exp. Fluids* **15**, 279–94 (1993).
[27] C. J. Veenman, M. J. T. Reinders, and E. Backer, "Resolving motion correspondence for densely moving points," *IEEE T. Pattern Anal.* **23**, 54–72 (2001).
[28] F. Bourgeois and J.-C. Lasalle, "An extension of the Munkres algorithm for the assignment problem to rectangular matrices," *Commun. ACM* **14**, 802–4 (1971).
[29] N. Mordant, E. Lévêque, and J.-F. Pinton, "Experimental and numerical study of the Lagrangian dynamics of high Reynolds turbulence," *New J. Phys.* **6**, 116 (2004).
[30] H. Xu, "Tracking Lagrangian trajectories in position-velocity space," *Meas. Sci. Technol.* **19**, 075105 (2008).
[31] G. A. Voth, G. Haller, and J. P. Gollub, "Experimental measurements of stretching fields in fluid mixing," *Phys. Rev. Lett.* **88**, 254501 (2002).
[32] M. Mathur, G. Haller, T. Peacock, J. E. Ruppert-Felsot, and H. L. Swinney, "Uncovering the Lagrangian skeleton of turbulence," *Phys. Rev. Lett.* **98**, 144502 (2007).
[33] N. T. Ouellette and J. P. Gollub, "Curvature fields, topology, and the dynamics of spatiotemporal chaos," *Phys. Rev. Lett.* **99**, 194502 (2007).
[34] N. T. Ouellette and J. P. Gollub, "Dynamic topology in spatiotemporal chaos," *Phys. Fluids* **20**, 064104 (2008).
[35] T. Dracos, "Particle tracking in three-dimensional space," in T. Dracos (ed.), "Three-dimensional velocity and vorticity measuring and image analysis techniques," pp. 129–52 (Kluwer Academic Publishers, Dordrecht, The Netherlands, 1996).

[36] K. Hoyer, M. Holzner, B. Lüthi, M. Guala, A. Liberzon, and W. Kinzelbach, "3D scanning particle tracking velocimetry," *Exp. Fluids* **39**, 923–34 (2005).

[37] J. Lu, J. P. Fugal, H. Nordsiek, E. W. Saw, R. A. Shaw, and W. Yang, "Lagrangian particle tracking in three dimensions via single-camera in-line digital holography," *New J. Phys.* **10**, 125013 (2008).

[38] J. P. Fugal, R. A. Shaw, E. W. Saw, and A. V. Sergeyev, "Airborne digital holographic system for cloud particle measurements," *Appl. Optics* **43**, 5987–95 (2004).

[39] M. P. Arroyo and K. D. Hinsch, "Recent developments of PIV towards 3D measurements," *Top. Appl. Phys.* **112**, 127–54 (2008).

[40] J. C. del Álamo, R. Meili, B. Alonso-Latorre, J. Rodriguez-Rodriguez, A. Aliseda, R. A. Firtel, and J. C. Lasheras, "Spatio-temporal analysis of eukaryotic cell motility by improved force cytometry," *Proc. Natl. Acad. Sci.* **104**, 13343–8 (2007).

[41] K.-Y. Chan, D. Stich, and G. A. Voth, "Real-time image compression for high-speed particle tracking," *Rev. Sci. Instr.* **78**, 023704 (2007).

8

Cellular automata models of granular flow

G. WILLIAM BAXTER

8.1 Introduction

The modeling and simulation of granular materials is important to our understanding of their behavior and the wealth of phenomena they exhibit [1–3]. Many phenomena and practical applications, such as the design of industrial processes, remain out of reach of traditional simulation methods due to the large numbers of grains involved.[1] A liter of fine sand, $m_{grain} \approx 0.1$ mg, may contain 10^7 grains. An industrial process such as mixing or hopper flow can easily involve 10^1 to 10^3 liters. Geophysical processes, such as sand dunes and earthquake faults, involve even larger numbers of grains. Traditional simulation techniques are currently unable to deal with such large numbers of grains; however, cellular automata models are able to simulate larger numbers of grains for longer times and show promise in the simulation of large, real-world granular flows.

Molecular dynamics, where Newton's laws are applied to individual grains and the resulting motion is determined from the forces, has made dramatic progress in the past two decades through improved techniques and more powerful computers. Molecular dynamics is particularly effective for gases of simple grains interacting through hard-sphere collisions. In principle, even the physics of complex grain interactions can be included. However, as the complexity of the interactions increases or the number of grains increases, the computational demands increase as well. As a result, molecular dynamics is primarily used for relatively small numbers of grains (10^2–10^5), whereas real granular flows often contain many more.

Cellular automata do not attempt to exactly model the physical interactions between grains. Instead, automata rules are constructed to imitate some behaviors

[1] For this chapter, *modeling* will refer to a simplified model, picture, or rule of how grains interact, whereas *simulation* will refer to the use of this model to study the large-scale temporal or spatial behavior of a granular material subjected to a specific set of driving forces and boundary conditions.

Experimental and Computational Techniques in Soft Condensed Matter Physics, ed. Jeffrey Olafsen. Published by Cambridge University Press.

of the grains. Simplifying the grain interaction makes it possible to simulate many more grains for longer periods of time. This allows the study of large-scale flows and processes. This is the essential trade-off of cellular automata models: reduced microscopic exactness in exchange for larger numbers of grains and longer timescales.

Over the past two decades, cellular automata models have been introduced to study many granular flow phenomena. One can loosely group these automata models into two types: surface flows where the behavior is dominated by avalanches and bulk flows where grain motion throughout the material is important. Following a brief history of cellular automata and a look at their key features, we will examine these two types of cellular automata models and how to optimize them. The chapter will conclude with a short discussion of future directions for grain flow cellular automata.

8.2 History

Cellular automata were invented in the 1940s by John von Neumann and Stanislaw Ulam [4]. In a traditional cellular automata, parameters of time, space, and value are all discrete. The automata operates on a lattice of sites, where each site has a discrete number of allowed values. This lattice can have any dimensionality. The automata rule determines how lattice site values change in time. The value of the lattice site at the next time step depends on the value at the current time as well as the values of neighboring sites. Traditionally, all sites on the lattice will update simultaneously.

Cellular automata became popular in the 1960s and 1970s as their behavior reminded people of other physical systems. Perhaps the most famous is John Horton Conway's *Life* which operates on a two-dimensional square lattice and looks like a bacterial colony [5]. Though using a simple rule, *Life* displays quite complex behavior and could be used for computation [6]. Cellular automata can be designed to resemble and model many different physical systems [7–11]. Automata have been especially successful in the modeling of fluids.

Cellular automata used for modeling fluids are known as lattice gas automata [12]. The first lattice gas automata used square lattices. One of the first successful lattice gas automata was the HPP lattice gas [13]. In the HPP lattice gas, up to four molecules or fluid particles are allowed at each lattice point and molecules have unit velocities directed toward one of the four nearest-neighbor sites. All sites are updated simultaneously following a *propagate and collide* procedure. First, all particles are moved one lattice site along their velocity vector. Sites with multiple particles experience collisions in which two particles with velocities directly toward each other are given new velocity vectors at right angles to the original directions. In 1986, Frisch *et al.* [14] introduced a lattice gas automata which used a two-dimensional regular triangular lattice. In this model, particles may have one of six unit velocities which move toward each of the six nearest neighbor sites. Each site on the lattice may hold up to six fluid particles with the constraint that each particle must have a unique velocity. Again, all sites are updated simultaneously using a *propagate and collide* procedure. Each particle moves to the next lattice site along its velocity vector and then collisions are processed at each site to determine the new particle velocities. They called this cellular automata the hexagonal lattice gas, but it has come to be known as the FHP lattice gas after its creators. The FHP lattice gas, unlike the HPP lattice gas, can be shown to reduce to the correct continuum equations [14, 15] for Newtonian fluids, the Navier–Stokes equations. Related lattice gas automata have been used to study fluid flows in many geometries [16, 17] and magnetohydrodynamics [18]. Some lattice gas automata have been applied directly to granular systems such as fluidized beds [19]; however, the behavior of granular media is often very different from that of fluids. As a result, cellular automata describing the flow of grains generally differ from lattice gas automata describing fluids, although they share many common elements.

Cellular automata models of granular materials use a simplified model of grain interactions. Instead of the rigorous physical modeling of grain interactions used in molecular dynamics [20–22], the use of a simplified model allows a cellular automata to follow many more grains for a longer time. It is not uncommon for cellular automata of granular systems to employ a random number generator. This may be used to determine the order in which the rule is applied to lattice sites or it may be part of the rule itself. Traditionally cellular automata are purely deterministic and their rules are not probabilistic; nevertheless, we will call the model a cellular automata even if a random number generator is used provided most of the other features of a cellular automata are present. There are many variations of granular cellular automata, but they can be grouped into two broad categories: surface flows or avalanches where the motion is primarily along the surface and bulk flows where grain movement at the surface and interior are treated the same.

8.3 Surface flow models

A cellular automata model of avalanches was first introduced by Bak, Tang, and Wiesenfeld (BTW) in 1986 [23, 24]. This simple model was originally intended to study self organized criticality (SOC) [25]. It may be applied to many different systems; but, conceptually, it is easiest to think of the model as representing a sandpile, and that was how the authors described it. In the one-dimensional BTW model, the lattice extends from $x = 0$ to N and the height of the pile is an integer $h(x)$. $z(x) = h(x) - h(x + 1)$ is the local difference in heights or effectively the

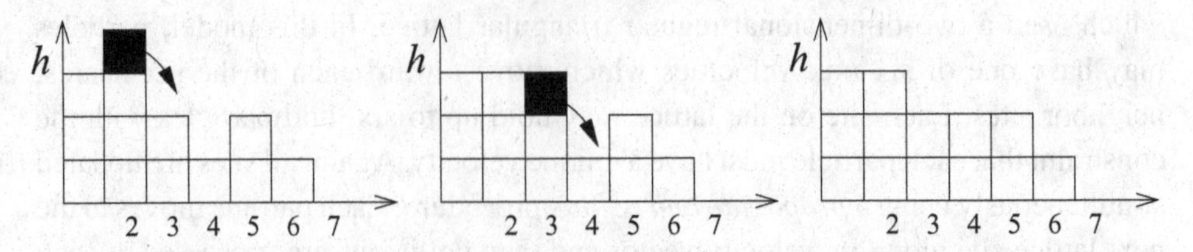

Figure 8.1 When a grain is added at a random lattice site, here $x = 2$, the local slope may exceed the critical, resulting in a series of avalanche events.

local slope. In the standard BTW model, grains are added to the pile at random sites x one at a time as $h(x) = h(x) + 1$ (Figure 8.1). Adding a grain to site x results in $z(x) \to z(x) + 1$ and $z(x-1) \to z(x-1) - 1$. This creates the possibility that the local slope will exceed its critical slope or *angle of repose*, z_c. In the BTW model, $z_c = 1$. If $z(x) > z_c$, then the grains rearrange through an avalanche event:

$$h(x) \to h(x) - 1 \tag{8.1}$$

$$h(x+1) \to h(x+1) + 1. \tag{8.2}$$

Equivalently, the avalanche event can be written in terms of the local slope:

$$z(x) \to z(x) - 2 \tag{8.3}$$

$$z(x \pm 1) \to z(x \pm 1) + 1. \tag{8.4}$$

A local avalanche at site x may cause neighboring sites to exceed their critical angles. The avalanches continue until no further motion occurs, at which time another grain may be added. In the original BTW model, boundary conditions are a fixed wall at $x = 0$ with no limit in height $z(0) = 0$ and an edge at $x = N$ where $h(N) = 0$. Avalanches only occur in the direction of increasing x, and over time the slope gradually evolves to a constant value of z_c. This gradual evolution toward a critical value is central to the concept of SOC, which may occur in many systems [25]. SOC was thought to explain the common presence of $1/f$ noise in physical systems and the BTW model's avalanche sizes and frequencies were extensively studied to probe the mechanisms of SOC [26–31].

The BTW model can be extended to a two-dimensional square lattice [23–25] with $0 < x \leq N$ and $0 < y \leq M$. $z(x, y)$ represents the local slope at a point on the lattice which, extending the one-dimensional case, might be related to height as $z(x, y) = 2h(x, y) - h(x-1, y), -h(x, y-1)$. The rule for adding a grain at site x, y as $h(x, y) \to h(x, y) + 1$ can be written in terms of slopes:

$$z(x, y) \to z(x, y) + 2 \tag{8.5}$$

$$z(x-1, y) \to z(x-1, y) - 1 \tag{8.6}$$

$$z(x, y-1) \to z(x, y-1) - 1. \tag{8.7}$$

Again avalanches occur when $z(x, y) > z_c$ and the avalanche rule in terms of the local slope becomes:

$$z(x, y) \to z(x, y) - 4 \tag{8.8}$$

$$z(x, y \pm 1) \to z(x, y \pm 1) + 1 \tag{8.9}$$

$$z(x \pm 1, y) \to z(x \pm 1, y) + 1. \tag{8.10}$$

Typically, $z = 0$ at fixed walls and $h = 0$ at edges.[2]

Despite the appealing simplicity of the BTW model, experimenters were quick to point out that it does a poor job of describing some features of real sandpiles. Simple BTW models tend to evolve toward a universal slope and a pure power law of avalanche sizes (hallmarks of SOC). Real piles of granular materials are generally not characterized by a single slope and their avalanches have a different power law in both size and lifetime [32,33]. In experiments, SOC was found under some conditions and not under others [33, 34]. Nevertheless, the model of spatial rearrangement of grains by avalanches was quickly adapted by other researchers to make it more realistic. Adaptations include the following:

- modifying the critical slopes
- multiple grain types
- rotations
- inertial effects.

While the interaction rules remain simple at the grain–grain level, these improved avalanche models have proven successful at capturing the behavior of real flows on larger scales.

Frette addressed the problem of poorly defined angles of repose by modifying the automata rules to allow two critical slopes [34,35]. Each time a lattice site was processed, one of these two critical slopes was chosen at random. In real sandpiles, he argued that varying critical slopes were caused by anisotropies in the grains themselves (varying size or shape), which affected the local packing. The automata rules needed to change in time if they were to capture changes in critical slope as

[2] Basic code implementing a similar rule is provided by Pöschel and Schwager in *Computational Granular Dynamics* [22].

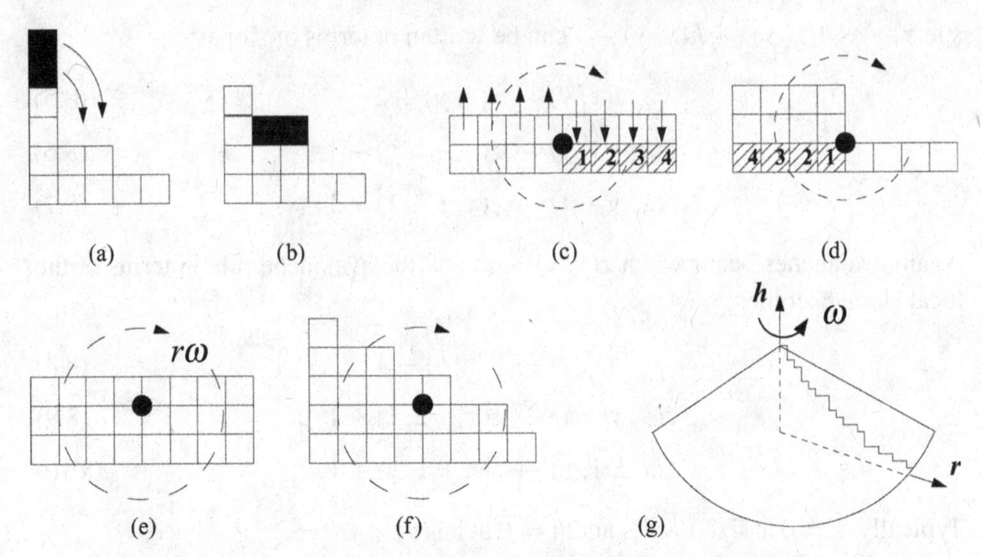

Figure 8.2 (a) and (b) illustrate the toppling rule of Prado and Olami. (c) and (d) illustrate the rotation method of Yanagita. (e) and (f) illustrate the solid-body rotation of Lai *et al.* (g) illustrates the centrifuging pile of Shinbrot *et al.* [44].

grains moved and the grain packing evolved. His model resulted in higher local slopes at the down-stream end of avalanches, reminiscent of height steps seen in real avalanches.

Alternately, the BTW model can be modified to have two species of grains with different frictional properties and therefore different critical angles. Cizeau *et al.* [36] studied sandpile formation with two grain types 1 and 2 and four angles θ_{ij} for the critical slope when grain type i was on top of grain type j. Depending on the relative sizes of these critical angles, they found either stratification or segregation, much as occurs in real sandpiles.

Prado and Olami [37] modified the BTW model to include inertia by using a rule where, whenever the slope at a site exceeds the threshold, two grains from that site topple onto the two downhill nearest neighbors as shown in Figures 8.2(a) and (b). This is no longer a nearest-neighbor model, but it could be extended to topplings of three or more grains. Moreover, they counted the number of times a grain at site x had moved, $n(x)$, during each avalanche and gradually decreased the critical slope at that site as $z_c(x) = z_c - \alpha n(x)$. This models inertia by making avalanches easier for grains that have already avalanched. The parameter α determines the magnitude of the inertial effect. The resulting avalanche behavior was considerably more complex [38]. Their results were in agreement with the experimental avalanche studies of Held *et al.* [33].

A variation on the inertia idea was used by LaMarche *et al.* to study chute flow [39]. In their model, both the number of grains in an avalanche and the distance they moved during an avalanche were determined by the local slope. They added white noise to the local slope as a means of incorporating diffusive motion. The resulting flows were quite complex and included shockwaves, ripples, and vortices. The authors noted that the model's parameters had to be carefully adjusted to produce results in agreement with a particular experimental flow – a situation common to automata models of granular materials.

Many researchers have addressed segregation and flow in rotating drums as studied experimentally by Hill *et al.* [40] and by Nakagawa *et al.* [41,42]. Lai *et al.* [43] used an automata with two grain types to study avalanches in a two-dimensional half-filled rotating cylinder. As shown in Figures 8.2(e) and (f), rotations were modeled by removing $r\omega$ grains from the bottom front of the grain pile (moving the rest of the grains down) and adding these to the bottom of the back of the pile (moving the rest of the back grains up). r here is the distance from the axis of rotation so more grains are moved at sites farther from the axis. This has the effect of raising grains at the back higher than at the front so that avalanches occur when the critical slopes are exceeded. Further, Lai *et al.* allowed a limited number of avalanche events N_a between each rotation event. If the number of avalanches required to bring all sites below the critical slope N_r was less than N_a, then the pile's slope was static; however, if $N_r > N_a$, then the pile's slope was dynamic. The disadvantage of this scheme was that the system had to be half filled.

Ktitarev and Wolf [45] used a similar two-dimensional model to Lai *et al.*, but resolved the filling limitation by using a cylindrical lattice where grains were arranged in circular rings. As a result, they could study stratification at any fill level.

Yanagita [46] used a simpler rotation mechanism to model three-dimensional flow in a rotating drum with two grain types with different critical slopes. As shown in Figures 8.2(c) and (d), rotations were modeled by removing the bottom layer of the grain pile from the front of the pile (moving the rest of the grains down), flipping it over, and adding it to the bottom layer of the back of the pile (moving the rest of the back grains up). This model was able to model both axial and radial segregation in the rotating drum.

Several authors have experimented with adding inertial effects to BTW-type surface-flow models. Nerone and Gabbanelli [47] used an unusual two-dimensional square lattice orientation on which they defined both a height and momentum m at each lattice site. Avalanches occur along diagonals. The effect of increasing momentum at a lattice site was to lower the critical slope for avalanches at that site as $z_c = z_0 - \alpha m(x, y)$ (a technique similar to that of Prado and Olami [37]).

BTW inspired models of grain flow have a lot in common with BCRE theory, named after its authors Bouchaud, Cates, Prakash, and Edwards [48]. In BCRE theory, a flowing granular material can be considered to be composed of two regimes: fast-moving grains along the surface and slower near-static grains in the bulk. This separation of time scales, where one region's grains move at a much faster time scale than another region's grains, is common in surface flows of granular materials. In BCRE theory, the two regions are treated separately. Similarly, BTW-inspired automata models focus on the avalanche behavior and the interior of the sandpile is treated as static or subjected to simple solid-body rotations. By focusing the calculation on the surface grains rather than the bulk, the computational complexity scales as order $\sqrt{N}$ for N grains.

Avalanches in certifuging systems, rotating about the vertical axis as shown in Figure 8.2(g), can also be simulated with BTW-type surface flow models. Shinbrot *et al.* studied rotating granular heaps [44]. In their model, which uses a cartesian coordinate system x, y, h, avalanches always occur in the direction of the local gradient of height. Moreover, both the number of grains moved and the distance at which they are deposited is determined by the local slope. A radial centrifugal force is imposed by adding a term to the local slope proportional to $r\omega^2$, where r is the distance to the axis of rotation and ω was the angular velocity of the system. Shinbrot and coworkers performed experiments using the same geometry and saw comparable static and dynamic patterns in both experiments and the cellular automata experiments. Yoon *et al.* [49] used a similar idea to study conical depressions in rotating containers, where the axis of the container coincided with the axis of rotation and both were tilted slightly from the vertical. Both the angle of tilt and the rotation rate could change during runs. Their simpler model assumes grains move only radially and thereby reduces to a one-dimensional BTW-type surface flow model. Nevertheless, the automata reproduces the kinks and surface shapes seen in the experiments.

8.4 Bulk flow models

The alternative to surface flow automata like the BTW-type models is to model flow in the bulk. This is typically done in one of three ways:

- probabilistic models (diffusing void models)
- momentum models
- energy and interaction models.

8.4.1 Probabilistic models

It is not unusual for cellular automata models to require the occasional use of a random number generator; however, some rules can be overwhelmingly dominated by probabilities. We refer to these as *probabilistic models*. Random walks have been used in modeling the flow of granular media since at least Litwiniszyn [50] in 1956. He used a two-dimensional random walk to model the gravity flow of a thin sheet of sand through a small bottom slot and compared these predictions with experimental measurements [51]. Later, Mullins [52] used a similar idea to relate the flow of voids on a two-dimensional square lattice to a two-dimensional diffusion equation, which he then solved for a variety of boundary conditions. While these were discrete models, they were not cellular automata.

The primary example of a probabilistic cellular automata model of granular media is that introduced by Caram and Hong in 1991 [53]. In their *diffusing void* model, sites on a two-dimensional triangular lattice were either grains, walls, or voids. Voids are introduced at the bottom of the lattice and diffuse upwards by randomly switching places with grains above them. This model was able to capture broad surface shapes and particle streamlines appropriate to flow around obstacles and through hoppers. A similar probabilistic cellular automata using a square lattice was introduced by Savage [54].

A more complex model was introduced by Fitt and Wilmott [55]. This model also used a square grid on which holes moved upward while grains moved downward but with two types of grains. The probability of a particle, located in one of the three sites above a hole, moving into a hole is determined by the size of the particle type and the hydrostatic pressure or height of grains above that site. This model allowed the authors to study segregation in gravity-driven hopper flows.

8.4.2 Momentum models

Cellular automata models, which implement momentum conservation, are the granular flow equivalents of FHP lattice gases. While FHP lattice gases conserve both momentum and energy in collisions, granular materials are dissipative so that energy is not conserved. These granular automata tend to use a regular triangular lattice such as that of the FHP lattice gas with up to seven particles per lattice site, one at rest, and one with a velocity toward each of the six nearest neighbor sites. The rules generally have a collision step and a propagation step. During the collision step, grains collide at each site according to a rule. Figure 8.3 shows one example.[3] Similar collision rules have been used by many researchers [56–58] with different probabilities, depending on the behavior being modeled. Note that some collisions result in more than one stationary particle, even though the model allows only one rest grain per site. Since only one rest particle is allowed per site, the extra rest particles are moved at random to the nearest lattice site with

[3] Basic code implementing a similar rule is provided by Pöschel and Schwager in *Computational Granular Dynamics* [22].

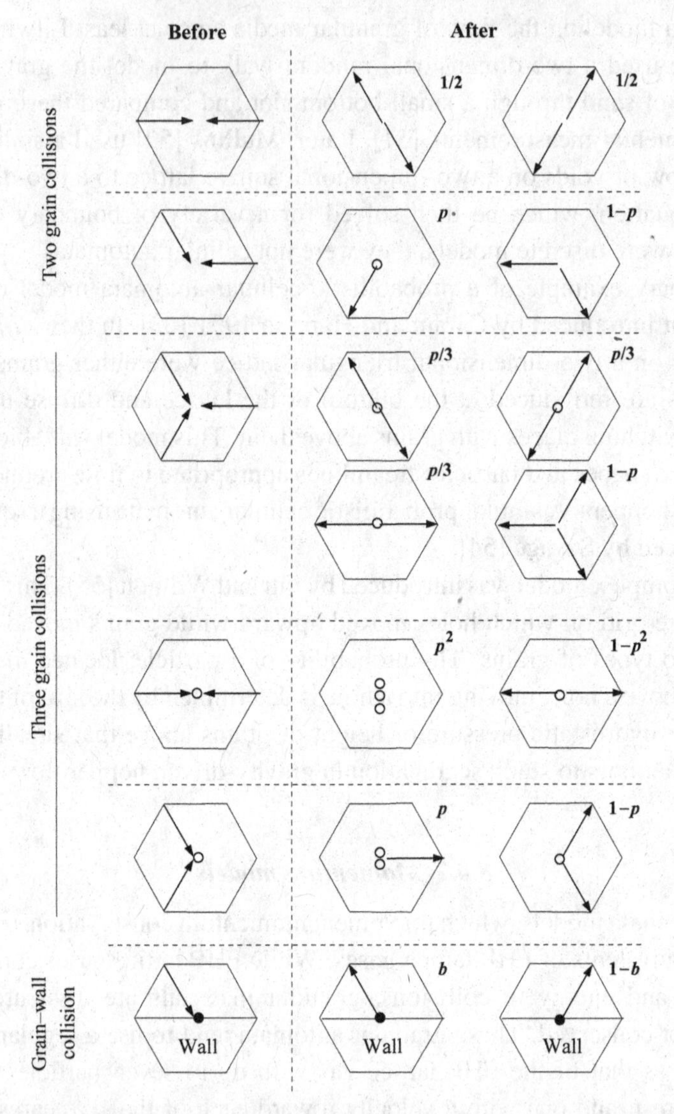

Figure 8.3 A possible nearest-neighbor collision rule for a triangular lattice. The incoming grains are shown in the left column and the outgoing grains in the right. When more than one output state is possible, the probability of each is shown with p being the probability of one of the output states. Moving grains are indicated with a vector; stationary grains are indicated with an open circle and wall points with a filled circle. Each diagram represents three similar cases, each rotated through 60°.

no rest grain. Some models [56, 57] implement gravity by setting rest particles into motion downward with probability g if the neighboring lattice site in that direction is empty. Walls are implemented though lattice sites that are not allowed to move. The probability b determines whether a grain reflects back from a rough wall. For a smooth wall, $b = 0$. This model has been successfully used by Alonso and

Herrmann to study the shape of sandpiles [56], and by Peng and Herrmann [57] and later Peng and Ohta [58] to study granular flow in a channel.

Károlyi and Kertész [59] used a triangular lattice with up to seven particles per site (one rest particle and one velocity in each direction) to study sandpile flows. They implemented different collision rules for the flowing surface layers and for the static interior of the pile. In the rapidly flowing region, the rule was momentum based, similar to that discussed above, with both elastic and inelastic collisions; but, in the slowly changing or static portion of the pile, the rule was nearly static with motion determined only by gravity and the presence of holes. They were able to build their piles grain by grain using this model, and watch the resulting avalanches. The results looked much like the behavior seen in experiments on sand [33] and rice [34], which had previously been simulated using BTW-like automata.

A more complicated automata was implemented by Kozicki and Tejchman to study flow in hoppers [60]. Each site on a 2d triangular lattice was allowed 13 particles: a rest particle and particles with velocities toward each of the nearest and next-nearest neighbors. The collision rule was motivated by momentum conservation, but there were two propagation steps. A look-up table was used to dramatically speed up the simulation. The results were compared with experiments on sand flow through thin hoppers, with very good agreement in surface shapes and flow fields.

8.4.3 Energy and interaction models

Automata rules can also be constructed using energy minimization. The assumption is that the motion of any individual grain will be that which lowers its local energy. This technique was used by Baxter and Behringer [61–63] in one of the earliest grain flow automata. Their automata modeled ellipsoidal-shaped grains, such as rice or grass seed, on a triangular lattice with one grain, wall, or hole per site. Grains were described by an orientation and a velocity as shown in Figure 8.4. There were three allowed orientations along the three primary basis vectors of the triangular lattice. There were seven allowed grain velocities which were unit velocities toward any of the nearest-neighbor sites, or the rest state, or zero velocity. The energy included contributions from gravity and both the motion and orientation of the grains and their nearest neighbors. Baxter and Behringer were able to model the flow of shaped grains in a hopper geometry and achieve good agreement with flows of *Kentucky Bluegrass* grass seed in a thin nearly two-dimensional hopper.

The motion of grains is determined by a local energy function, which includes both a gravitational interaction and interactions with nearest-neighbor grains as shown in Figure 8.4. For a grain at site i at iteration t, the energy is:

$$E_{i,t} = E_{gravity} + \Sigma_{nn} E_{inter}, \tag{8.11}$$

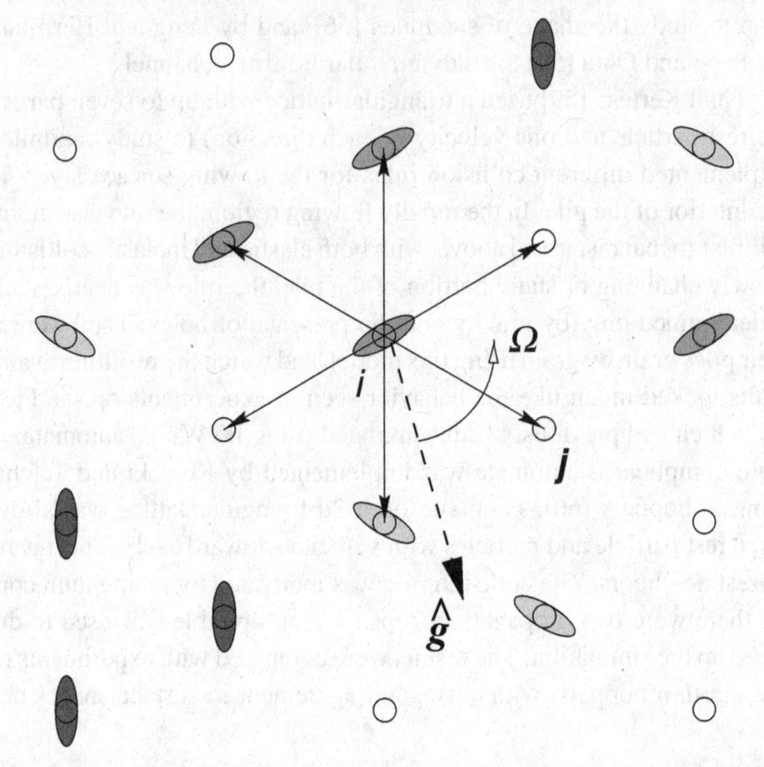

Figure 8.4 A portion of a regular triagonal lattice showing holes and grains. The shading indicates the three allowed grain orientations. Also shown are the directions of motion of a grain i toward its six nearest neighbors, three of which are occupied. The interaction energy E_{inter} is computed separately for each nearest neighbor site such as j. 'Down' is determined by the gravity vector $\hat{g}$, which rotates during the simulation.

where the sum is over nearest neighbors:

$$E_{gravity} = \begin{cases} \Delta * (\hat{d}_{ij} \cdot \hat{g}) & \text{if an adj. site below is vacant} \\ & \quad \& \ \hat{v}_i \text{ is not toward it.} \\ 0.0 & \text{if there are no empty holes below.} \end{cases} \tag{8.12}$$

Here Δ is a scale factor determining the strength of the gravitational interaction. $\hat{v}_{i,t}$ is the velocity at site i and the next iteration t, $\hat{d}_{ij}$ is a vector from site i to the empty site j located below i, and $\hat{g}$ is a unit vector along the direction of gravity at iteration t. If there are more than one vacant site below, $E_{gravity}$ is computed for each and the largest value is used. The same automata can be extended to study the motion of shaped grains in a rotating two-dimensional cylinder by allowing the gravity vector to rotate in time as shown in Figure 8.4.

The interaction term E_{inter} models the effects of shearing, dilation, and relative grain orientation between a grain and its nearest neighbors:

$$E_{inter} = F^{\alpha}_{shear} F^{\beta}_{orient} F^{\gamma}_{dilation}. \tag{8.13}$$

The exponents α, β, and γ make it possible to vary the strengths of the interactions. The shearing factor F_{shear} increases the local energy of grain i if it is moving in a different direction from its neighbor:

$$F_{shear} = |\hat{v}_j - \hat{v}_i| + 0.1. \tag{8.14}$$

$\hat{v}_i$ is a possible velocity vector for grain i at the next iteration and $\hat{v}_j$ is the velocity vector of the neighboring grain after the previous iteration. The orientation factor reduces the energy if shaped grains are aligned while moving:

$$F_{orient} = \begin{cases} S \leq 1 & \text{if both grains are aligned along} \\ & \text{one of their velocity vectors} \\ 1.0 & \text{otherwise.} \end{cases} \tag{8.15}$$

Note that $S = 1$ or $\beta = 0$ correspond to spherical grains. Finally, the dilation factor ensures that grains will only interact when they are in contact:

$$F_{dilation} = \begin{cases} \frac{1}{d} & \text{if } d \leq \frac{1}{1-C} \\ 0.2 & \text{otherwise.} \end{cases} \tag{8.16}$$

Here, d is the mean spacing (time-averaged) of the grains and $0 \leq C < 1$. We have studied the case $C = 0.2$ corresponding to slightly cohesive grains. ($C = 0$ corresponds to perfectly noncohesive grains.)

The automata uses a two-step rule consisting of an *interaction step* followed by a *propagation step*. The interaction step is where each grain may select a new orientation and direction of motion which minimizes its local energy. However, the interaction rule includes an explicit density coupling such that the local density surrounding a grain determines the probability that the grain will be allowed to change its orientation and direction of motion. This density coupling is implemented by choosing a random number between 0 and 1 for each grain. If this number is less than the density coupling function at this grain's local density, then the grain is allowed to change its orientation and direction of motion; otherwise, no changes are allowed. Several possible density coupling functions are shown in Figure 8.5. If a grain is allowed to change its orientation and direction of motion, then these are chosen to minimize its local energy $E_{i,t}$. This set of calculations is done simultaneously for all sites. Following the interaction step, all grains are moved one site along their direction of motion during the propagation step.

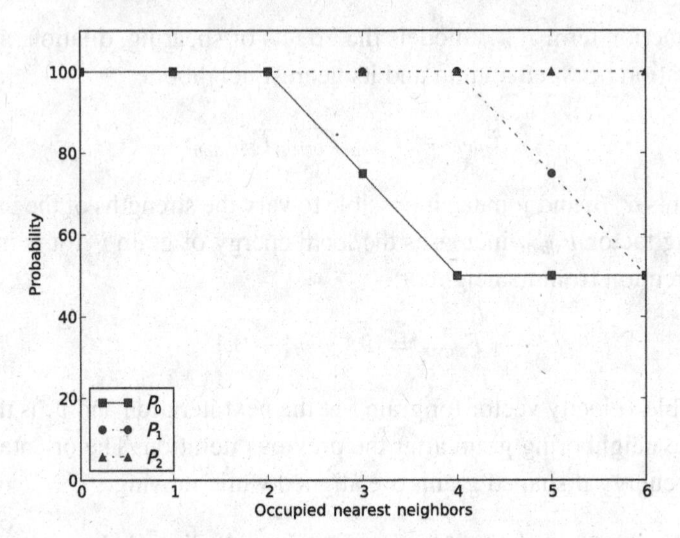

Figure 8.5 Probability that a grain will be allowed to move based on the local density. Several possible density coupling functions are shown.

This automata has been used to study the flow of shaped grains in a hopper geometry with results very similar to the orientations, flow patterns, and density waves seen in grass seed or rice [61–63]. By tagging otherwise identical shaped grains with a color, blue or yellow, and allowing the gravity vector to slowly rotate, this model can simulate mixing of two colors of rice in a two-dimensional rotating container. This is similar to the mixing experiments of Metcalfe *et al.* [64], but the automata allows the grain shape to be varied from round to ellipsoidal.

The mixing simulation is initialized by placing wall points at every lattice site located greater than a distance R from the lattice center. As in the physical experiment, grains are placed at lattice points within the walls up to an initial fill height. Grains to the left of the vertical center line have one color, while grains to the right of the vertical center line have the the other color. These grains are given random initial orientations. The remaining lattice sites above the initial fill height are left vacant. An example of the initial state is shown in Figures 8.6(a) and (b).

The container is incrementally rotated by rotating the direction of gravity in the $E_{gravity}$ calculation. After each change in the gravity vector, any avalanches that result are allowed to complete before the gravity vector is changed again. Sample results showing the evolution of mixing and grain alignment after one, four, and nine revolutions are shown in Figure 8.6. The data run shown in Figure 8.6 used a lattice with 65536 sites on which is built a container with a diameter of 250 lattice spacings, which is 70% filled with shaped grains. The gravity vector was rotated in 30° increments with 1500 iterations between each increment. (Parameters: $\Delta = 3.0$, $S = 0.01$ corresponding to ellipsoidal grains, $C = 0.2$, $\alpha = 3$, $\beta = 2$, $\gamma = 1$, and

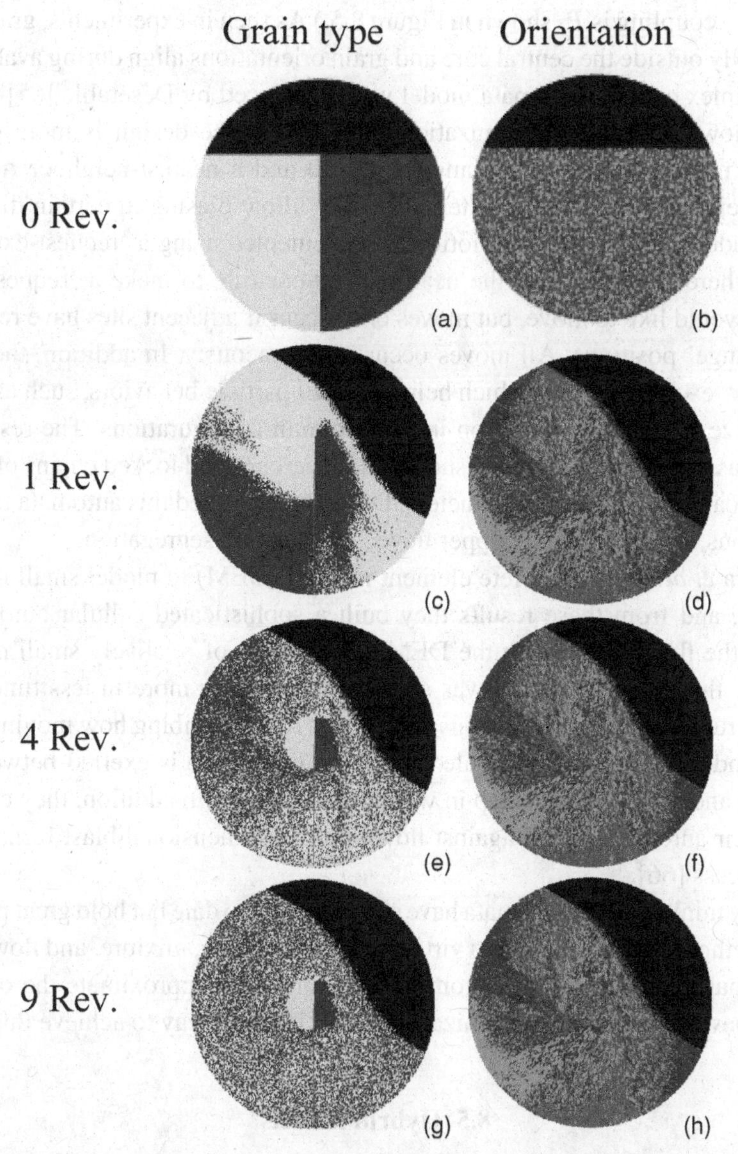

Figure 8.6 Cellular automata states from a run with shaped grains using parameter values given in the text. (a), (c), (e), and (g) are the grain types after 0, 1, 4, and 9 revolutions with shade indicating grain type. (b), (d), (f), and (h) are the corresponding grain orientations after 0, 1, 4, and 9 revolutions with shade indicating the grain orientations from Figure 8.4. Notice that the grain types mix quickly outside the central core, but grain orientations align during the avalanches.

the density coupling is P_2 shown in Figure 8.5.) As seen in experiments, grain types mix quickly outside the central core and grain orientations align during avalanches.

A complex cellular automata model was introduced by Désérable [65] in 2002 which allows an energy-minimization rule, though the design is more general. The automata uses a regular triangular lattice and a nearest-neighbor rule. The nearest-neighbor rule uses an external field to allow biasing in certain directions and excluded sites like walls. Motion is implemented using a 'request-exchange' model, where each site uses the nearest-neighbor rule to make a 'request' as to where it would like to move, but moves only occur if adjacent sites have requested to 'exchange' positions. All moves occur simultaneously. In addition, the model allows for 'exclusion rules' which help to model particle behaviors, such as roughness or size, by inhibiting motion in certain grain configurations. The result is to make it easier to form structures, such as stable arches and locked chains of grains. Through careful choices of parameters, Désérable has used this automata to model suspensions, sedimentation, hopper flows, mixing, and segregation.

Katsura *et al.* used a discrete element method (DEM) to model small numbers of grains, and from these results they built a sophisticated cellular automata to describe the flow [66]. While the DEM was capable of relatively small numbers of grains, the automata model was able to model many more in less time. Their resulting rule had three components: a collision rule describing how moving grains collide and interact, a contact rule describing how force is exerted between rest particles, and a propagation step in which grains move. In addition, they carefully tested their automata model against flow in a two-dimensional blast furnace with good success [66].

Energy minimization automata have seen little use to date but hold great promise. These methods can be adapted to virtually any grain type, mixture, and flow geometry. As has been shown, the automata rule must only approximate the observed grain behavior and energy minimization is one obvious way to achieve this.

8.5 Hybrid models

These models mix continuous parameters, such as those found in molecular dynamics models with discrete parameters commonly found in cellular automata. As such, they are not traditional cellular automata, but they attempt to capture some of the speed of automata while retaining one or more continuous variables. Hybrid models are faster than molecular dynamics models. The presence of continuous variables can guarantee the numerical stability of the calculation, offer additional flexibility, and lower statistical noise in the measurements [10]. The most common two-dimensional model, introduced by Gutt and Haff [67], used continuous x and y, but overlayed a grid on this space and allowed only a single particle per grid site. This made the collision book-keeping easier since a particle could only collide with those in neighboring lattice sites.

A variation was employed by Jasti and Higgs to study granular lubrication [68]. They used a square lattice with one grain per site and continuous velocities.

Baldassarri *et al.* [69] studied cooling in a granular gas using a two-dimensional triangular lattice on which grains were allowed continuous velocities but only discrete positions (lattice sites). Their model explicitly conserved both linear and angular momentum and included dissipation. By confining grains to lattice sites, they obtained much of the efficiency advantages of automata while preserving the continuous velocities necessary to carefully study internal energy and temperature.

8.6 Optimizations

Evaluation of the rule becomes more computationally intensive as more sites are involved. The most efficient rules use only the nearest-neighbor sites. Depending on the complexity and the number of sites involved in the rule, it is sometimes possible to implement the rule through a *lookup table* [9, 70], where the new state is determined by reading a value from a multidimensional array. Once built, a lookup table requires no significant calculations for the remainder of the simulation. The resulting efficiency means that larger lattices and more iterations are possible in less time. A simple automata with a nearest-neighbor rule on a triangular lattice, one grain or hole or wall per site, and seven unit velocities would require a lookup table of $9^7 = 4.7$ megabytes. This is sufficiently small that the time required to build it may quickly be recouped in its use. More complex rules may require much larger lookup tables. For example, the energy minimization rule of Baxter and Behringer [61] has 23 possible states for each lattice site (three possible grain orientations with seven possible velocities as well as a wall or a hole), and the new state is determined by the old state and its six nearest neighbors. This requires a lookup table of $23^7 = 3.4$ gigabytes (without invoking any symmetries). These are within the current memory ranges found in modern desktop computers; however, some grain automata rules are sufficiently complex that their lookup table would be too large to be practical. The lookup table makes sense if the number of entries in the table is much less than the number of times the rule would be evaluated during a run. Alternatively, the table can be created during the course of the run so that each configuration of local states need be evaluated only once.

It is also worth considering whether a simpler rule would produce usable results. For example, there have been many different BTW-type grain automata applied to sand piles [34–37, 39] and flows in rotating cylinders [43, 45–47, 71]. The goal in any automata model is to find the simplest, and therefore computationally fastest, set of rules that describe the behavior of the granular material sufficiently well for a particular geometry and grain type. As yet, there has been no head-to-head comparison of the strengths and weaknesses of these different models.

8.7 The future

Many interesting processes are three dimensional so cellular automata models need to be fully three dimensional. This is already being done with the surface flow models, but little has been done with bulk flow models. It is certainly possible to create three-dimensional momentum rules as has been done for lattice gas fluids [72]. In addition, it is not hard to imagine how to create energy or interaction rules for three-dimensional lattices.

Much of the industrial design of granular processes still uses 'trial and error' rather than simulation, but cellular automata simulation is now producing results sufficiently close to real flows that this could change. It is important that automata models, as with any computer simulation, always be proofed against experiments; and many of the models mentioned above have been carefully compared with real flows to demonstrate their similarities and limitations. We are, as yet, unaware of the use of grain automata in the design of any industrial processes, though this will be a logical step as grain automata continue to improve.

Finally, it is astonishing that the simple models of grain interactions found in cellular automata can produce such close approximations to the behavior of real granular materials. This may cause us to rethink how grain–grain interactions lead to the formation of large-scale behavior and which interactions are most important. The results may some day lead to both an enhanced understanding of the nature of granular materials and improved theories of their behavior.

Acknowledgments

The author would like to thank J. A. McCausland for useful comments on this chapter.

References

[1] H. M. Jaeger, S. R. Nagel, and R. P. Behringer, "Granular solids, liquids, and gases," *Reviews of Modern Physics* **68**, 1259–73 (1996).
[2] H. M. Jaeger, S. R. Nagel, and R. P. Behringer, "The physics of granular materials," *Physics Today* **49**, 32–38 (1996).
[3] I. Aranson and L. Tsimring, "Patters and collective behavior in granular media: theoretical concepts," *Reviews of Modern Physics* **78**, 641–92 (2006).

[4] J. von Neumann, *Theory on Self-Reproducing Automata* (University of Illinois Press, 1966).
[5] M. Gardner, "Mathematical games: the fantastic combinations of John Conway's new solitary game 'Life'," *Scientific American* **223**, 120–3 (1970).
[6] R. T. Wainwright, "Life is universal!" in *Winter Simulation Conference: Proceedings of the 7th conference on Winter simulation*, volume 2, pp. 449–59 (1974).
[7] T. Toffoli and N. Margolus, *Cellular Automata Machines: A New Environment for Modeling* (MIT Press, 1987).
[8] L. B. Kier, P. G. Seybold, and C.-K. Cheng, *Modeling Chemical Systems Using Cellular Automata* (Springer, 2005).
[9] S. Wolfram, *A New Kind of Science* (Wolfram Media, Champaign, IL, 2002).
[10] B. Chopard and M. Droz, *Cellular automata modeling of physical systems* (Cambridge University Press, Cambridge, 1998).
[11] D. Désérable, P. Dupont, M. Hellou, and S. Kamali-Bernard, "Cellular automata models for complex matter," *PaCT 2007, LCNS*, pp. 385–400 (2007).
[12] S. Wolfram, "Cellular automata fluids 1: Basic theory," *J. Stat. Phys.* **45**, 471–72 (1986).
[13] J. Hardy, Y. Pomeau, and O. de Pazzis, "Time evolution of a two-dimensional classical lattice system," *Phys. Rev. Lett.* **31**, 276–79 (1973).
[14] U. Frisch, B. Hasslacher, and Y. Pomeau, "Lattice-gas automata for the Navier–Stokes equation," *Phys. Rev. Lett.* **56**, 1505–8 (1986).
[15] H. Chen, S. Chen, and W. H. Matthaeus, "Recovery of the Navier–Stokes equations using a lattice-gas Boltzmann method," *Phys. Rev. A* **45**, R5339–R5342 (1992).
[16] S. Chen and G. D. Doolen, "Lattice Boltzmann method for fluid flows," *Annual Review of Fluid Mechanics*, vol. 30, pp. 329–64 (1998).
[17] S. Chen, G. D. Doolen, and W. H. Matthaeus, "Lattice gas automata for simple and complex fluids," *Journal of Statistical Physics* **64**, 1133–62 (1991).
[18] H. Chen and W. H. Matthaeus, "New cellular automaton model for magnetohydrodynamics," *Phys. Rev. Lett.* **58**, 1845–8 (1987).
[19] B. G. M. van Wachem, A. F. Bakker, J. C. Schouten, M. W. Heemels, and S. W. de Leeuw, "Simulation of fluidized beds with lattice gas cellular automata," *Journal of Computational Physics* **135**, 1–7 (1997).
[20] H. J. Herrmann, "Simulation of granular media," *Physica A* **191**, 263–76 (1992).
[21] N. V. Brilliantov and T. Pöschel, *Kinetic Theory of Granular Gases* (Oxford Graduate Texts, New York, 2004).
[22] T. Pöschel and T. Schwager, *Computational Granular Dynamics – Models and Algorithms* (Springer-Verlag, 2005).
[23] P. Bak, C. Tang, and K. Wiesenfeld, "Self-organized criticality: An explanation of the 1/f noise," *Phys. Rev. Lett.* **59**, 381–4 (1987).
[24] P. Bak, C. Tang, and K. Wiesenfeld, "Self-organized criticality," *Phys. Rev. A* **38**, 364–74 (1988).
[25] P. Bak, *How nature works : the science of self-organized criticality* (Copernicus, 1996).
[26] K. Wiesenfeld, J. Theiler, and B. McNamara, "Self-organized criticality in a deterministic automaton," *Phys. Rev. Lett.* **65**, 949–52 (1990).
[27] S. Lubeck, N. Rajewsky, and D. E. Wolf, "A deterministic sandpile automaton revisted," *The European Physical Journal B* **13**, 715–21 (2000).
[28] L. P. Kadanoff, S. R. Nagel, L. Wu, and S. min Zhou, "Scaling and universality in avalanches," *Phys. Rev. A* **39**, 6524–37 (1989).

[29] E. Goles, "Sand pile automata," *Annales De L'Inst. Heri Poincare, Section A* **56**, 75–90 (1992).
[30] D. Dhar, "Self-organized critical state of sandpile automaton models," *Phys. Rev. Lett.* **64**, 1613–16 (1990).
[31] D. Dhar, "Sandpiles and self-organized criticality," *Physica A: Statistical Mechanics and its Applications* **186**, 82–7 (1992).
[32] H. M. Jaeger, C. Liu, and S. R. Nagel, "Relaxation at the angle of repose," *Phys. Rev. Lett.* **62**, 40–3 (1989).
[33] G. A. Held, D. H. Solina, H. Solina, D. T. Keane, W. J. Haag, P. M. Horn, and G. Grinstein, "Experimental study of critical-mass fluctuations in an evolving sandpile," *Phys. Rev. Lett.* **65**, 1120–23 (1990).
[34] V. Frette, "Sandpile models with dynamically varying critical slopes," *Phys. Rev. Lett.* **70**, 2762–5 (1993).
[35] K. Christensen, A. Corral, V. Frette, J. Feder, and T. Jøssang, "Tracer dispersion in a self-organized critical system," *Phys. Rev. Lett.* **77**, 107–10 (1996).
[36] P. Cizeau, H. Makse, and H. E. Stanley, "Mechanisms of granular spontaneous stratification and segregation in two-dimensional silos," *Phys. Rev. E* (1998), arXiv:cond-mat/9809430v1.
[37] C. Prado and Z. Olami, "Inertia and break of self-organized criticality in sandpile cellular-automata models," *Phys. Rev. A* **45**, 665–9 (1992).
[38] S. T. R. Pinho, C. P. C. Prado, and S. R. Salinas, "Complex behavior in one-dimensional sandpile models," *Phys. Rev. E* **55**, 2159–65 (1997).
[39] K. R. LaMarche, S. L. Conway, B. J. Glasser, and T. Shinbrot, "Cellular automata model of gravity-driven granular flows," *Granular Matter* **9**, 219–29 (2007).
[40] K. M. Hill, A. Caprihan, and J. Kakalios, "Bulk segregation in rotated granular material measured by magnetic resonance imaging," *Phys. Rev. Lett.* **78**, 50–3 (1997).
[41] M. Nakagawa, "Axial segregation of granular flows in a horizontal rotating cylinder," *Chemical Engineering Science* **49**, 2540–44 (1994).
[42] K. Yamane, M. Nakagawa, S. A. Altobelli, T. Tanaka, and Y. Tsuji, "Steady particulate flows in a horizontal rotating cylinder," *Physics of Fluids* **10**, 1419–27 (1998).
[43] P.-Y. Lai, L.-C. Jia, and C. K. Chan, "Friction induced segregation of a granular binary mixture in a rotating drum," *Phys. Rev. Lett.* **79**, 4994–7 (1997).
[44] T. Shinbrot, N. H. duong, M. Hettenbach, and L. Kwan, "Coexisting static and flowing regions in a centrifuging granular heap," *Granular Matter* **9**, 295–307 (2007).
[45] D. V. Ktitarev and D. E. Wolf, "Stratification of granular materials in a rotating drum: Cellular automata modeling," *Granular Matter* **1**, 141–4 (1998).
[46] T. Yanagita, "Three-dimensional cellular automaton model of segregation of granular materials in a rotating cylinder," *Phys. Rev. Lett.* **82**, 3488–91 (1999).
[47] N. Nerone and S. Gabbanelli, "Surface fluctuations and the inertia effect in sandpiles," *Granular Matter* **3**, 117–20 (2001).
[48] J.-P. Bouchaud, M. E. Cates, J. R. Prakash, and S. F. Edwards, "Hysteresis and metastability in a continuum sandpile model," *Phys. Rev. Lett.* **74**, 1982–5 (1995).
[49] S. Yoon, B.-h. Eom, J. Lee, and I. Yu, "Circular kinks on the surface of granular material rotated in a tilted spinning bucket," *Phys. Rev. Lett.* **82**, 4639–42 (1999).
[50] J. Litwiniszyn, "Application of the equation of stochastic processes to mechanics of loose bodies," *Archivuum Mechaniki Stosowanej* **8**, 393–411 (1956).
[51] J. Litwiniszyn, "Statistical methods in the mechanics of granular bodies," *Rheologica Acta* **1**, 146–50 (1958).

[52] W. W. Mullins, "Stochastic theory of particle flow under gravity," *Journal of Applied Physics* **43**, 665–78 (1972).
[53] H. Caram and D. C. Hong, "Random-walk approach to granular flows," *Phys. Rev. Lett.* **67**, 828–31 (1991).
[54] S. B. Savage, "Disorder, diffusion and structure formation in granular flows," in D. Bideau and A. Hansen (eds.), *Disorder and Granular Media*, p. 264 (North-Holland, 1993).
[55] A. D. Fitt and P. Wilmott, "Cellular-automaton model for segregation of a two-species granular flow," *Phys. Rev. A* **45**, 2383–8 (1992).
[56] J. J. Alonso and H. J. Herrmann, "Shape of the tail of a two-dimensional sandpile," *Phys. Rev. Lett.* **76**, 4911–14 (1996).
[57] G. Peng and H. J. Herrmann, "Density waves and 1/f density fluctuations in granular flow," *Phys. Rev. E* **51**, 1745–56 (1995).
[58] G. Peng and T. Ohta, "Velocity and density profiles of granular flow in channels using a lattice gas automaton," *Phys. Rev. E* **55**, 6811–20 (1997).
[59] A. Károlyi and J. Kertész, "Lattice-gas model of avalanches in a granular pile," *Phys. Rev. E* **57**, 852–6 (1998).
[60] J. Kozicki and J. Tejchman, "Application of a cellular automaton to simulations of granular flow in silos," *Granular Matter* **7**, 45–54 (2005).
[61] G. W. Baxter and R. P. Behringer, "Cellular automata models of granular flow," *Phys. Rev. A* **42**, 1017–20 (1990).
[62] G. W. Baxter and R. P. Behringer, "Cellular automata models for the flow of granular materials," *Physica D* **51**, 465–71 (1991).
[63] G. W. Baxter and R. P. Behringer, "Density effects in cellular automata models of granular materials," in R. P. Behringer and J. T. Jenkins (eds.), *Powders and Grains '97,* Proceedings of the Third International Conference on Powders and Grains (A. A. Balkema, 1997).
[64] G. Metcalfe, T. Shinbrot, J. J. McCarthy, and J. M. Ottino, "Avalanche mixing of granular solids," *Nature* **374**, 39–41 (1995).
[65] D. Désérable, "A versatile two-dimensional cellular automata network for granular flow," *SIAM Journal of Applied Mathematics* **62**, 1414–36 (2002).
[66] N. Katsura, T. Mitsuoka, A. Shimosaka, Y. Shirakawa, and J. Hidaka, "Development of new simulation method for flow behavior of granular materials in a blast furnace using cellular automaton," *ISIJ International* **45**, 1396–405 (2005).
[67] G. Gutt and P. Haff, "An automated model of granular materials," in *Proceedings of the Fifth Distributed Memory Computing Conference, 1990,* pp. 522–9 (1990).
[68] V. K. Jasti and C. F. H. III, "A lattice-based cellular automata modeling approach for granular flow lubrication," *Transactions of the ASME* **128**, 358–64 (2006).
[69] A. Baldassarri, U. Marini Bettolo Marconi, and A. Puglisi, "Cooling of a lattice granular fluid as an ordering process," *Phys. Rev. E* **65**, 051301 (2002).
[70] S. Wolfram, "Statistical mechanics of cellular automata," *Reviews of Modern Physics* **55**, 601–44 (1983).
[71] T. Shinbrot, M. Zeggio, and F. J. Muzzio, "Computational approaches to granular segregation in tumbling blenders," *Powder Technology* **116**, 224–31 (2001).
[72] U. Frisch, D. d'Humieres, B. Hasslacher, P. Lallemand, Y. Pomeau, and J. P. Rivet, "Lattice gas hydrodynamics in two and three dimensions," in *Modern Approaches to Large Nonlinear Physical Systems* (Sante Fe Institute, 1986).

9

Photoelastic materials

BRIAN UTTER

9.1 Introduction

The behavior of dense granular materials, which consist of large collections of individual grains, is an example of a complex system. Despite the relative simplicity of the constituents, the large number of frictional contacts leads to indeterminacy, history dependance, and jamming. We still lack a general set of macroscopic equations to describe their flow. A continuum description of the relevant state variables is desirable, and early studies in soil mechanics focused on characterizing bulk stress/strain relationships and failure. However, it was determined through experiments using photoelastic materials [1–3] that forces transmitted through granular assemblies are carried through an inhomogeneous network of stress chains in which the majority of force is carried through chains of particles comprising a minority of grains (e.g. Figure 9.1(b)). The creation and failure of these chains are central to the fluctuations that can dominate in measurements of dense, granular systems [4].

To visualize internal stresses, these experiments used grains composed of photoelastic materials, which exhibit stress-induced birefringence. When placed between crossed polarizers, in a polariscope, the intensity of transmitted light varies with the local principal stress difference, allowing visualization of the internal stresses in the system. Regions of differential stress appear as a series of bright and dark fringes. The resulting pattern offers both an immediate insight into the spatial stress distribution and the opportunity to measure quantitative local force data in the sample.

Photoelasticity has historically been used in models of mechanical parts to measure locations of high stress to avoid failure, particularly for parts that fail due to concentrated stresses that occur at sharp discontinuities in the design which are

Experimental and Computational Techniques in Soft Condensed Matter Physics, ed. Jeffrey Olafsen.
Published by Cambridge University Press.

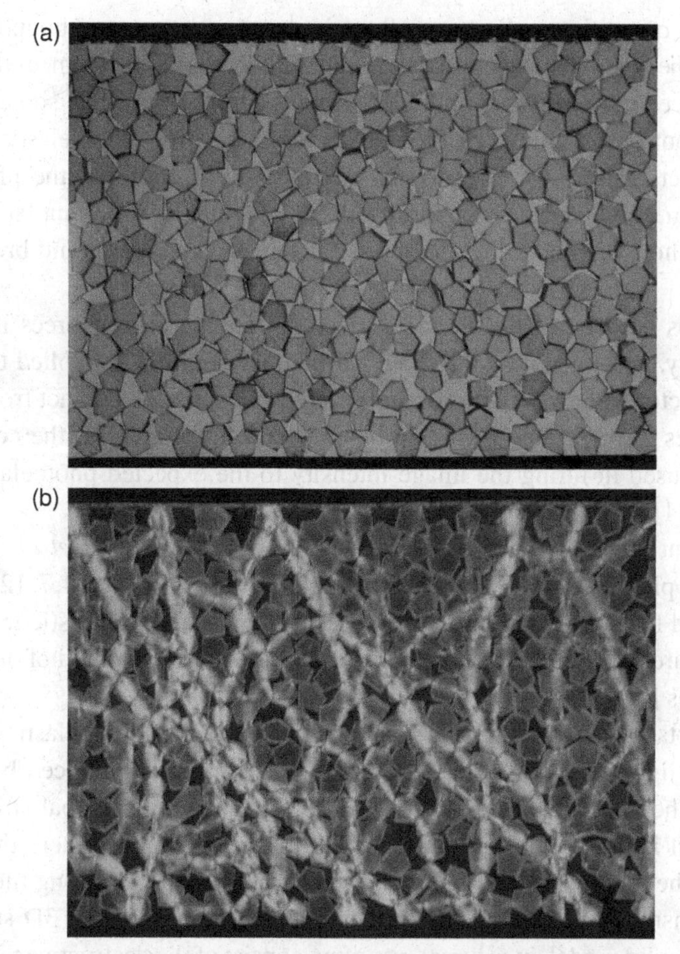

Figure 9.1 Images of a sheared granular material. (a) The analyzer is removed to observe the grains without the photoelastic effect. (b) The same granular assembly is shown in which the analyzer has been inserted to image the stress chains. In each image, the bottom surface is moving to the left relative to the top surface.

difficult to model, either due to geometry or complicated loading [5]. There has been a resurgence in their use, though, as constituents in experiments in which one desires to know the sometimes complicated internal stresses in complex systems, most notably in granular materials, the focus of this chapter.

Figure 9.1(a) shows the top view of a 2D granular assembly of pentagonal photoelastic grains in which the bottom surface is moving left to impose a shear. Figure 9.1(b) shows the same granular assembly placed between crossed polarizers in order to visualize the internal stresses. Bright lines indicate the force- or stress-chains through which the majority of the pressure is transmitted through the experiment.

Although quantities such as overall strain, mean pressure, and boundary forces are often experimentally accessible, it is generally not possible to know the internal forces between constituents. While these forces can be inferred in some cases, for example from deformations of droplets in emulsions [6], photoelastic materials offer a direct measure of the internal stress state. In addition, the photoelastic response can be calibrated to measure mean pressure in the granular assembly, which can show dramatic fluctuations as the stress-chains form and break during flow [7].

The stress field in an elastic body, given particular contact forces imposed at the boundary, was known by the time that photoelasticity was applied to granular systems. Vector force information can be determined at each contact from photoelastic images through the "inverse problem," in which forces at the contacts are parameters used in fitting the image intensity to the expected photoelastic fringe pattern [8–11].

Significant advances have been made recently by Behringer *et al.* in making quantitative photoelastic measurements in granular experiments [4, 7, 12–14]. This has included the gradient-squared method of calibrating photoelastic images with mean pressures [4, 7] and an implementation of the inverse method of determining vector forces [12–14].

In attempts to visualize 3D stress patterns in an extended photoelastic object, the transmitted light intensity is determined by the phase shifts induced by different regions of the experiment due to local differences in the principal stress tensor. Therefore, it is not possible to know from a particular image where the stress is located or the magnitude and orientation of principal stresses along the direction of light transmission, making 3D measurements difficult. When 3D stresses are desired in a solid model, measurements often consist of the destructive technique in which stresses in an elastic solid are frozen-in during the experiment and the sample is machined to image 2D slices [15]. More recent attempts based on correlations in scattered light [16] and integrated, or tomographic, photoelasticity [17, 18] have enabled nondestructive measurements of 3D stress fields in limited situations.

While straightforward interpretation of photoelastic data is limited to two-dimensional systems, particle tracking allows the ability to fully describe both the kinematics and forces at the grain scale, since all grains can be kept in view during an experimental run [19–21].

Increasingly powerful numerical methods and discrete element simulations have reduced the need for experimental tests on stresses in systems traditionally studied with photoelastic models. However, there is ongoing research in related areas, such as the minimization of photoelastic defects in fiber optics, use of photoelastic models for effects of stresses in osteology, and integrated photoelasticity mentioned above. In addition, while quantitative results were historically determined by a trained expert assigning a fringe order to the centers of photoelastic fringes, digital photoelasticity has resulted in both improved accuracy and automation [15,22,23].

9.2 Theory

In a birefringent material, there is a different refractive index, and therefore speed of propagation, for two orthogonal linear polarizations of light, such that they exit the material with a relative phase difference. These orthogonal directions are referred to as fast and slow axes. Birefringence is also referred to as double refraction, since a single beam of light entering a birefringent material will split into two refracted beams (the ordinary and extraodinary rays) that exit the material.

In a photoelastic material, the amount of birefringence at each point in the sample is set by the local principal stresses. For polymeric materials, this is often due to the local alignment of polymer chains or, for glasses, details of the dielectric response. As a result, many transparent plastics and glasses are, to some degree, photoelastic [24].

In the simplest experimental setup, a 2D photoelastic material can be placed between plane polarizers, that is in a plane polariscope, illuminated by a light source and viewed from the opposite side. The first polarizer is typically referred to as the "polarizer" and the latter, oriented at 90° relative to the first, is the "analyzer." In this crossed-polarizer geometry, polarized light from the first polarizer that is undisturbed by the photoelastic material will be blocked by the analyzer. However, stressing a photoelastic material induces a local birefringence and therefore alters the polarization of light from the polarizer, which allows light to pass through the analyzer. This is the basic operation of a dark-field polariscope, in which the image is dark in the absence of a photoelastic material and stresses appear as bright regions. By rotating the analyzer to the same orientation as the polarizer, one can create a bright-field polariscope, in which stressed regions appear as dark fringes on a bright background.

In a plane polariscope, there is an additional condition that leads to extinction of light. If a local stress is induced in the photoelastic medium such that the principal stress axes are aligned with the orientations of the polarizer and analyzer, no light is allowed to pass. That is, if the polarizer allows light which is oriented entirely along the fast or slow axis, it retains this polarization through the birefringent material and is blocked by the analyzer. In these cases, a dark band will be produced, regardless of level of stress on the system. This limitation can be avoided by using crossed circular polarizers, or a circular polariscope, as shown in Figure 9.2 for which this dark band is absent.

A circular polarizer is a linear polarizer with a fixed-birefringence, quarter wave plate attached to it such that its fast axis is at 45° relative to the polarizer. The

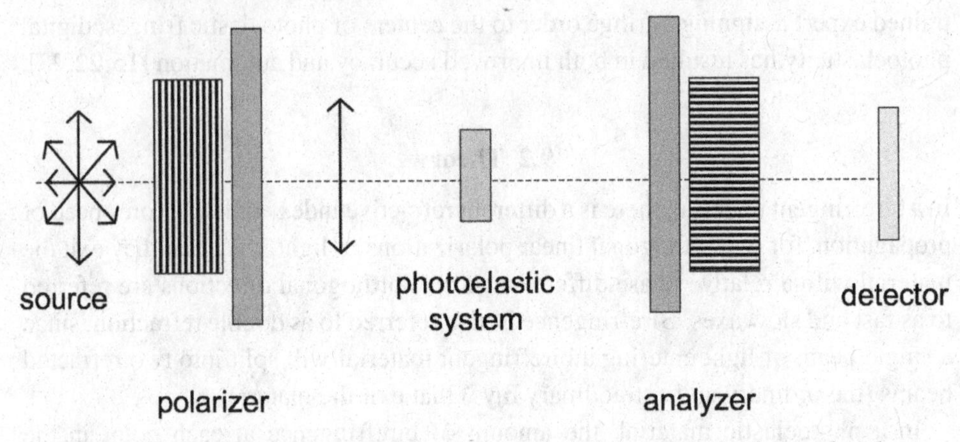

Figure 9.2 Schematic of a circular polariscope. A light source is observed at the detector to study a photoelastic system observed between crossed polarizers. The polarizer and analyzer are circular polarizers, consisting of a linear polarizer with a quarter-wave plate.

linearly polarized light can be viewed as equal components along the fast and slow axes, initially in phase:

$$\begin{aligned}\vec{E}_{fast} &\propto \cos(2\pi z/\lambda)\hat{x}\\ \vec{E}_{slow} &\propto \cos(2\pi z/\lambda)\hat{y}\end{aligned} \tag{9.1}$$

for light traveling in the z direction and the fast axis oriented in the $\hat{x}$ direction. The slow axis is retarded by a quarter of the wavelength of light relative to the fast axis:

$$\vec{E}_{slow} \propto \cos\left(2\pi\left(\frac{z}{\lambda} - \frac{1}{4}\right)\right)\hat{y} \propto \sin\left(\frac{2\pi z}{\lambda}\right)\hat{y}. \tag{9.2}$$

The result, $\vec{E}_{fast} + \vec{E}_{slow}$, is a polarization that rotates in a circle in the (x, y) plane.

Circularly polarized light is right- or left-handed, based on the sense in which the polarization rotates in the (x, y) plane, and it is not possible to determine the orientation of the polarizer from the light that is transmitted. That is, it is nondirectional, and unlike the plane polariscope the orientation of the experiment relative to the polarizer is therefore not important. The second quarter-wave plate restores the circular light to a plane polarized light which enters the analyzer as with the plane polariscope. The analyzer should be the opposite-handed polarization compared to the polarizer for a dark field image. Strictly speaking, the quarter-wave plate is specific to a particular wavelength of incident light, so monochromatic light is often used in precise photoelastic measurements. However, in most cases, the use of white light does not alter the results significantly [25].

The speed of light through a photoelastic material, v, is dependent on whether the polarization of light is parallel or perpendicular to the local principal stress orientation [5]. The optical path difference, or the phase difference for light polarized parallel or perpendicular to the local stress orientation, is:

$$\alpha = \frac{2\pi(\phi_{n_\parallel} - \phi_{n_\perp})t}{\lambda} = \frac{2\pi C\sigma t}{\lambda}, \tag{9.3}$$

where ϕ are the phase differences for light parallel and perpendicular to the local polarization axis, t is the thickness or optical path length of the light, C is the stress-optic coefficient, $\sigma \equiv \sigma_1 - \sigma_2$ is the difference in local principal stresses, and λ is the wavelength of light. Similar expressions can be written in terms of strain or written in terms of a retardation length $\Delta = \lambda\alpha/2\pi$ or relative retardation $\delta \equiv \Delta/\lambda = \alpha/2\pi$, which correspond to the relative phase shift of the two rays in terms of distance or fraction of a wavelength respectively [25].

The principal stresses are axes of maximum and minimum direct stresses and can be either extension or compression. These stresses act in mutually perpendicular planes. The phase difference for light polarized along these principal stresses (with the waves oscillating in these perpendicular principal planes) depends on the local shear stress, $(\sigma_1 - \sigma_2)/2$ [25].

For a material with fixed birefringence, leading to a fixed-phase difference between the two polarizations of light α, the intensity of light exiting a dark-field polariscope is given by:

$$I = I_0 \sin^2\frac{\alpha}{2} = I_0 \sin^2\frac{\pi\Delta}{\lambda} = I_0 \sin^2 \delta\pi, \tag{9.4}$$

where I is the measured light intensity and I_0 is the incident light intensity. In this case, in the absence of a photoelastic response, the light is extinguished, but if the relative phase shift is $\lambda/2$, a bright spot will be observed.

For photoelastic materials then:

$$I(x, y) = I_0 \sin^2\left(\frac{\pi(\sigma_2 - \sigma_1)tC}{\lambda}\right), \tag{9.5}$$

where σ_1 and σ_2 are local planar, principal stresses at position (x, y). The stress optic coefficient C is material dependent and typically relatively insensitive to the wavelength of light [25].

Regions with the same transmitted intensity are referred to as isochromatics and indicate locations in the sample with the same prinicipal stress difference. As the difference in principal stresses changes in a sample, this will be visualized as a series of bright and dark bands referred to as fringes. The degree to which each of these bands of light are retarded, as a fraction of the wavelength of light, is

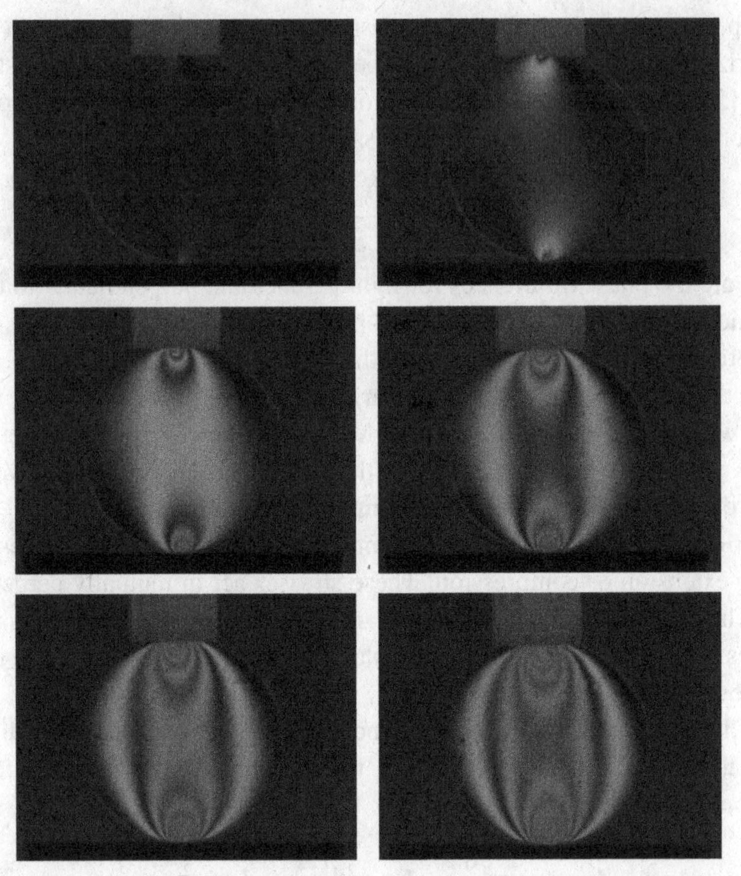

Figure 9.3 Photoelastic disk (PSM-4, diameter = 3 cm) with compressional force of approximately 0, 0.3, 1.0, 2.0, 2.9, and 3.8 N, illuminated by white light.

referred to as the fringe order. For instance, the centers of dark and bright bands have fringe orders n and $n + 1/2$, where n is any integer, with larger fringe orders associated with larger principal stress differences. Since light of different colors will be extinguished at different locations, monochromatic light is often used, though white light will often times produce a similar resulting image. An example is shown in Figure 9.3. In this case, different isochromatics will appear as different colors. In grayscale, the primary effect is that higher-order fringes close to the contact force are blurred as one color is extinguished while another is transmitted.

The relative retardation δ can take any value, however, expressing $\delta = n + \delta^*$ where n is an integer and δ^* is the fractional part $0 < \delta^* < 1$:

$$I(\delta) = I_0 \sin^2(\pi(n \pm \delta^*)) = I_0 \sin^2(\pi\delta^*). \tag{9.6}$$

Therefore, only the fractional part of the relative retardation is found directly by measuring the light intensity. That is, the integral part must be determined separately and does not result directly from the intensity. Historically, this would be done visually based on the experience of a photoelastician. The quantity δ, or sometimes n, is referred to as the fringe order.

While it is often sufficient to identify the location or number of fringes, it is possible to make accurate measurements of the fractional fringe order using the well-established Tardy method, a compensator, or fringe multiplication [26]. Additionally, it is possible to use two images taken with two different wavelengths of light to determine fringe order automatically. In this method, using Equation 9.3, $\delta = \alpha/2\pi$, and the fact that the sample thickness t and stress σ is the same for both images:

$$\frac{\delta_1 \lambda_1}{C_1} = \frac{\delta_2 \lambda_2}{C_2}. \tag{9.7}$$

Since $C_1 \approx C_2$ for different wavelengths of light, $\delta_1 \lambda_1 \approx \delta_2 \lambda_2$, which provides an additional constraint to determine the integer portion of the fringe order [26].

Lines along which the principal stress directions are identical are called isoclinics. A plane polariscope is able to show a particular isoclinic, since it produces a dark line for local stresses for which one principal stress orientation is along the axis of the polarizer. This particular isoclinic is the dark band mentioned above that is absent in a circular polariscope. The measured intensity in a plane polariscope is:

$$I = I_0 \sin^2 \frac{\pi \Delta}{\lambda} \sin^2 2(\alpha - w), \tag{9.8}$$

where $\alpha - w$ is the difference between the orientation of the principal stress axes in the sample and the polarization axes of the polariscope. The quantity α is also referred to as the isoclinic angle. By rotating the optical elements in the polariscope, or phase-stepping, it is possible to observe isoclinics as dark bands moving across the sample and to determine both the local isoclinic angle and the relative retardation δ [23]. Therefore, isochromatics and isoclinics allow determination of principal stress difference and principal stress direction respectively.

9.3 Experimental methods

While many materials are photoelastic, there are some which are designed to have a large stress optic coefficient, for instance, those produced by Vishay [27]. There are a range of Young's moduli for available materials (from 1 to 3000 MPa), which are supplied in sheets and can easily be cut or machined, using razor blades, rotating punches for circles, or end mills on a CNC mill for other shapes. Images for this article were taken using PSM-4, a fairly soft material. The stiffer materials also exhibit a strong photoelastic effect, but frequently retain "frozen in" stresses for many hours after being compressed, while the softer materials lose their induced birefringence almost immediately, making them ideal for dynamic experiments. PSM-4 has a Young's modulus of 4.8 MPa, a Poisson ratio of 0.4–0.5, a strain optic coefficient of $C \approx 0.002 MPa^{-1}$, and a static friction coefficient of 0.85 [27], [13]. In experiments, to avoid grains sticking to surfaces of the experiment, one can use a dry lubricant such as baking powder [19].

It is possible that in cutting grains, some may be left with similar frozen-in stresses, particularly if they are heated during cutting, not cut cleanly, or not lubricated when punched. These grains will appear bright even in the absence of external forces. While it is often a small fraction, which may be discarded, it is possible to anneal the grains at a moderate temperature to remove the frozen-in stresses, though grains may become somewhat discolored in the process. There are also liquid materials available that cure to photoelastic solids, and which can be used to cast other shapes. However, creating grains with flat surfaces and no bubbles can be difficult.

Detailed calibration of the force F (in Newtons) versus displacement x (in mm) at a single contact yields $F \propto x^{3/2}$ for $x > 250\ \mu$m, or forces larger than approximately 0.05 N. The photoelastic response is detectable at $\delta \approx 150\ \mu$m [14]. Under gravitational loading, the material is sensitive enough that forces due to the "hydrostatic" pressure are observable for 5–7 mm diameter grains more than ten layers below the surface.

One way to highlight changes in the force network is to use difference images, in which sequential images are subtracted from each other and rescaled, as shown in Figure 9.4. Regions which remain unchanged between successive pictures are observed as grey; white lines indicate stress chains that formed, and black lines indicate chains that disappeared. In the left image of Figure 9.4, a difference image is shown for planar shear, similar to Figure 9.1. In the right image, a difference image is shown for a free-surface avalanching flow. In this case, stresses near the free surface, which are nearly imperceptible in the original images, are emphasized. The disadvantage is that in applications with significant rearrangement of grains, changes in the stress network are obscured by changes in the positions of the grains. This is particularly evident in the free surface flow on the right of Figure 9.4.

A peculiar fact about these materials is that for a given applied force, the photoelastic response (that is the fringe pattern) is to lowest order, independent of the thickness of the grains [25]. A thicker sample will have a stress and path length which are inversely proportional to each other. The only constraint is that the grains must have a diameter larger than their thickness to avoid flipping. In the results presented here, sheets of 3 mm thick material are used to create grains of 5–7 mm diameter.

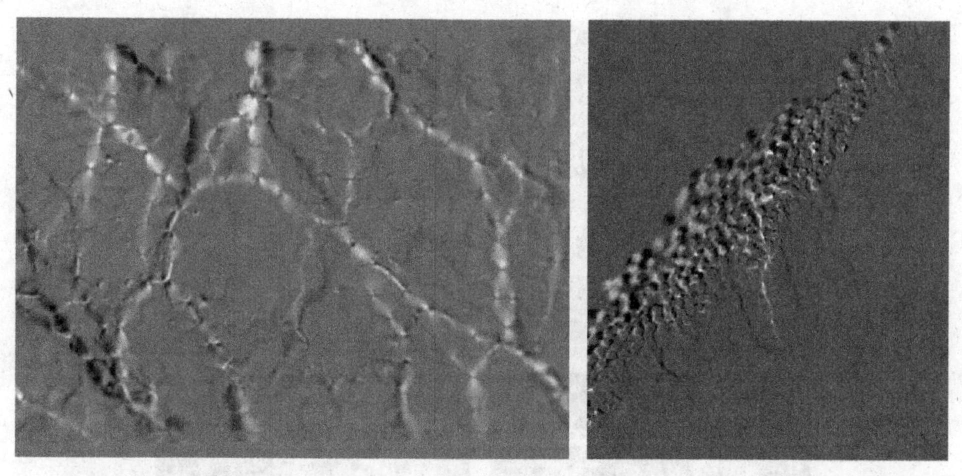

Figure 9.4 Difference images, in which sequential pictures are subtracted to highlight changes in the force network, are shown for (left) planar shear (e.g. in Figure 9.1) and (right) a free-surface avalanching flow.

Experimentally, for a diffuse light polariscope, the light source should be relatively uniform, providing reasonably parallel light.[1] (It is also possible to illuminate the sample using a point light source at the focus of a large lens in a lens polariscope geometry [25].) Polarizers of sufficient quality can be purchased as large sheets of polarizing film from a variety of sources. Their primary measure of quality is the degree to which they extinguish light when crossed. The polarizer can be placed directly on the light source and the analyzer can be placed on the lens of the camera used for data collection. If desired, a second camera without an analyzer in place can simultaneously record images for particle tracking.

Experiments can be constructed using cast acrylic. Note that acrylic material, particularly when extruded, will act as a weak linear polarizer and can compromise the quality of images. Additionally, scratches on either the polarizer or experimental apparatus are typically more apparent when viewed with crossed polarizers, so care should be taken to avoid scratches.

9.4 Analysis

9.4.1 G-squared calibration

While photoelastic experiments will readily allow visualization of stresses, quantitative data can be obtained by calibrating the photoelastic response of the system to a known applied pressure [4]. An example of such calibration images is shown

[1] The LD Illuminators from Lumined are one example of suitable light sources. They are CCFL lamps with a flicker rate of over 40 kHz, making them practical also for high-speed photography. For slower frame rates, similar illuminators by S&S X-ray Products may be useful.

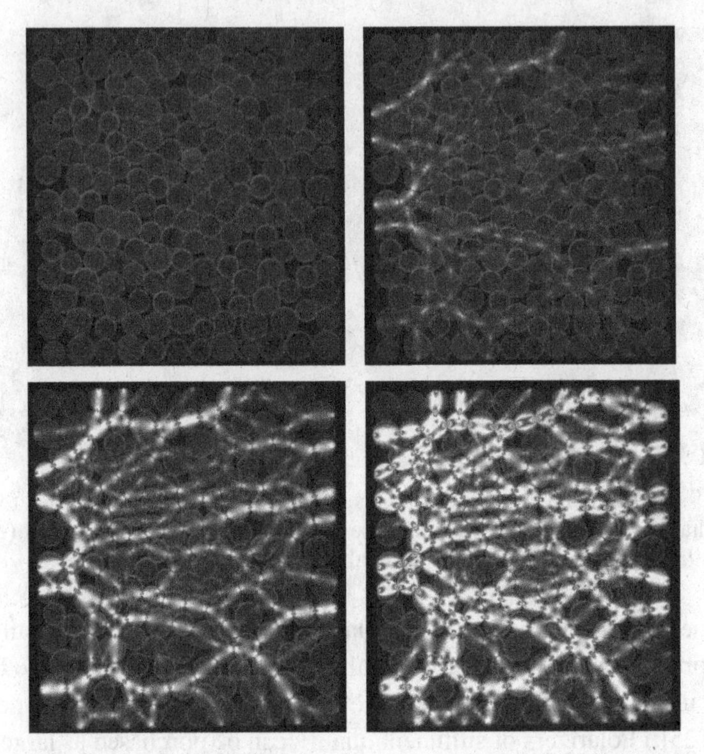

Figure 9.5 Images used for calibrating the g-squared technique [29]. The increasing force on the piston is 0.04, 0.25, 0.78, and 1.73 N.

in Figure 9.5. In this figure, a piston on the left is compressed with a force gauge and the photoelastic response is calibrated to the force reading.

With larger mean pressure, there are more grains that allow light to pass through the analyzer. The intensity of the image roughly corresponds to the overall pressure. In addition, the number of fringes increases monotonically with the force at the contact. For instance, in Figure 9.5 grains with larger forces are those with a fringe structure of light and dark bands observed within the grain. Correspondingly, the average gradient in image intensity increases. The scale of the local forces can then be characterized by the average square gradient or "G-squared" ($\langle G^2 \rangle$) at a pixel (i, j) written in terms of the intensity or pixel values of its neighbors. In a particular region of space, $\langle G^2 \rangle$ correlates with the mean pressure in the sample, that is the trace of the stress tensor [28]:

$$\langle G^2 \rangle \propto \sum_{i,j} \Bigg((I_{i-1,j} - I_{i+1,j})^2 + \left(\frac{I_{i-1,j-1} - I_{i+1,j+1}}{\sqrt{2}} \right)^2 + (I_{i,j-1} - I_{i,j+1})^2 + \left(\frac{I_{i+1,j-1} - I_{i-1,j+1}}{\sqrt{2}} \right)^2 \Bigg), \tag{9.9}$$

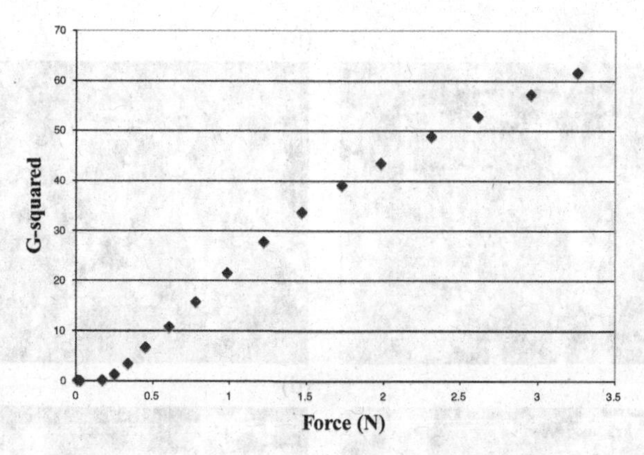

Figure 9.6 Calibration curve for Figure 9.5 [29].

where the sum is taken over all pixels in the region of interest and the average $\langle G^2 \rangle$ per pixel is used.

An example of the calibration curve for the images in Figure 9.5 is shown in Figure 9.6. At small forces, the photoelastic effect is difficult to detect and the background noise due to the edges of the grains dominates. At the upper limit, one has to resolve the fringes, typically with at least 10–20 pixels across a grain diameter, but this is rarely a limitation in situations of interest. While not strictly linear, the calibration is monotonic and sufficiently linear in the region of interest (forces of 0.25–1.5N) to allow a scale for pressure measurements.

Additionally, since $\langle G^2 \rangle$ corresponds to force analogous to a spring of effective constant k diplaced a distance x ($F \sim kx$), it is also possible to calibrate the energy stored in the system. In this case, the energy applied to the piston, calculated as the force times the displacement of the piston, correlates well with the square of $\langle G^2 \rangle$ (energy $= \frac{1}{2}kx^2 \propto \langle G^2 \rangle^2$) [29].

9.4.2 *The inverse problem*

Photoelastic data can be used to identify contact forces by assigning a fringe order to the images and using the stress optic coefficient to determine the stress. Early methods proposed using fringe orders from two or more experimental data points to calculate forces at the boundary, including the overdetermined system in which multiple points from a photoelastic image, are used in a nonlinear fit [8–10, 30]. This is referred to as the inverse problem and has recently been implemented in experimental systems by Behringer *et al.* [12–14].

The measured photelastic images are fit using the plane elastic solution, assuming that deformations are not substantial. The force components at the boundaries are treated as fitting parameters, which are determined through a nonlinear fit. For

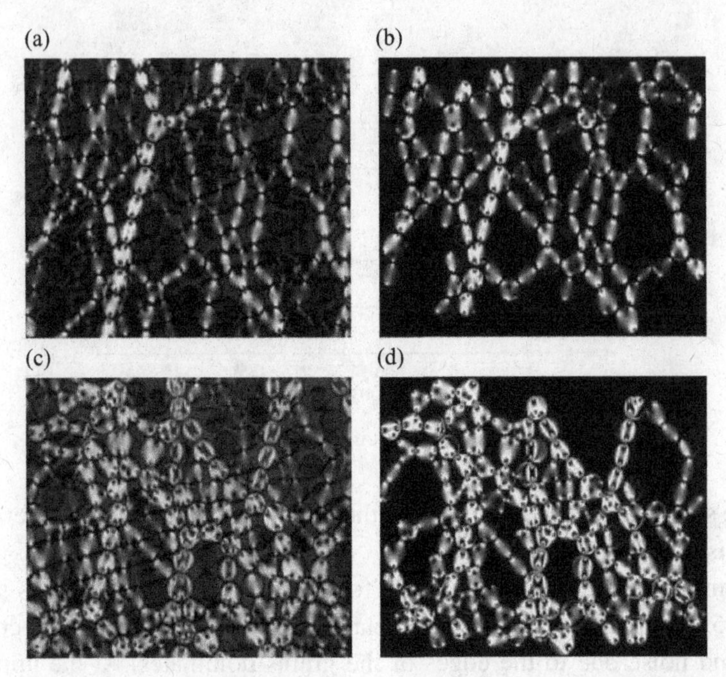

Figure 9.7 Experimental images (a) and (c) with the corresponding theoretical images (b) and (d) expected based on contact forces determined through the inverse method. (Reprinted from Figure 9.2 of [12] by permission from Macmillan Publishers Ltd, Nature **435**, 1079 (2005), copyright 2005.)

adequate results, it is necessary to also know the locations of the grains and contacts, which is most easily obtained with a second camera without the analyzer in place. Differences between the experimental fringe pattern and a calculated pattern, given a known set of forces are minimized to find the best fit for the forces. The $\langle G^2 \rangle$ measurement is one way to estimate the initial forces for the fit. In addition, a minimum $\langle G^2 \rangle$ threshold can be used to verify that a force contact exists, rather than simply a "rattler," which does not exert significant force on its nearest neighbors. These are difficult to distinguish using nearest neighbor distances from nonphotoelastic images alone. This results in vector forces at each contact rather than simply a mean measure of the local pressure.

Examples of the use and validity of the inverse problem can be found in [12–14]. It is possible, upon fitting experimental data with the inverse method to find vector forces at each contact, to then use the calculated forces to reconstruct a photoelastic image by solving elasticity equations to find the theoretical stress in each grain. The result can be compared with the experimental image as a validation of the method. An example is shown in Figure 9.7, reproduced from [12], with experimental images (a), (c) and corresponding theoretical images (b), (d). While a completely automated process is not easy to accomplish, given the nonlinear fit inherent in the method, experimental images, along with images reconstructed using contact forces found through the fit, appear remarkably similar.

9.5 Photoelastic measurements in granular media

As mentioned earlier, photoelastic measurements were the first to reveal the inhomogeneous network of stress chains (or force chains) which transmit force through a granular medium [1–3]. More recently, photoelasticity has played a prominent role in elucidating the properties of these force chains and the behavior of granular matter. Below, some of the recent findings from photoelastic granular measurements are summarized.

A number of photoelastic granular studies focus on the transmission of forces through a static granular packing. Geng *et al.* found that ordered packings can lead to strong forces along crystallographic directions, while disorder leads to a broadened, diffusive force transmission [28]. These results produce visual evidence for similar conclusions observed in experiments at the boundaries of three-dimensional packings [31]. It is also possible to use sheets of photoelastic material at the boundaries of such an experiment to measure boundary forces in 3D experiments [32].

Even in granular systems that appear similar, for instance with two qualitatively similar piles of grains, the construction history is found by Geng *et al.* to have a significant effect on the force structure of the pile as evidenced by photoelastic measurements [33, 34]. A pile created from a point source typically exhibits a dip in pressure below the center of the pile due to the avalanche-induced force network which establishes stress chains that transmit force along the sides of the pile [33, 34]. This stress dip was found by Zuriguel *et al.* to be accentuated with elongated particles [35, 36].

Through the inverse problem, Majmudar and Behringer determined vector forces at each contact, thereby allowing a bulk measurement of the probability distribution of forces [12]. They also used photoelasticity to determine the presence or absence of contacts and make measurements of scaling laws for contact number and pressure versus packing fraction near jamming [13]. They have similarly applied this technique to dynamics, such as pure shear [37].

Under shear, stress chains are found to align along the principal stress direction and fail due to buckling [38]. These chains have been proposed to be strong in compression, but weak to transverse forces through a buckling instability [39]; therefore, experiments that are able to access the force network can offer significant insight. This buckling is fundamental to the behavior of granular materials, particularly at the threshold of jamming, where grains may become frustrated and immobile or suddenly flow with instantaneous changes in the force network. Photoelastic experiments also allow visualization of the stress network during transient situations [21], where the anisotropy of the force network is proposed to lead to weakening observed when the direction of shear is reversed [40]. In this case, the strong force network established during shear establishes a force network that cannot oppose significant shear in the reverse direction.

These force fluctuations are found to dominate in granular dynamics [4]. Photoelastic experiments by Howell *et al.* revealed a diverging force chain length scale similar to a critical transition as the critical packing fraction is approached from above [7, 19]. In dense flows, the mean (time-averaged) stress in a Couette shear also displays a perhaps surprising logarithmic dependence on the shear rate [41].

Other properties of the flow, such as diffusion, are also affected by the force network. For instance, diffusivity is observed to be lowest along directions corresponding to the orientation of the observed stress chains [20].

Additional applications of photoelastic granular materials include the failure of granular soils [42], visualization of the load transfer profiles due to explosive loading in granular aggregates [43], measurement of various fabric tensors during cyclic shear [44], and characterization of the jamming transition through responses of the force network and resulting flow to imposed vibrations [45]. Shear in an anolog to geological faults is also studied using photoelastic techniques [46]. In these experiments, small stress drops are observed to occur due to localized events punctuated by occasional, large system-spanning events.

It should be noted that it is possible to exploit the somewhat weaker photoelastic response in other materials. Lesniewska *et al.* used glass beads in an indexed matched fluid to measure internal stresses and particle displacements [47].

As interest in the jamming transition in granular materials continues, photoelastic experiments are sure to contribute insight into the complicated, dynamic force network that exists during the jamming/unjamming transition.

9.6 Conclusions

Photoelastic materials have been receiving increased attention in recent years as the building blocks of complex systems, such as granular materials. Due to their stress-induced birefringence, when placed between crossed polarizers, they allow visualization of internal stresses within 2D experiments – information that is inaccessible in most other experiments. In granular systems, early photoelastic experiments were the first to reveal that stresses are carried by an inhomogeneous network of force chains. Subsequent experiments continue to probe the real-time stress response of these complex systems.

While they provide immediate, full-field information about the relative magnitude and location of stresses in the system, it is also possible to gather quantitative data, either through calibration of a gradient-squared method using images at known pressures or using nonlinear fitting to solve the fill inverse problem to determine contact forces based on measured photoelastic fringes. Additionally, photoelastic materials are easily machined into a variety of shapes and sizes, which allows for a large range of experimental systems.

Although implementing the full inverse problem involves a significant amount of work, acquiring qualitative or average results in simple geometries requires little more than polarizers, a light source, and a camera. In this sense, photoelastic materials offer an easy way to visualize forces in systems for which stress information is typically difficult to access.

Acknowledgments

I would like to thank R. P. Behringer, with whom I learned many of the experimental techniques presented here.

References

[1] P. Dantu, "Etude statistique des forces intergranulaires dans un milieu pulverulent," *Géotechnique* **18**, 50–5 (1968).
[2] A. Drescher and G. de Josselin de Jong, "Photoelastic verification of a mechanical model for the flow of a granular material," *Géotechnique* **18**, 50–5 (1968).
[3] T. Travers, D. Bideau, A. Gervois, J. P. Troadec, and J. C. Messager, "Uniaxial compression effects on 2d mixtures of 'hard' and 'soft' cylindars," *J. Phys. A: Math. Gen.* **19**, L1033–L1038 (1986).
[4] D. W. Howell, R. P. Behringer, and C. T. Veje, "Fluctuations in granular media," *Chaos* **9**, 559–72 (1999).
[5] M. M. Frocht, *Photoelasticity*, vol. 1 (John Wiley and Sons, New York, 1941).
[6] J. Brujić, S. F. Edwards, D. V. Grinev, I. Hopkinson, D. Brujić, and H. A. Makse, "3d bulk measurements of the force distribution in a compressed emulsion system," *Faraday Discuss.* **123**, 207–20 (2003).
[7] D. Howell, R. P. Behringer, and C. Veje, "Stress fluctuations in a 2d granular couette experiment: a continuous transition," *Physical Review Letters* **82**, 5241–44 (1999).
[8] R. J. Sanford and J. W. Dally, "A general method for determining mixed-mode stress intensity factors from isochromatic fringe patterns," *Engineering Fracture Mechanics* **11**, 621–33 (1979).
[9] A. J. Durelli and D. Wu, "Use of coefficients of influence to solve some inverse problems in plane elasticity," *Transactions of the ASME* **50**, 288–96 (1983).
[10] A. Shukla and H. Nigam, "A numerical-experimental analysis of the contact stress problem," *J. Strain Anal.* **20**, 241–5 (1985).
[11] S. Paikowsky and K. J. DiRocco, "Image analysis for interparticle contact modeling," in *Conference on Digital Image Processing: Techniques and Applications in Civil Engineering*, ASCE Publication No. 236 (1993).

[12] T. S. Majmudar and R. P. Behringer, "Contact force measurements and stress-induced anisotropy in granular materials," *Nature* **435**, 1079–82 (2005).
[13] T. S. Majmudar, M. Sperl, S. Luding, and R. P. Behringer, "Jamming transition in granular systems," *Physical Review Letters* **98**, 058001 (2007).
[14] T. S. Majmudar, M. Sperl, S. Luding, and R. P. Behringer, "The jamming transition in granular systems – supplementary information," available at www.aip.org/pubservs/epaps.html, document number E-PRLTAO-98-020705.
[15] T. Y. Chen, *Selected Papers on Photoelasticity* (SPIE Optical Engineering Press, Bellingham, WA, 1999).
[16] J.-C. Dupré, "Photoelastic analysis of a three-dimensional specimen by optical slicing and digital image processing," *Experimental Mechanics* **18**, 393–7 (1968).
[17] H. Aben, L. Ainola, and J. Anton, "Integrated photoelasticity for nondestructive residual stress measurements in glass," *Optics and Lasers in Engineering* **33**, 49–64 (2000).
[18] L. Ainola and H. Aben, "On the optical theory of photoelastic tomography," *J. Opt. Soc. Am. A* **21**, 1093–101 (2004).
[19] C. T. Veje, D. W. Howell, and R. P. Behringer, "Kinematics of a two-dimensional granular couette experiment at the transition to shearing," *Physical Review E* **59**, 739–45 (1999).
[20] B. Utter and R. Behringer, "Self-diffusion in dense granular shear flows," *Physical Review E* **69**, 031308 (2004).
[21] B. Utter and R. Behringer, "Self-diffusion in dense granular shear flows," *European Physical Journal E* **14**, 373–80 (2004).
[22] A. Ajovalasit, S. Barone, and G. Petrucci, "Towards rgb photoelasticity – full-field automated photoelasticity in white-light," *Experimental Mechanics* **35**, 193–200 (1995).
[23] A. Ajovalasit, S. Barone, and G. Petrucci, "A review of automated methods for the collection and analysis of photoelastic data," *Journal of Strain Analysis* **33**, 75–91 (1998).
[24] V. Galiatsatos, "Refractive index, stress-optical coefficient, and optical configuration parameter," in J. E. Mark (ed.), *Physical Properties of Polymers*, pp. 823–56 (Springer Science, New York, 2007).
[25] R. B. Heywood, *Designing by Photoelasticity* (Chapman and Hall, London, 1955).
[26] T. Y. Chen, "Digital determination of photoelastic birefringence using two wavelengths," *Experimental Mechanics* **37**, 232–6 (1997).
[27] Vishay Intertechnology, Inc. (Malvern, PA).
[28] J. F. Geng, G. Reydellet, E. Clement, and R. P. Behringer, "Green's function measurements of force transmission in 2d granular materials," *Physica D* **182**, 274–303 (2003).
[29] B. Utter and R. P. Behringer, unpublished.
[30] M. M. Frocht, *Photoelasticity*, vol. 2 (John Wiley and Sons, New York, 1948).
[31] N. W. Mueggenburg, H. M. Jaeger, and S. R. Nagel, "Stress transmission through three-dimensional ordered granular arrays," *Physical Review E* **66**, 031304 (2002).
[32] E. I. Corwin, H. M. Jaeger, and S. R. Nagel, "Structural signature of jamming in granular media," *Nature* **435**, 1075–8 (2005).
[33] J. F. Geng, E. Longhi, R. P. Behringer, and D. Howell, "Memory in two-dimensional heap experiments," *Physical Review E* **64**, 060301 (2001).
[34] J. F. Geng, D. Howell, E. Longhi, R. P. Behringer, G. Reydellet, L. Vanel, E. Clement, and S. Luding, "Footprints in sand: The response of a granular material to local perturbations," *Physical Review Letters* **87**, 035506 (2001).

[35] I. Zuriguel, T. Mullin, and R. Arevalo, "The role of particle shape on the stress distribution in a sandpile," *Proc. R. Soc. London. Series A – Mathematical, Physical and Engineering Sciences* **464**, 99–116 (2008).
[36] I. Zuriguel, T. Mullin, and R. Arevalo, "Stress dip under a two-dimensional semipile of grains," *Physical Review E* **77**, 061307 (2008).
[37] J. Zhang, T. Majmudar, and R. P. Behringer, "Force chains in a two-dimensional granular pure shear experiment," *Chaos* **18**, 041107 (2008).
[38] M. Oda and H. Kazama, "Microstructure of shear bands and its relation to the mechanisms of dilatancy and failure of dense granular soils," *Geotechnique* **48**, 465–81 (1998).
[39] M. Cates, J. Wittmer, J.-P. Bouchaud, and P. Claudin, "Jamming, force chains, and fragile matter," *Physical Review Letters* **81**, 1841–4 (1998).
[40] M. Toiya, J. Stambaugh, and W. Losert, "Transient and oscillatory granular shear flow," *Physical Review Letters* **93**, 088001 (2004).
[41] R. Hartley and R. Behringer, "Logarithmic rate dependence of force networks in sheared granular materials," *Nature* **421**, 928–31 (2003).
[42] R. Wan, P. Guo, and M. Al-Mamun, "Behaviour of granular materials in relation to their fabric dependencies," *Soils and Foundations* **45**, 77–86 (2005).
[43] R. J. Sanford and J. W. Dally, "Dynamic photoelastic studies of wave-propagation in antigranulocytes media," *Optics and Lasers in Engineering* **14**, 165–84 (1979).
[44] R. J. Sanford and J. W. Dally, "Experimental investigation of fabric-stress relations in antigranulocytes materials," *Mechanics of Materials* **11**, 87–106 (1991).
[45] D. Amon, E. Hoppmann, and B. Utter, unpublished.
[46] K. E. Daniels and N. W. Hayman, "Force chains in seismogenic faults visualized with photoelastic granular shear experiments," *Journal of Geophysical Research* **113**, B11411 (2008).
[47] D. Lesniewska and D. M. Wood, "Observations of stresses and strains in a granular material," *Journal of Engineering Mechanics – ASCE* **135**, 1038–54 (2009).

10

Image acquisition and analysis in soft condensed matter

JEFFREY S. OLAFSEN

10.1 Introduction

Fast and affordable computing power, especially in the form of personal computers and workstations, has enabled the expansion of the study of soft condensed matter physics over the last two decades. The use of computing power to not only analyze image data but acquire it from high-resolution and high-speed digital sources has also made many significant investigations in soft condensed matter experiments accessible to students while still pursuing their undergraduate degrees. Unlike students just a generation before, contemporary undergraduates are well versed in the use of computing power, even operating systems such as Linux, when they first arrive on campus. Because of outreach initiatives such as REU programs, intentions to recruit future graduate students, and an increasingly competitive trend in the careers of undergraduate majors, the opportunities to engage undergraduates in research has flourished over nearly the same period of time.

Granular systems, with simple hard sphere interactions and inter-particle friction, tend to be investigated in experiments that are tabletop in scale. As a subset of soft condensed matter systems, the macroscopic nature of granular physics makes the systems conceptually accessible to students as early as the sophomore or junior year of their baccalaureate careers. This no way trivializes the investigations or minimizes the advances in knowledge that a properly trained undergraduate can contribute to the larger scientific community when mentored well. Indeed, as most undergraduate physics programs do not typically offer courses in contemporary advanced topics, such an nonequilibrium thermodynamics or soft condensed matter physics, the ability to have students pursue journal-publishable research efforts, while still enrolled in more classical undergraduate physics courses, is a transformational opportunity for their educational development.

Experimental and Computational Techniques in Soft Condensed Matter Physics, ed. Jeffrey Olafsen.
Published by Cambridge University Press.

This chapter will discuss the topics of image acquisition and analysis for soft condensed matter systems in general before detailing the specific example of techniques for particle tracking in granular gas systems. Finally, the chapter will focus on the author's experience in mentoring undergraduate research and the specifics of introducing students to the tools of the trade. The intent is to present the material in such a manner that one might use the general outline of undergraduate research presented by the author with the specifics of any of the soft condensed matter techniques offered in this book. Also, the framework for involving an undergraduate student researcher in journal-publishable research will be discussed with the view that the chapter should be read by both the student and the faculty mentor, as well as any graduate students or postdoctoral fellows who will be involved in the mentoring of undergraduate students. It is by no means the only manner in which to involve undergraduates in a research laboratory in soft condensed matter physics, but it is one that has worked extremely well for the author.

10.2 Image acquisition

While there have been investigations of granular systems with what might be termed more exotic image acquisition systems, such as X-ray [1] and magnetic resonance imaging (MRI) [2–4], or novel light sources and spectroscopic techniques [5,6], the primary image acquisition in granular gas systems, and soft condensed matter systems more broadly, is through high-resolution and high-speed digital camera technology. The CCD (charge-coupled device) sensor that makes such digital imaging possible was recognized with the 2009 Nobel Prize in Physics awarded in part to Willard S. Boyle and George E. Smith [7]. The use of such digital optical acquisition relies on using primarily visible light sources. There are cameras that are designed or enhanced with optics that are optimized to operate in particular wavelength ranges such as infrared (IR), and most off-the-shelf CCD cameras are sensitive in the near-IR spectrum in addition to the visible. Several different companies exist that will design and construct image acquisition systems as intermediaries to original equipment manufacturers (OEMs). These intermediate companies are beneficial in building pre-packaged systems that are more turn-key and accessible for undergraduate students with limited backgrounds in interfacing computer components. However, even such platforms as LabVIEW have recently made the development of the image acquisition portion of granular gas investigations accessible to relatively inexperienced undergraduates. The primary consequence of the limitation of optical imaging to the visible spectrum is that most granular investigations are thereby constrained to being 1D or 2D in design [8,35] (although one can pursue a 3D investigation in a dilute particle density limit [36,37] or with a tracer particle technique similar to MRI [38,39]). While some applications have been able to develop 3D imaging through the use of multiple orthogonal cameras [40] or novel uses of index matching and illumination [41–43], the typical use of high-speed and high-resolution digital imaging is primarily for the investigations of systems in fewer than three dimensions.

As computers have developed faster and faster processors and bus speeds, the data transfer rates have allowed computer handshaking through quicker framegrabbers. In the past, the data transfer rates typically were expended to acquire either high-resolution images (with a large number of pixels) at a relatively slow frame rate, or a smaller image with fewer pixels at a relatively fast frame rate. The increased reliance on progressive scan technology (where the camera image is written in sequential lines rather than in an interlaced manner of half images for even and odd lines [44]) allows for newer cameras that are able to be more flexible in how the data transfer rate is used. Lines of progressive images may be cut to improve the camera frame rate, resulting in a camera that may have several modes that exchange image size for speed. There are both expensive and inexpensive models of such cameras. Measurements of mean field properties of granular systems, such as density and position correlations, are typically made utilizing images with high resolution over large spatial scales of the granular media. Corresponding "microscopic" or local measurements that are particularly sensitive to small windows of time, such as collision times or velocity fluctuations, are made with high-speed cameras. In the case of velocity measurements that are not capable of measuring the entire system, ensemble statistical techniques are used to sample the system and describe it in a statistically meaningful manner [10–12, 16–18, 22, 25, 31, 35, 42, 43, 45]. In larger scale images, PIV techniques, similar to that described in Chapter 7, can be used for particle tracking over larger volumes. In this chapter, the specific example that will be discussed was designed to be able to track the motion of a granular bath as well as a tracer particle in a finite volume, so there were no particles either entering or leaving the sample volume. This aspect of the particular experiment that will be discussed is relatively unique [13, 15, 21, 23, 26, 46], and the selection of a sample volume that is a subset of the entire system is more common in order to obtain data away from the experimental boundaries. The details of the experiment used for the specific example discussed in this chapter are available from references [47, 48]. A broader and more fundamental discussion of image analysis beyond this specific example is also available [49].

10.3 Lighting

Unlike chemical film from traditional optical cameras, the CCD element of a digital camera can be operated at a variety of frame rates. As discussed above, the frame rate is often limited by the speed at which the image can be triggered and electronically processed. Separate from the frame rate of the camera is the iris and electronic shutter (or both), which allow the user to adjust the amount of light that is acquired within the trigger window of the CCD element. While available CCD cameras have improved the higher end frame rates, such cameras require an increasingly large amount of light in order to properly stimulate the elements of the CCD (or CMOS) sensor to create the electrical response that is processed by the camera's electronics to produce the image. The faster the camera frame rate, the brighter the image source in general that is needed to illuminate the region of interest of the camera. Lenses may have manual or electronic irises and a lens is typically chosen based on the experimental application to image the proper field of view (dimension) and depth of field, as discussed in Chapter 1. A complication of some CCD cameras is what are intended to be helpful features, such as automatic gain and white level adjustment, which may need to be disabled or placed in a manual mode so they do not undo adjustments made by the user in attempting to set the light level correctly.

Particle tracking through image analysis with such software languages as the interactive data language (IDL) or MATLAB rely on the contrast between the illumination of the items of interest in the cell and the undesired portions of the image. In images relying on backscattered illumination, the particles are identified by a bright spot reflecting off the particle against the background of the experimental cell see Figure 10.1, although many experiments use the reverse process of backlighting an experimental region and tracking the particles as the dark regions (shadows) against the bright background [10, 12, 14, 26, 28–30, 34, 35, 50]. The processing of both types of lighting is essentially the same – whereas reflected light produces peaks in the image, backlighting produces minima in the image. In both practices, it is still the contrast between the particle and the background that is used to identify the location of the particles. Care should be taken with high-speed cameras and light sources running off AC sources by which it is possible to see 60 or 120 Hz artifacts in the illumination intensity.

The ability of a CCD camera to resolve light intensity is denoted by the bit-level of the camera. Most inexpensive off-the-shelf cameras are typically 8-bit, which means the camera has 2^8 levels of light intensity that may be resolved. The z-direction of a three-dimensional rendering of the image (pixel intensity), such as in Figure 10.2, would have values between 0 (darkest pixels) and 255 (brightest pixels) for a complete range of 256 levels. A 10-bit camera would have 1024 levels, and so on. The production of an image with too much illumination (such as the left image in Figure 10.1) will produce mesas in the pixel intensity – regions for which several neighboring pixels are all saturated to their maximum pixel value (255 for an 8-bit camera). This is typically a poor image for particle tracking, which works best when each particle is illuminated with a single sub-maximum peak. In

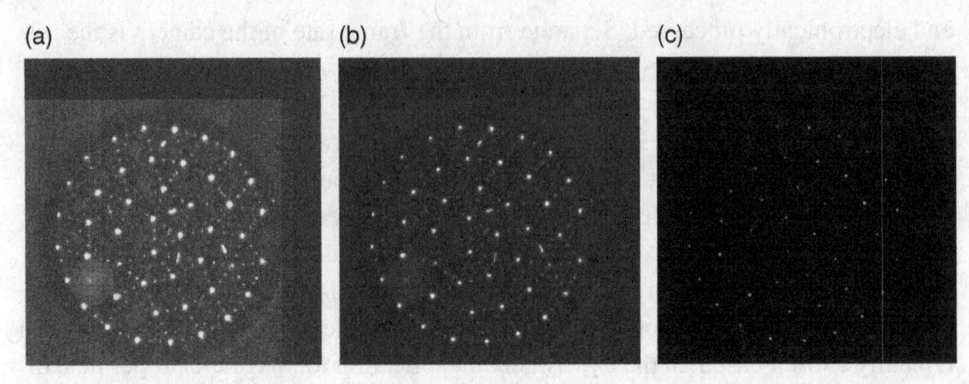

Figure 10.1 The experiment that will be used as the specific example discussed in this chapter. Viewed from above, the dynamics of a tracer particle (lower left of each image) are tracked as the lower dimer layer is shaken in the vertical direction. (a) Illumination with too much light that produces secondary reflections on the edge of the cell (lower right) as well as the connecting rods and the spheres of the metallic dimers, (b) a lower level of lighting that produces one bright spot on each of the particles that are desired to be tracked, and (c) an insufficient amount of light that fails to illuminate all of the particles within the cell.

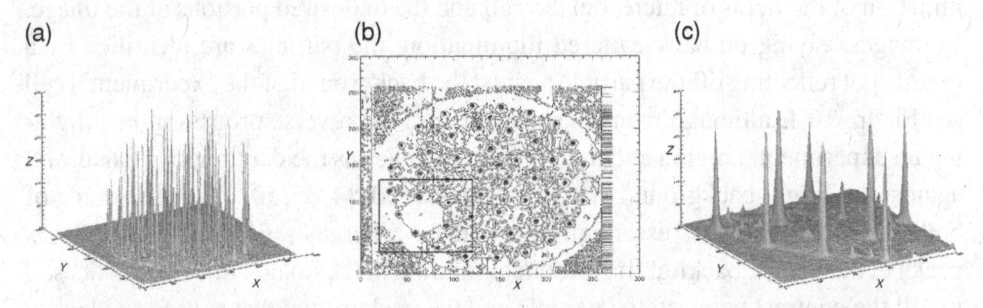

Figure 10.2 Two-dimensional images can be rendered as three-dimensional surfaces where the x and y are dimensions of the image and the z value is the pixel intensity at each element in the image as obtained from the CCD array. (a) A three-dimensional surface representing the image with proper lighting from Figure 10.1, (b) a contour plot of the same image in which the tracer particle is clearly evident, (c) an expanded view of the lower left portion of the image in which the tracer particle is located. Much of the rest of this example will focus on this lower left corner of the image.

other words, a single pixel on the particle that has a higher intensity than any other is optimum for tracking. In addition, such saturated images are not useful for a refinement, to be discussed later, that allows for the identification of the particle position to sub-pixel resolution.

One will note a gradient in the light from one side of the image to the other in the left panel of Figure 10.2. Since the camera in this example is positioned directly above the experiment being imaged, any light source will be at an angle to the experimental cell. That angle results in the gentle gradient observed in the surface plot of the image. While a second light could be added opposite the first, solving the one problem introduces a new one, namely two spots of reflected light off of each particle that one wishes to track. Some experiments employ a circular light source closer to the surface of the cell to solve this issue [21]. The reliance on the use of light reflection can produce circles or halos in this instance rather than a single spot on the particle. Light sources positioned too closely to the horizontal also may introduce variations of lighting in this example and similar experiments, in that nearby particles may cast shadows on their neighbors, further causing problems for the methods of peak detection and particle identification discussed here.

10.4 Particle identification

One helpful pedagogical technique, especially in training undergraduates to understand and develop their own particle tracking algorithms, is to ask them to reflect on how they themselves interpret an image by eye and know the location their brain is assigning as the particle's position. Encouraging this sort of self-reflective method allows us to appeal to the student's own intuition in later tailoring their software to issues specific to their particular experiment, which almost always results in modifications to techniques being presented even in this illustration.

The specific example of image analysis that is being presented here now turns from image preparation (lighting) to image manipulation. However, as different techniques of image modification are introduced, it is wise to keep in mind that as one tries to optimize the image for each of the different corrections discussed, it is often necessary to return to the issue of lighting and contrast and to take new images in order to optimize the particle tracking results. There is a "knack" to this sort of feedback process between the lighting and the image analysis which is best understood experientially. This is often an investment of time that can be quite significant (and frustrating to inexperienced students, but is absolutely necessary in order to rely on the veracity of results of subsequent analysis), mainly to remove inadvertent reflections and other artifacts that can corrupt the experimental data. Particularly when training undergraduates, this is a learning experience that cannot be stressed enough.

Before moving on to other image manipulations to begin to find the particle positions, it is often helpful, to minimize unintended side effects, to remove extraneous background information in the image. This can be done by simply zeroing out the pixel values below a certain value. Subsequent image manipulations will end up mixing values in adjoining and neighboring pixels, so removing noise and unnecessary background light can aid in removing information that may result in unintended artifacts. At this point, however, it should be noted that any image

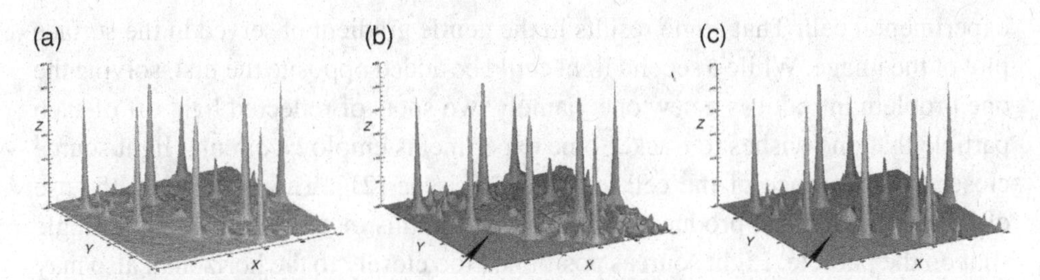

Figure 10.3 Removing background noise and secondary reflections. The original subimage (a), corresponding to the right panel of Figure 10.2, is filtered (b) by zeroing out pixel values below a particular threshold. It is possible to zero out too much (c) and remove information of interest – in this case, the contour of the tracer particle (arrows).

algorithm process should always maintain a copy of the original image so that there is always a baseline measure to reference in order to quantify the effectiveness of the particle identification algorithm [51].

The fundamental basis for particle identification in this specific example is a peak detection of the pixel intensities, the single light source producing a spot of light reflecting off each particle. (In a backlit experiment, this peak would be the minimum/darkest point detection, or the blacklit image could simply be inverted to produce a peaked image similar to the images in Figure 10.3.) If each pixel in a CCD array were truly isolated from its neighbors, the peaks produced in such images would be more pristine. However, due to crosstalk and noise, the peaks corresponding to the bright spots are more complicated, as discussed in Chapter 7.

10.4.1 Resolution

One could conceivably extract the isolated peaks and achieve the particle localization by passing each peak through a parabolic or Gaussian fitting routine. The precision of such a technique would necessitate having a larger reflected light spot in order to provide a suitable number of weighted points for the fit to converge in a robust manner [49–52]. In addition, such particle-by-particle fittings require in some sense the ability to separate out the particles without yet identifying where the particles are actually located. This is an issue if the particles are small and in contact (which is an effect minimized by illuminating only the center region of each sphere with light, thus separating the peaks of light by a significant fraction of the particle radius, and adjusting the resolution of the image so that each complete sphere is rendered over several more pixels than the number illuminated with the bright spot). What is presented here is computationally quicker and can extract the particle location with only a few (~10) pixels of illumination per particle. However,

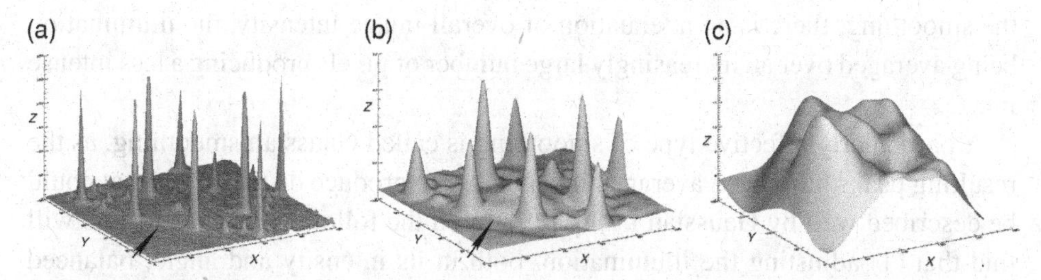

Figure 10.4 The effect of smoothing the image. The raw image surface plot (a) is smoothed over a small number of nearest neighbor pixels (b) and detrimentally smoothed with a radius that is too large, much of the detail of the image being lost (c). The arrows indicate the edge of the tracer contour.

because the procedure is not a fit *per se*, it is beneficial to "sculpt" the peaks before using them to identify the particles. Because the noise and lighting may not result in a single largest pixel in a peak, but perhaps two or three, one should first smooth the peak to ensure there is a single maximum to the peak [49]. There is a fine balance between having a significant number of illuminated pixels per particle and the shape of the illumination peak and illuminating the particle with too much light, resulting in a mesa of pixels all with an intensity of maximum value.

10.4.2 Smoothing

In order to identify a single peak per particle from the bright spot reflecting off each particle, it is beneficial to smooth the image, dulling it in a uniform manner that eliminates noise or spikes in the illumination that could produce artifacts such as two maxima. The simplest sort of smoothing is to reassign the pixel intensity at each location within the image to be the average of the pixel and its eight nearest neighbors. This can be thought of as smoothing over each pixel with a radius equal to one pixel. Averaging over larger numbers of nearest neighbors is, of course, possible, but smoothing over too large a radius can be detrimental.

Figure 10.4 is a demonstration of an unsmoothed image (a), previously seen in Figure 10.3, that is smoothed over a relatively small radius (b) and also an example of smoothing with an extreme radius (c). Note that in the center panel of Figure 10.4, the small smoothing radius does not result in loss of any of the structure within the image. The low intensity circle corresponding to the nonmetallic tracer particle is still evident in the smoothed image. Only the smallest structure and tiniest peaks are washed away by the smoothing. By eye, the reader will be able to identify each of the raw peaks (a) corresponding to the smooth peaks in (c). However, if the radius of surrounding pixels is increased too much, many of the details in the image's structure are averaged away. No matter what the radius of the smoothing, there is an attenuation of overall image intensity, the illumination being averaged over an increasingly large number of pixels producing a less intense image.

A particularly effective type of smoothing is called Gaussian smoothing, as the resulting peaks have been averaged in a manner to produce distributions that could be described well by Gaussian fitting [49]. With the following four steps one will find that (1) adjusting the illumination, both in its intensity and angle, balanced against (2) the field of view, as well as (3) how many pixels at the center of each sphere are illuminated, and (4) the quality of the peaks after smoothing the image comprise the first iterative loop of image acquisition and analysis.

Caution should be taken when smoothing a particle that is too close to the edge of an image. If the radius of smoothing extends beyond the boundary of the image, an artifact can be introduced from the lack of pixel intensities with which to average beyond the boundary of the image. This can adversely affect the weighting of the center of the illumination and introduce an error in the particle location.

10.4.3 Peak detection

Now that the peaks in the image have been properly sculpted, simple peak detection may be used to identify the location of each particle. By shaping the peaks in the image intensity, local maxima can be obtained in a straightforward manner, often with something as simple as comparing the value of each pixel with its eight nearest neighbors. The MATLAB and IDL software platforms, having been developed with powerful compilers specifically to handle image data, are capable of doing this in a straightforward manner with limited computational structure.

The example presented here was chosen because it demonstrates the need to discriminate between one of two types of particles. The highly reflective metallic particles produce nice peaks in the pixel intensity, which are easily located once the image has been sculpted. However, the duller plastic particle offers the opportunity to discuss another type of particle detection before moving on to the further refinement of sub-pixel resolution of the particle locations. The center of the larger plastic particle can be seen to produce a much smaller peak in Figures 10.4(a) and (b). This small peak is situated within the much duller and larger circle of the ball itself. Simply increasing the illumination, as shown in Figure 10.1, does not help, creating a mesa in the bright spot of the particle and never resulting in an illumination of the particle that is brighter than the metallic particles. However, the size of the dull sphere is larger, albeit dimmer. Again, particularly with an undergraduate student, one might ask: "How does one differentiate the plastic particle from the metallic one?" The trick is folding into the algorithm a way of teaching the software to do what the brain does almost instantly: differentiating shapes and sizes within an image [49, 51].

10.4.4 Convolutions

Just as the brain quickly differentiates the two types of particles, the particle identification should locate the plastic particle in this example with a different method from the metallic particles. Here, the concept of a convolution, a concept that students often find opaque from introductory mathematics [53], can be of great assistance.

In a two-dimensional image, one can think of a convolution as an overlap between the image, I, and a mask of the same geometrical shape, M. The convolution in effect passes the mask over the entire two-dimensional image and at each pixel location calculates the overlap between the image value and the mask value of size (m, n) at each location (i, j):

$$C_{i,j} = \frac{1}{W} \sum_{k=-m/2}^{m/2} \sum_{l=-n/2}^{n/2} I((i+k),(j+l))M(k,l), \qquad (10.1)$$

where the convolution is normalized by a weighting W, which can simply be the sum of the pixel values of the mask, thus removing the magnitude of the mask from result. Figure 10.5 is a demonstration of using a mask for the larger, dull tracer particle to determine its location within the image. Since the convolution is a geometric calculation, it is highly dependent upon the use of a mask of the appropriate size of the particle being identified. Note that although the mask is larger than the smaller particles in the lower layer of the experiment, the convolution generates a nonzero background as the mask passes over these particles. This background can be somewhat reduced by choosing a mask that is an annulus rather than a filled disk, thus reducing the overlap between the mask and the smaller particles (see Figure 10.6). As can be noted from the equation above, if the mask is hollowed out as with the annulus, it reduces the value of the convolution at the centers of the smaller particles, the mask having zero weight at those pixel locations. The mask then weights the edge detection of the larger, duller particle, but nonetheless still properly identifies its center. The presentation here is similar to a kernel-based Hough transform [10, 12, 49, 50] and can also be used to better detect distorted or partial objects, particularly near an image edge.

Since the identification in this case is based more on the size of the particle rather than its peak intensity, it is often helpful to sculpt the image in a different fashion

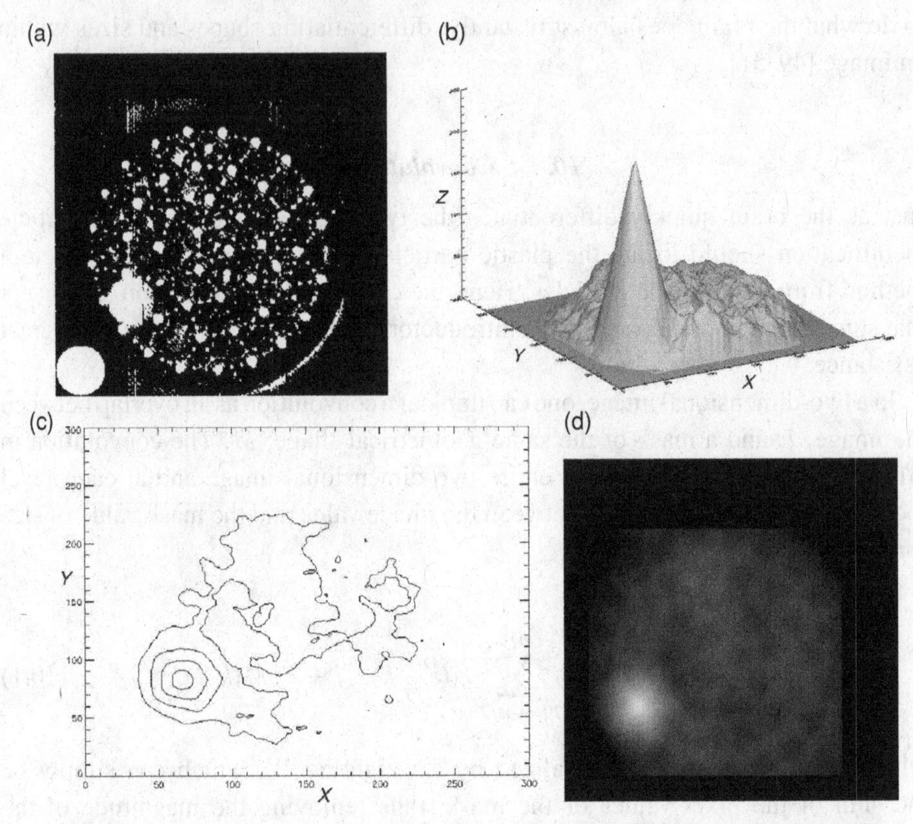

Figure 10.5 Convolution to locate the larger, dimmer particle. The sculpted image (a) with each pixel set to either 0 or the maximum intensity to create a sharp edge to aid in the convolution with the test mask (lower left inset of (a)). A surface plot (b) and contour plot (c) of the result of the image convolution (d). The result of the convolution produces a single definite peak at the center of what was the larger, dimmer particle in Figure 10.1. The pentagonal shape evident in the convolution (d) and the contour plot of the convolution (c) is due to the five metal particles that encircle the larger Delrin ball in the image (a).

than described in Section 4.2. As the convolution is simply looking to identify, geometrically, the position in the image with maximum overlap with the test mask, the pixel variations in the original image are not necessarily helpful. As shown in Figure 10.5, the image may be modified in an extreme fashion, first passing over the image and setting each pixel to either its minimum (0) or maximum (bit level – 1, that is 255 for an eight-bit image) values. Whereas peak detection was not assisted by mesas in the pixel intensities, the convolution can get a better calculation of the center of the larger particle in this manner. However, since this method of particle identification is based on the particle's size, the construction of a mask of appropriate radius is important.

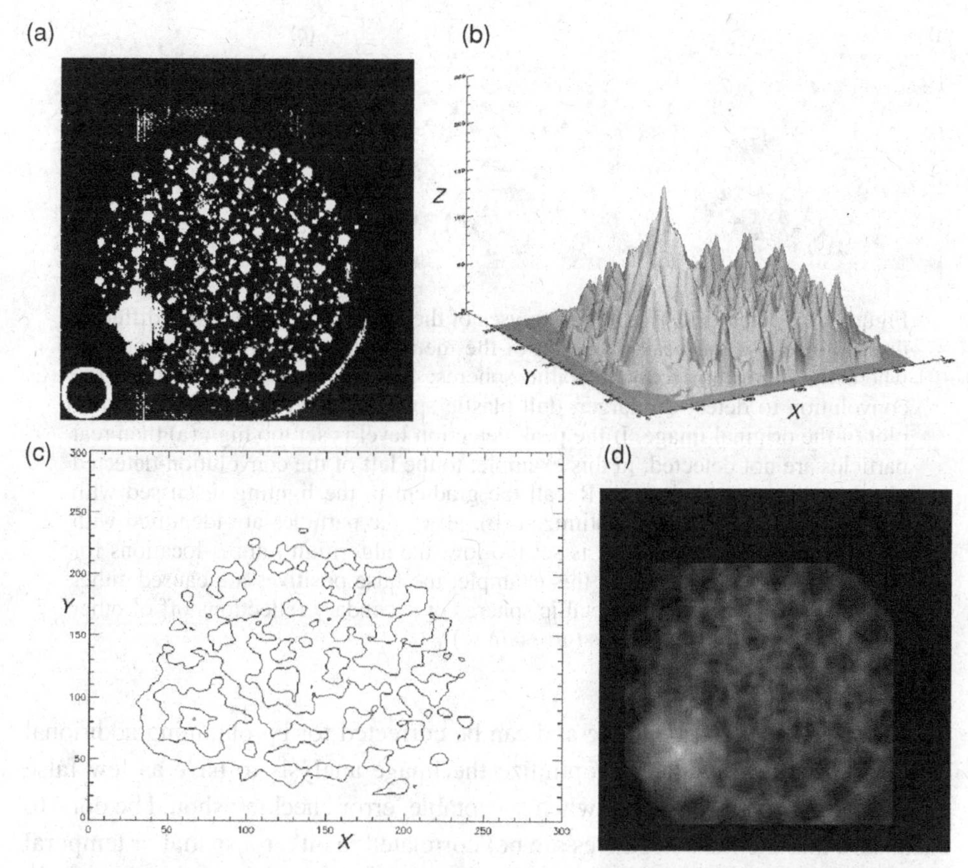

Figure 10.6 Comparable results to those shown in Figure 10.5 for a convolution with a ring mask instead of a filled circular mask. The sculpted image (a) is shown with an inset of the hollow ring mask used for this convolution. The results for the convolution are shown as surface plot of the convolution intensity (b), a contour plot (c), and the actual convolution (d). The high intensity center determined from the convolution is more pronounced here relative to the noisy background. The peak is the location of the center of the larger, dimmer plastic tracer riding on top of the smaller, metallic lattice (see Figure 10.1).

10.4.5 Error checking

Now that all of the particles have been potentially identified, it is important to pause and examine the results at this intermediate level before proceeding with a refinement of the particle positions. The caveat of "nominal" identification reflects that the results need to be checked for both false positives (objects that were identified as particles that were not present) and false negatives (objects that were within the image that were not identified). An ideal analysis would have neither false positives nor false negatives. However, in many statistical experiments, where data are obtained through ensemble averaging, some small percentage of false negatives

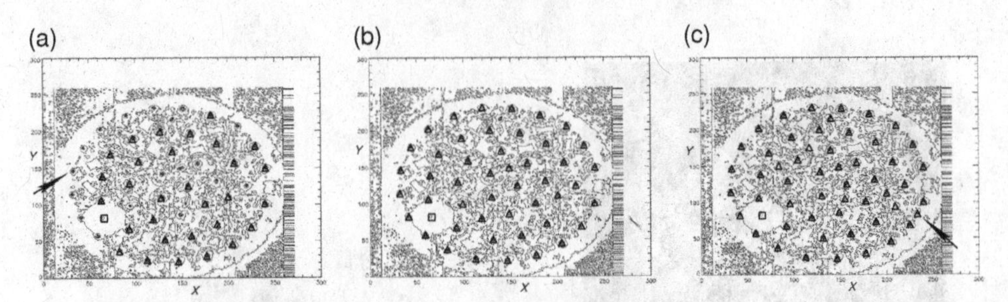

Figure 10.7 The results for three analyses of the same image data with a different threshold set for the peak detection of the metallic spheres. Triangles represent where the algorithm found metallic spheres; the square is the peak from the convolution to detect the larger, dull plastic sphere plotted on top of a contour plot of the original image. If the peak detection level is set too high (a) then real particles are not detected, in this example, to the left of the convolution-detected plastic sphere (arrow in (a)). Recall the gradient in the lighting discussed with Figure 10.2. If the level is optimized (b), all of the particles are identified with no false positives. If the level is set too low, the algorithm returns locations for particles that are incorrect. In this example, the false positives are caused either by the rods connecting the metallic spheres or secondary reflections off of other locations on one of the spheres (arrow in (c)).

(missing particles) is acceptable and can be corrected for by obtaining additional data. Care should be taken to optimize the image analysis to have as few false negatives as possible, and even when acceptable, error checking should be done to make sure that the false negatives are not correlated in either a spatial or temporal fashion, implying that there is a "gap" in the observations, due to a shadowing on one side of the cell, for example see Figure 10.7. The presence of false positives cannot be tolerated as the data are corrupted for any further use until and unless the false positives are removed or otherwise filtered from the data analysis.

The error checking at this point in the image analysis process can typically be thought of as a separate iterative loop similar to steps (1)–(4) discussed in Section 4.2. However, if care was not taken with the illumination, the optimization process may require using the results of the error checking to return to the adjustment of lighting and acquire new images. One typically learns in this process to take only small sets of images until the lighting, smoothing, and peak detection have been optimized.

The discrete elements in the CCD array of the digital camera force a spatial truncation on an image, and the image analysis methods discussed so far only allow the particle location to be determined to the nearest pixel. If the purpose of the image analysis is particle tracking, this introduces a least measure that must be addressed. Imagine an object whose center moves only a fraction of a pixel between two frames. The methods described here would locate the particle center in the same pixel in both frames, leading to the incorrect result that the object's displacement was zero. One might think this is simply the smallest resolution of the particle location, and a reflection of the minimum sensitivity of the measurement. However, there is a refinement that allows for the sub-pixel determination of the particle location.

10.4.6 Sub-pixel resolution

In particle tracking, what is desirable is measurement of the smooth transition of the object being tracked across the field of view. To recapture this from the image data, it is common to not only identify the particle center by locating the pixel with the peak illumination, but also to examine the surrounding pixels to refine the particle center, allowing the particle's dynamics to be captured as the peak traverses a pixel element.

To accomplish this refinement, the information surrounding the peak illumination is used. Given that the illumination should be consistent from the source as the particle is tracked, then the illumination can be thought of in terms of the extended Gaussian shape of the peak and not simply its peak value. This shape is what transverses the field of view from one pixel to another. The pixel of peak illumination simply finds the rough location of the center. However, if the illumination is considered as the whole of the three-dimensional shape, then that information can be used to refine the position, removing the pixilation introduced by only tracking the discrete pixel where the peak of illumination is located. This refinement can be considered a "center of light" calculation, akin to a center of mass location of the object, where the consistency of the illumination is relied upon to calculate the refined center (X, Y) of the illumination as the entirety of the shape moves across several pixels. The pixel intensity I and distances x_p and y_p from the peak pixel are used to weight the refined center in the direction of the local gradients of illumination out to a distance r away from the pixel of peak illumination:

$$X = \frac{\sum_{p=0}^{r} x_p I(x_p, y_p)}{\sum_{p=0}^{r} I(x_p, y_p)}, \qquad Y = \frac{\sum_{p=0}^{r} y_p I(x_p, y_p)}{\sum_{p=0}^{r} I(x_p, y_p)}, \tag{10.2}$$

where a sample refinement using the above calculation is shown in Figure 10.8. Note that while the pixel at the (x, y) location of the peak illumination does not contribute to the numerators in the equations above (as it is a distance 0 from the peak), it does weight the denominator, playing the same role that inertia would in a center of mass calculation (see Figure 10.9).

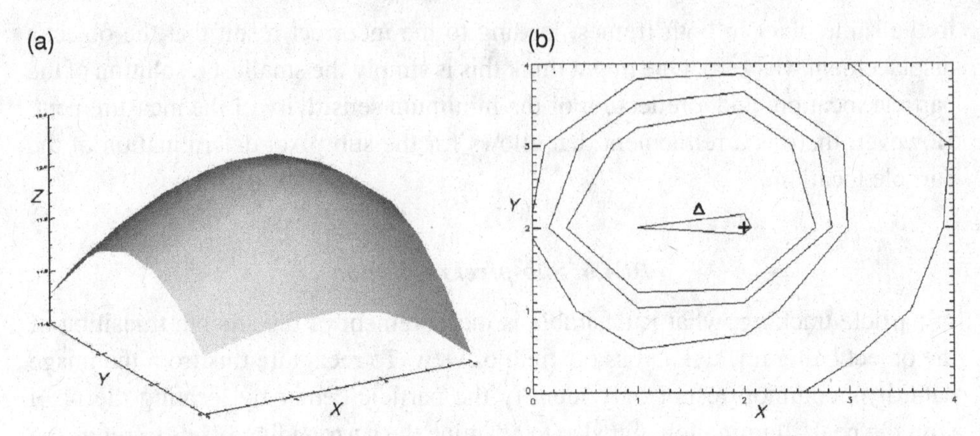

Figure 10.8 Expanded view in surface plot of a single peak in the pixel intensity of the sample image (a). Contour image (b) of the same peak. The plus symbol denoted the pixel with the highest illumination value. The triangle symbol denotes the result of the sub-pixel refinement of the peak location.

Care should be taken to error check the sub-pixel resolution. Too small a radius in the calculation can result in sub-pixel localizations that are still tightly tied to the integer pixel values. Too large a radius may introduce artifacts from secondary reflections and illumination that are time and space dependent in the image. One quick method of error checking a large amount of data is to perform a *modulo* function of the sub-pixel location with integer 1 to remove the leading integer value and either plot the resulting fractional locations or produce a histogram of the values. A sufficiently large data set of properly valued sub-pixel locations will produce a uniform plot or histogram, there being no bias in the system to, on average, favor any of the fractional pixel locations. Here again, it is worth noting that it is assumed that the lighting is as uniform as possible. Biases in such a plot toward the integer values indicate a residual pixilation in the values. While varying some of the parameters discussed in Section 4 can improve the results, one should keep in mind that changing the illumination to a more diffuse, less-localized type is sometimes necessary to remove the pixilation completely, thus beginning the entire process all over again with a new set of image data. Experience is the best teacher in this process, and a sample tracking set up has been used in the author's laboratory to introduce new students to this process.

10.4.7 Momentum filtering

Chapter 7 discusses the use of image locations to produce particle tracks of tracers in soft condensed matter systems. However, if the purpose of image tracking is to extract information about the velocities of particles *between* collisions, there is

21	34	36	20	14
41	118	146	45	19
52	200	211	48	17
35	141	81	27	15
23	26	25	16	12

Figure 10.9 Example of the pixel values centered on the pixel with the highest intensity denoting a peak. A cursory view of the pixel values would indicate that the corresponding "center of light" would refine the peak up and to the left. The quantitative result of averaging over the numerical values shown for a radius of one pixel from the maximum value of 211 would result in the refined pixel location moving to the left by 0.38 pixels and up by 0.22 in agreement with the triangle's location in Figure 10.8. The pixel values are taken from the original image, not the sculpted image, so that the refinement is tied to the actual lighting in the original image. A similar, but more refined, result can be calculated for extending the calculation to a radius of two pixels from the maximum.

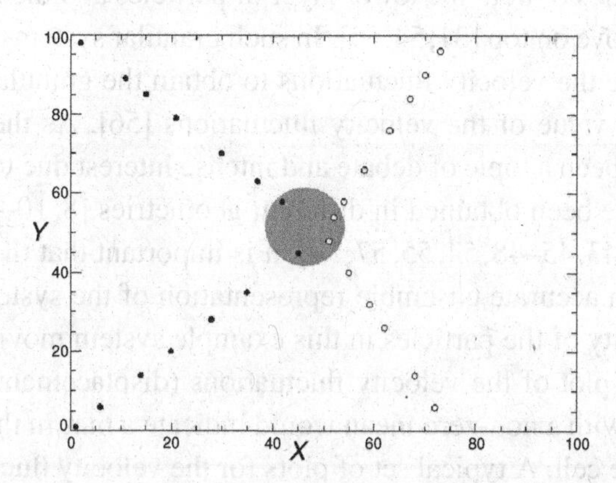

Figure 10.10 A set of sample data from a dilute cell experiment showing the trajectories of two particles before and after collision.

an additional filtering that must be performed on the particle locations. Since the capture of images is a discrete process, one should be careful to remove errant data (again, a criterion that denies false positives.)

In Figure 10.10, we show an example of two particle tracks with a collision in the middle of the image, where momentum was exchanged between the two particles. It is not possible, due to the discrete nature of the image capture to know where (exactly) the collision took place. As such, the four positions within the grey circle are suspect. In which interval did the collision take place? If the purpose of the analysis is to examine the velocities of the particles between collisions, then the displacements on either side of these four suspect particle positions should be removed from the ensemble due the uncertainty of the moment of collision. A calculation of the velocities (displacements per frame) in each of the trajectories can be used to calculate a change in momentum along the particle paths to filter out the collisions.

10.5 Data analysis

While many of the refinements discussed in the previous section are meant to optimize the analysis process, small residual effects may sometimes remain even after the process in the previous sections is followed. This is due to small biases in the particle locations that may not be evident in small subsets of the analysis. Once the complete data set is analyzed, it is important to scrutinize the results at least one more time.

In the example data set discussed in most of this chapter, the aim was to examine the velocity statistics of both the lower layer of particles and the upper layer tracer that was free to move on top [31,54,55]. In such granular systems, particle tracking is used to measure the velocity fluctuations to obtain the granular temperature, or root mean square value of the velocity fluctuations [56]. As the investigation of such systems has been a topic of debate and intense interest due to the wide variety of results that have been obtained in different geometries [8, 10–13, 15–18, 22, 23, 25–27, 31, 33, 39, 43, 45–48, 54, 55, 57, 58], it is important that the data analysis be proven to yield an accurate ensemble representation of the system. Since there is an equal probability of the particles in this example system moving in either the x or y directions, a plot of the velocity fluctuations (displacements of the particles between frames) with a non-zero mean would indicate a bias in the system, such as a tilt in the sample cell. A typical set of plots for the velocity fluctuations as both a diagram in the xy plane and a histogram of fluctuations is shown in Figure 10.11.

In the case of the convolution example, pixilation may occur due to either the sub-pixel refinement, similar to the situation shown above, but also due to the incorrect sizing of the mask for the convolution procedure. Figure 10.12 demonstrates a pixilation in the velocity fluctuations (particle displacements) in the xy plane for the tracer particle due to a slight under-sizing of the mask, which leads to an incorrect identification of the pixel about which the sub-pixel refinement is then processed.

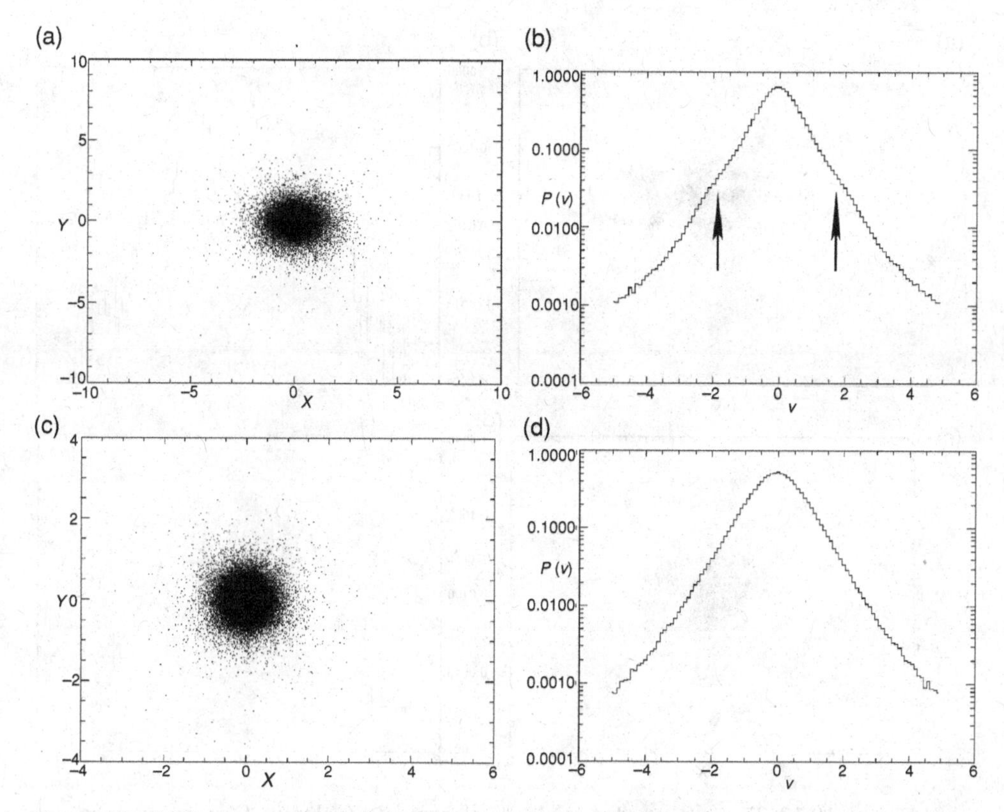

Figure 10.11 The measured particle displacements ((a) and (c)) about the zero mean value in the xy plane for the lower layer particles and a histogram ((b) and (d)) of the displacement probabilities in the x direction, respectively. Since the frame rate of the camera is a constant, the particle displacements are related to the velocity fluctuations by that constant rate. The shoulders on the distribution correspond to a residual bias of pixilation around the integer pixel locations ((a) and (b)). After a slight adjustment of the smoothing radius as well as the sub-pixel refinement radius, the robust behavior of the fluctuations is obtained in the absence of pixilation ((c) and (d)).

As stated earlier, this entire process is an iterative one, and a technique which is best conveyed experientially in the lab with a sample set of data, or sample experiment for a new student to use to learn in a hands-on manner. Once the fundamental process is understood, a student is better equipped to know how to prove the robust nature of their data acquisition and data analysis in a novel experimental manifestation, with a better appreciation of what is a real dynamic behavior and what is an undesired artifact from their data-processing algorithms. The author has had significant success in teaching this general procedure to undergraduate students who are then able to take their general understanding of the ideas of image acquisition and analysis and apply them to a new experimental investigation.

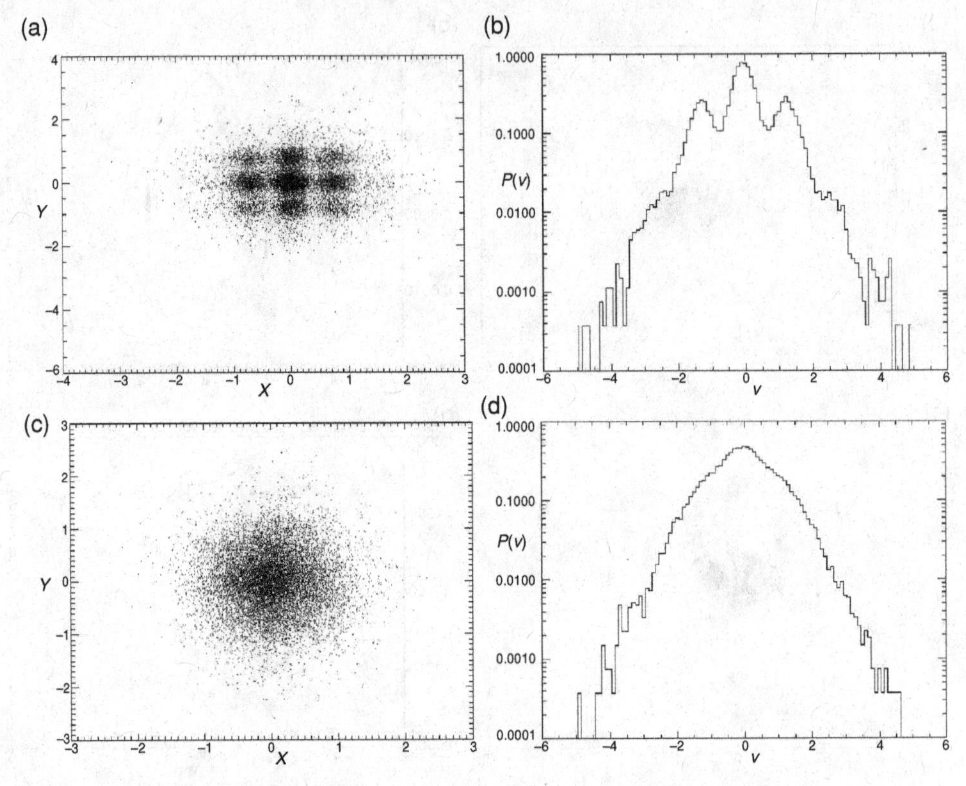

Figure 10.12 The measured particle displacements (velocity fluctuations, see Figure 10.11) about the zero mean value in the xy plane for the upper layer Delrin particle ((a) and (c)) and a histogram ((b) and (d)) of the fluctuation probabilities in the x direction, respectively. The multiple peaks on the distribution correspond to a residual bias of pixilation around the integer pixel locations ((a) and (b)). After a slight adjustment of the mask radius, the robust behavior of the fluctuations is obtained in the absence of pixilation ((c) and (d)).

10.6 Incorporating undergraduates into soft condensed matter research

The tabletop nature of many soft condensed matter investigations makes them accessible to students early in their development as scientists, even at the undergraduate level and soon after their completion of the introductory physics sequence. Unlike large experiments with numerous personnel, each of whom are given a subtask of the experiment, soft condensed matter systems and their smaller scale allow a student to learn about a wide variety of experimental and computational techniques. A tabletop experiment can expose a student to a suitable sampling of theory, computation, data acquisition and analysis, design and machining, electronics, automation, and troubleshooting. The experience can aid a student in learning about their own strengths and weaknesses, helping to guide their decision about what sub-discipline of physics holds both their interest and their aptitude.

The advent of the personal computer and undergraduates' familiarity with the technology – computing or otherwise – will aid in the incorporation of their efforts into physics in general, and soft condensed matter systems more specifically. As many departments do not yet offer classes in soft condensed matter systems, research opportunities in the field allow undergraduates to incorporate the science of these modern endeavors into their undergraduate experience. The benefits are not limited to the undergraduates themselves. More focus has been made of late on the professional development of both postdoctoral fellows and graduate students as science mentors [59]. Undergraduate researchers in a soft condensed matter laboratory can quickly be trained to the point where they are seen as contributing colleagues within the research group. There are important differences, however, that the author would like to highlight between the successful integration of undergraduates into a thriving research laboratory and simply "helping out."

While REU experiences are fantastic opportunities for students to visit other universities and colleges during the summer months and be exposed to research and resources not available at their home institution, the typical REU experience lasts on the order of ten weeks during the summer. Typically, the student is between their sophomore and junior year (a rising junior) or their junior and senior year (a rising senior) when they engage in such a research experience. The limited duration of the REU puts tremendous pressure on someone who is typically still trying to make the transition from what is taught in a lecture class out of a textbook to understanding how to make a contribution to a modern research endeavor.

The author has developed what is a more prosperous model of incorporating undergraduates into a research setting. The involvement of undergraduates involves a significant contribution to the mentoring relationship – more than is typical for a graduate student or postdoctoral fellow. Also, the supervision of an undergraduate student means accepting an alternative timeline to the advancement of a research effort from the expectation that would be placed upon a graduate student or postdoctoral fellow. A typical postdoctoral fellow arrives in a research laboratory already trained, hired for a period of time that is on average two years, and is on a career path to establishing their credentials for a more permanent position in either academia or industry. Graduate students may arrive with a wide diversity of undergraduate backgrounds, many with little more than a set of lecture setting classes to their physics background, but with a more open-ended time period of five to seven years to develop their skills as they work toward their degree.

Undergraduates represent a third distinct group: their course of study is not as open-ended as a graduate student, with a typical bachelor's degree still taking four (to sometimes five) years. However, unlike either the postdoctoral fellow or the graduate student, their aims in pursuing research are broader. Some students are simply looking for a summer job, some are trying to understand what a career in physics entails, and others even come with a true passion to make their mark on a topic that truly interests them. Investing in undergraduates as researchers is to everyone's advantage: schools are increasingly challenged to identify and recruit driven, capable graduate students, and all graduate students start out as undergraduates somewhere. Hence, the investment made in developing an undergraduate researcher impacts the entire scientific community, even (and especially) if that undergraduate researcher pursues their graduate degree at another institution. As an undergraduate has not yet made the commitment of either a graduate student or postdoctoral fellow, they do require additional care and mentoring, lest they be overwhelmed too quickly as they assimilate the differences between working in the laboratory setting and simply learning in the classroom environment.

Unlike a postdoctoral fellow, who may be teaching a class as part of their postgraduate position, or a graduate student who is progressing from taking classes to developing their thesis topic, an undergraduate student should be primarily focused on their academic classes in each of their four years, pursuing their degree in order to develop both the diversity of their education as well as to build the proper theoretical foundation of their major. As such, the expectations for an undergraduate researcher should be flexible during the academic year. Here, the author discusses a model that has been developed over several years that has proven to work well.

10.6.1 The two-year plan/three-year plan

A science student in chemistry, physics, biology, or other such field is usually involved in their introductory major sequence during the freshman year of college, as well as in many cases, learning the social and self-sufficiency skills of living on their own for the first time in a new environment. In terms of soft condensed matter systems, it is in the best interest of the mentor (whether graduate student, postdoctoral fellow, or faculty member) to wait until after this first year to consider involving an undergraduate student in research. Simply put, there is too much that is in flux in a student's undergraduate development in that first year for them to be making any long-term commitments. For anything beyond an REU, incorporating an undergraduate into a research group in a meaningful way should be looked on as exactly that: a long-term commitment.

This commitment may raise the immediate question as to why one should mentor an undergraduate. This question is a valid one. As mentioned above, the implications go far beyond simply exposing someone to an opportunity to do research. Incorporating undergraduates in research is a way in which to develop future graduate students, postdoctoral fellows, and scientists to be recruited in academia and industry. As such, doing research with undergraduates is not something that is just to be pursued at institutions that do not offer graduate degrees. However, if the investment of time is not something that you can make as a mentor, then the question may best be answered that you should not mentor an undergraduate. It is true that, in an equal amount of time, a graduate student or postdoctoral fellow is going to be more productive from a research standpoint, and, thus, a more profitable investment of a faculty supervisor's time. That is not the only matter to consider. Graduate students and postdoctoral fellows were previously undergraduates themselves, and given the difficult trends in the sciences [60–62] are persons to be cultivated for their future contributions and not to be taken for granted. Additionally, the faculty member's development is not the only professional maturity to be considered. The graduate students and postdoctoral fellows in a research lab are all working towards being scientists themselves, and in their development need to learn how to manage and supervise students as well.

In the spring of either a student's freshman or sophomore year, depending upon their background and any college credit they may have been able to earn while still in high school, the undergraduate is finally prepared to make the commitment to their major in such a way that their potential for finishing what they have started can accurately be assessed. During the spring of that academic year, students often begin to seek opportunities to learn about the career paths that their major will offer. Their introductory major courses now completed, and supplemental courses such as introductory computer science and general education classes out of the way, the typical major curriculum is designed to immerse the student in the details of their major over the final two to three years of their undergraduate degree. During this spring semester, a student's focus will remain fixed on success in the classroom, as everything from scholarship to their familial assessments of their progress rely on maintaining as high a grade point average as possible. However, the interest in doing research may be cultivated via conversations with faculty about research opportunities in the laboratory for the coming summer months. These meetings are most fruitful when they include a frank assessment of the student's background and abilities (available via both conversation with the student and reviewing their academic transcript) and matching their interests and abilities to an available project in the research laboratory. Within every research lab there is a wide variety of projects that exist as either fully developed investigations, pilot projects, or efforts that are little more than a question in the faculty member's mind in need of background research. Every lab is different, yet with a little effort it is possible to find an endeavor that both the research group and the undergraduate will benefit from investigating.

These spring semester conversations should be used by the faculty member and the student to express what they are each hoping to gain from their collaboration. Expectations and lab practices should be made clear, including a thorough review of lab safety. It is important to remember that for all of their enthusiasm, undergraduates lack experiences that graduate students, postdoctoral fellows, and faculty members often take for granted. How to involve a student in a manner that is both safe for them and the other members of the laboratory (and for the equipment in the lab) is an important consideration. During this lead up to the first summer, the student can be given papers to read, even if what they digest of the content is limited, the experience should foster questions on the part of the student, and lead the faculty member to help identify those skills that can be developed part time during the preceeding academic semester. Since the topics of data acquisition and analysis discussed in the first part of this chapter are typically handled through a software platform such as MATLAB or IDL, it is an excellent time to give the student the opportunity to familiarize themselves with these packages, and both have manuals or have associated texts that lend themselves well to use as tutorials. Sample data from a prior investigation can be given to the student to develop an understanding of the skills laid out in this and other chapters in this volume.

During the first summer, the student should be given a project with a definite goal, and in soft condensed matter systems this can take a wide variety of manifestations depending on the student's abilities and the laboratory's needs. Everything from writing a small bit of code to prototyping a new experiment can give the student a better understanding of the demands of doing research, as well as the members of the research group the opportunity to mentor, encourage, and guide a student to a goal which will contain both failures and successes along the way. With undergraduates, one may find that after the first project has been completed, and the student has come to a better operational understanding of either the work done in that research group, specifically, or their major discipline, in general, that they decide they want to work on something different. In this regard, REU programs are excellent short-term programs for such experiences. Because one of the possible outcomes of the first summer is that a student decides they want to do something else, it can be decided not to invest the time in mentoring undergraduates. If committed to supervising undergraduate research, this must simply be accepted as a potential outcome.

The discussion here will continue, however, under the assumption that the first project has left the student eager to do more. Hence, there often exists within the first summer the time to begin to develop the student's own investigation, whether this is a new experiment, a different set of experiments on an existing apparatus, or development of the background information on a project that has not yet been pursued in the research group at all. Again, the particular strengths and interests of the student and mentor should come together to chart the course the research endeavor will take. In the author's laboratory, the first summer is often an opportunity for the undergraduate student to apply the training on image acquisition and analysis to a pilot project of interest within the lab. The undergraduate is closely supervised to make sure that frustration does not arise from insurmountable roadblocks in the investigation. The author has discovered that this is an important stage of the undergraduate's development in the laboratory.

Due to the larger amount of experience and knowledge, a faculty member, postdoctoral fellow, or graduate student may be able to see in advance the pitfalls to the student's approach to a problem. There is an almost natural reflex in these situations, to help the student by saving them the time of going down an avenue that will not pay off. It is the author's belief that this reflex should be suppressed, to a degree. Allow the student to develop their own experiences, which may mean failure, as experience often does. Point the undergraduate student in the correct general direction and do not let them languish in one failure after another for too long, but do not give them the answer, or part of their professional development will be subverted. By all means, continue to ensure their safety and the safety of the equipment in the lab by training and supervision. However, as one supervises more and more undergraduates, one develops a library of common guesses that students make about how best to solve a problem. When the avenue is correct, encourage them by complimenting their intuition. When the avenue is not correct, commiserate with them that they are not the first or only student to make that bad guess. From time to time, these common mistakes will even become signposts for a mentor about a student's development, helping to gauge how well they are developing their problem-solving skills. If a student begins to languish, offer the benefits of experience, but try to temper the desire to simply give them the answer. Use the same pedagogical skills developed in the classroom to ask questions to answer their questions, rather than simply supplying them with the answers. This process is not as completely negative as the cautions above might imply. The author has often been impressed by some of the novel techniques that his undergraduate colleagues have developed to solve a problem of image analysis from a fresh perspective. The author has also noted an effect of critical mass in the research group, when three or four well-trained undergraduates will begin to develop their own skills of interacting with each other to help solve issues with their image analysis algorithms. As with the similar observations in classroom experiences [62], these peer-related teaching moments in a research setting are a boon to the learning environment.

This is obviously a time-consuming process and one reason why it is not pursued by everyone and may not be for everyone. Remember that journal articles are only one product of a research laboratory. The other significant product is the professional development of the lab's members [60]. Both of these outputs benefit the larger scientific community. No matter how busy a lab is, there is always more to do than there is the time to investigate; ancillary or side projects of developed research efforts are good fodder for an undergraduate project, especially if they are being mentored by a postdoctoral fellow or graduate student responsible for the primary investigation to which the project is attached. By the end of the first summer, the synergy between the mentor and the student should allow for the identification of the particular project that the student will pursue. Depending upon the obstacles, the end of the first summer may actually result in some data acquisition on the student's project, but it should at least achieve the design of the experiment. This minimal goal for the end of the first summer is important because when the next academic year begins, the student's focus will shift back to their classroom requirements. While full-time students may not have the ability to devote significant investments of time in the research lab during the academic year, it is important to keep the momentum going from the first summer.

If the first summer resulted in only training and the background understanding for the experiment, the academic year that follows it should focus on the design and construction of the experiment. The guidance of a student in the design of machined parts is an important skill, but one that should be handled through several iterations of the drawings and conversations with the mentor about the pitfalls of the design, rather than the actual physical construction of incorrect or badly designed machined parts. Many institutions are fortunate to have extremely skilled and experienced machinists who have the patience to work in an advisory role with less-experienced students. The academic year should continue with moving the design and modification of the project forward in anticipation of hitting the ground running in the second summer, taking data and doing the necessary analysis to prove the robust nature of the project's results.

If the first summer resulted in the actual construction of a new experiment, or ownership of a previously constructed experiment by the undergraduate, then the academic year following the first summer experience is related but somewhat different. There may be modifications necessary to the existing experiment to allow for the particular study on which the student will focus or the adjustment or removal of part of the experiment that is resulting in an undesirable artifact in the data. If the student has progressed to this point before the end of the summer, then the reduced time available during the academic year is valuable time to allow the student to explore the phase space accessible by the experiment, and in the author's experience this has resulted in several serendipitous discoveries. Even working in the lab for as little as an hour each day is often enough to maintain momentum, albeit at a slower pace, during the academic year. Ultimately, the benefit of this reduced effort is to avoid losing time in a subsequent summer to a repetition of the learning curve.

In either course, the second summer should focus on the acquisition of the pristine data from the experiment, optimized for the apparatus and techniques, and careful to avoid experimental artifacts with which the student may be less familiar. While the student is more experienced in the second summer, and may be able to work for longer periods of time with less supervision, it is important to monitor their progress so that they are not going down fruitless avenues of exploration. This second summer is important as it will typically occur between the junior and senior year, although an advanced undergraduate may be between their sophomore and junior year. An important goal for the end of the second summer is for the student to have collected sufficient data with which to proceed with their analysis and authorship of the results to be published in a journal. The academic year after the second summer is an excellent period of time to pass such a manuscript back and forth between the student and the mentor, and may also serve as part of the undergraduate's senior thesis if they are involved in such an endeavor.

Some students either arrive with enough advanced preparation or end up pursuing their undergraduate degree over a longer period of time such that a third academic year and a third summer of association with the laboratory is an option. In such cases, the student may now be able to work as independently as an experienced graduate student would, with a nearly complete understanding of the investigation. In such instances, a second paper may be obtained from further data acquisition and analysis, and having the undergraduate present results at a professional meeting should also be encouraged. The additional year may be used to the benefit of the research group by having the undergraduate student hand the investigation over to a graduate student who will have the time to devote to a longer study of the system. Finally, the third year may be used to encourage and develop the undergraduate's own mentoring skills by having them supervise a new undergraduate student researcher who will assume the responsibility for continuing the investigation.

10.6.2 *Student outcomes*

If an undergraduate student and a mentor in a research laboratory are going to make the deep investment of time, the outcome of the effort should be at least one refereed journal article. However, if the student is not working on something that will result in a journal article over this period of time, it should only occur because both the student and the faculty member were clear that the goal was some sort of hands-on experience that was desired by the student, such as learning how to program, learning lab protocol, etc. However, such experiences can be accomplished over the course of a single summer, and again, REU programs are an excellent option for these goals. For the rubric described here, the goal should be a journal-publishable paper. Every school can offer research courses, and universities can encourage students to work on a senior thesis, but a refereed journal article is the "coin of the realm" that reflects the standard of peer review by the community. While the mentor may invest more time in writing the paper, the student should be encouraged to write the first version of the paper, if for no other reason than to ensure that the mentor can understand both what the undergraduate researcher has assimilated as well as to discern any misunderstandings between how the faculty member thought the experiment was being pursued and what the student actually did.

The author has been fortunate to supervise several undergraduates in research that resulted in refereed journal articles [33, 47, 48, 63–65]. In each case, the students joined the lab after either their sophomore year, or in certain cases where they already had advanced college credit for the introductory physics sequence, after their freshman year. The students began with a set of tutorials to be exposed to the data acquisition and analysis software packages in order to understand the issues discussed in optimizing image analysis as outlined in the example. After learning the basics of image acquisition and analysis with a test system, each of the students began a design phase of their research project, which included tailoring the general image analysis algorithms to their particular research application. In some cases, the modifications included clusters of individual micron-sized particles [63], tracking a single particle over a continuous set of 100 000+ images to investigate a dynamical system [33], and tracking pairs of connected particles [64] or longer structures with a set number of particles in the structure [65]. The students would often improve upon the general tracking algorithms with innovations specific to their needs. For instance, in a couple of the latter cases [33, 65], unlike granular gas experiments, where an ensemble sample could be obtained via several measurements, the demands of the particular investigations demanded the tracking of a single particle or a set number of particles in every frame. Variations in lighting would sometimes result in a failure to identify the single particle or the set number of particles in every frame. The students would introduce their own innovations to the basic tracking programs, allowing the software algorithms to self-correct by looping through images with variations of the tracking parameters until the proper number of particles were properly identified and accurately refined. More recently, in the work associated with the sample data shown in this chapter, a hybrid approach of two types of particle tracking were implemented by the student to track two different species of particles in a single experiment [47, 48]. In each of these examples, the students moved on to successful graduate programs with research skills and a better understanding of the role of research in their career paths, which did not necessarily continue in the same scientific discipline as their undergraduate work. Nevertheless, their professional development was advanced through the experiences.

Acknowledgments

The author would like to thank the summer sabbatical program at Baylor University for partial support of the time to plan and write this chapter. The author would also like to thank Kristin Combs, Ben Bammes, Kevin Kohlstedt, Sarah Feldt, Jesse Atwell, and all of the other undergraduates he has had the pleasure of mentoring over the years.

References

[1] G. W. Baxter and R. P. Behringer, "Pattern formation in flowing sand," *Physical Review Letters* **62**, 2825 (1989).

[2] M. Nakagawa, S. A. Altobelli, A. Caprihan, E. Fukushima, and E.-K. Jeong, "Non-invasive measurements of granular flows by magnetic resonance imaging," *Experiments in Fluids* **16**, 54 (1993).

[3] E. E. Ehrichs, H. M. Jaeger, G. S. Karczmar, J. B. Knight, V. Y. Kuperman, and S. R. Nagel, "Granular convection observed by magnetic resonance imaging," *Science* **267**, 1632 (1995).

[4] V. Y. Kuperman, E. E. Ehrichs, H. M. Jaeger, and G. S. Karczmar, "A new technique for differentiating between diffusion and flow in granular media using magnetic resonance imaging," *Review of Scientific Instruments* **66**, 4350 (1995).

[5] J. R. Royer, E. I. Corwin, P. J. Eng, and H. M. Jaeger, "Gas mediated impact dynamics in fine-grained granular materials," *Physical Review Letters* **99**, 038003 (2007).

[6] N. Menon and D. J. Durian, "Diffusing-wave spectroscopy of dynamics in a three-dimensional granular flow," *Science* **275**, 1920 (1997).

[7] W. S. Boyle and G. E. Smith, "Charge coupled semiconductor devices," *Bell System Technical Journal* **49**, 587–93, April 1970; W. S. Boyle and G. E. Smith, "A new approach to MIS device structures," *IEEE Spectrum* 18–27, July 1971.

[8] P. Pieranski, J. Malecki, W. Kuczynski, and K. Wjociechowski, "A hard-disc system, an experimental model," *Philosophical Magazine* **37**, 107 (1978).

[9] E. Clement and J. Rajchenbach, "Fluidization in a bidimensional powder," *Europhysics Letters* **16**, 133 (1991).

[10] S. Warr, G. T. H. Jacques, and J. M. Huntley, "Tracking the translational and rotational motion of granular particles: use of high speed photography and image processing," *Powder Technology* **81**, 41 (1994).

[11] V. V. R. Natarajan, M. L. Hunt, and E. D. Taylor, "Local measurements of velocity fluctuations and diffusion coefficients for a granular material flow," *Journal of Fluid Mechanics* **304**, 1 (1995).

[12] S. Warr, J. M. Huntley, and G. T. H. Jacques, "Fluidization of a two-dimensional granular system: experimental study and scaling behavior." *Physical Review E* **52**, 5583 (1995).

[13] D. V. Khakhar, J. J. McCarthy, T. Shinbrot, and J. M. Ottino, "Transverse flow and mixing of granular materials in a rotating cylinder," *Physics of Fluids* **9**, 31 (1997).

[14] C. T. Veje and P. Dimon, "Two-dimensional granular flow in a small angle funnel," *Physical Review E* **54**, 4329 (1996).

[15] L. Labous, A. D. Rosato, and R. N. Dave, "Measurements of collisional properties of spheres using high-speed video analysis," *Physical Review E* **56**, 5717 (1997).

[16] J. S. Olafsen and J. S. Urbach, "Clustering, order, and collapse in a driven granular monolayer," *Physical Review Letters* **81**, 4369 (1998).
[17] D. W. Howell, R. P. Behringer, and C. T. Veje, "Fluctuations in granular media," *Chaos* **9**, 559 (1999).
[18] W. Losert, D. G. W. Cooper, J. Delour, A. Kudrolli, and J. P. Gollub, "Velocity statistics in excited granular media," *Chaos* **9**, 682 (1999).
[19] L. S. Tsimring, R. Ramaswamy, and P. Sherman, "Dynamics of a shallow fluidized bed," *Physical Review E* **60**, 7126 (1999).
[20] I. S. Aranson, V. A. Kalatsky, G. W. Crabtree, W.-K. Kwok, V. M. Vinokur, and U. Welp, "Electrostatically driven granular media: phase transitions and coarsening," *Physical Review Letters* **84**, 3306 (2000).
[21] M. A. Scherer, K. Kötter, M. Markus, E. Goles, and I. Rehberg, "Swirling granular solidlike clusters," *Physical Review E* **61**, 4069 (2000).
[22] A. Kudrolli and J. Henry, "Non-Gaussian velocity distributions in excited granular matter in the absence of clustering," *Physical Review E* **62**, R1489 (2000).
[23] B. Painter and R. P. Behringer, "Dynamics of two-particle granular collisions on a surface," *Physical Review E* **62**, 2380 (2000).
[24] G. Strassburger and I. Rehberg, "Crystallization in a horizontally vibrated monolayer of spheres," *Physical Review E* **62**, 2517 (2000).
[25] F. Rouyer and N. Menon, "Velocity fluctuations in a homogeneous 2D granular gas in steady state," *Physical Review Letters* **85**, 3676 (2000).
[26] E. C. Rericha, C. Bizon, M. D. Shattuck, and H. L. Swinney, "Shocks in supersonic sand," *Physical Review Letters* **88**, 014302 (2002).
[27] A. Prevost, D. A. Egolf, and J. S. Urbach, "Forcing and velocity correlations in a vibrated granular monolayer," *Physical Review Letters* **89**, 084301 (2002).
[28] B. A. Gryzbowski, J. A. Wiles, and G. M. Whitesides, "Dynamic self-assembly of rings of charged metallic spheres," *Physical Review Letters* **90**, 083903 (2003).
[29] S. Aumaître, T. Schnautz, C. A. Kruelle, and I. Rehberg, "Granular phase transition as a precondition for segregation," *Physical Review Letters* **90**, 114302 (2003).
[30] J. Rajchenbach, "Dense, rapid flows of inelastic grains under gravity," *Physical Review Letters* **90**, 144302 (2003).
[31] G. W. Baxter and J. S. Olafsen, "Kinetics: Gaussian statistics in granular gases," *Nature* **425**, 680 (2003).
[32] J. Stambaugh, D. P. Lathrop, E. Ott, and W. Losert, "Pattern formation in a monolayer of magnetic spheres," *Physical Review E* **68**, 026207 (2003).
[33] S. Feldt and J. S. Olafsen, "Inelastic gravitational billiards," *Physical Review Letters* **94**, 224102 (2005).
[34] T. W. Martin, R. D. Wildman, G. K. Hargrave, J. M. Huntley, and N. Halliwell, "Capturing gas and particle motion in an idealized gas-granular flow," *Powder Technology* **155**, 175 (2005).
[35] P. M. Reis, R. A. Ingale, and M. D. Shattuck, "Crystallization of a quasi-two-dimensional granular fluid," *Physical Review Letters* **96**, 258001 (2006).
[36] E. Falcon, R. Wunenburger, P. Evesque, S. Fauve, C. Chabot, Y. Garrabos, and D. Beysens, "Cluster formation in a granular medium fluidized by vibrations in low gravity," *Physical Review Letters* **83**, 440 (1999).
[37] E. Falcon, S. Fauve, and C. Laroche, "Cluster formation, pressure and density measurements in a granular medium fluidized by vibrations," *European Physical Journal B* **9**, 183 (1999).
[38] R. D. Wildman, J. M. Huntley, J.-P. Hansen, D. J. Parker, and D. A. Allen, "Single-particle motion in three-dimensional vibrofluidized granular beds," *Physical Review E* **62**, 3836 (2000).

[39] R. D. Wildman, J. M. Huntley, and D. J. Parker, "Granular temperature profiles in three-dimensional vibrofluidized granular beds," *Physical Review E* **63**, 061311 (2001).
[40] G. Maimon, A. D. Straw, and M. H. Dickinson, "A simple vision-based algorithm for decision making in flying drosophila," *Current Biology* **18**, 464 (2008).
[41] S. Siavoshi, A. V. Orpe, and A. Kudrolli, "Friction of a slider on a granular layer: nonmonotonic thickness dependence and effect of boundary conditions," *Physical Review E* **73**, 010301 (2003).
[42] J.-C. Tsai, G. A. Voth, and J. P. Gollub, "Internal granular dynamics, shear-induced crystallization, and compaction steps," *Physical Review Letters* **91**, 064301 (2003).
[43] A. V. Orpe and A. Kudrolli, "Velocity correlations in dense granular flows observed with internal imaging," *Physical Review Letters* **98**, 238001 (2007).
[44] I. Taketoshi, T. Hiroshi, B. Kunio, I. Yoshio, W. Kazuhiro, and K. Yuzuru, "Progressive-scan CCD camera and motion picture recording system for high resolution visualized image capturing," *Journal of the Visualization Society of Japan* **19**, 61 (1999).
[45] I. S. Aranson and J. S. Olafsen, "Velocity fluctuations in electrostatically driven granular media," *Physical Review E* **66**, 061302 (2002).
[46] M. Shattuck, R. A. Ingale, and P. M. Reis, "Granular thermodynamics," *AIP Conference Proceedings* **1145**, 43 (2009).
[47] K. Combs and J. S. Olafsen, "Energy injection in a non-equilibrium granular gas experiment," *AIP Conference Proceedings* **1145**, 997 (2009).
[48] K. Combs, J. S. Olafsen, A. Burdeau, and P. Viot, "Thermostatistics of a single particle on a granular dimer lattice: Influence of defects," *Physical Review E* **78**, 042301 (2008).
[49] T. Pavlidis, "Image analysis," *Annual Review of Computer Science* **3**, 121 (1988).
[50] R. D. Wildman, J. M. Huntley, and J.-P. Hansen, "Self-diffusion in a two-dimensional vibrofluidized bed," *Physical Review E* **60**, 7066 (1999).
[51] Y. J. Zhang, "A survey on evaluation methods for image segmentation," *Pattern Recognition* **29**, 1335 (1996).
[52] S. Ott and J. Mann, "An experimental investigation of the relative diffusion of particle pairs in three-dimensional flow," *Journal of Fluid Mechanics* **422**, 207 (2000).
[53] G. Arfken, *Mathematical Methods for Physicists*, 3rd edition (Academic Press, New York, 1985).
[54] G. W. Baxter and J. S. Olafsen, "Experimental evidence for molecular chaos in granular gases," *Physical Review Letters* **99**, 028001 (2007).
[55] G. W. Baxter and J. S. Olafsen, "The temperature of a vibrated granular gas," *Granular Matter* **9**, 135 (2007).
[56] H. M. Jaegar, S. Nagel, and R. P. Behringer, "Granular solids, liquids and gases," *Reviews of Modern Physics* **68**, 1259 (1996).
[57] J. S. Olafsen and J. S. Urbach, "Velocity distributions and density fluctuations in a granular gas," *Physical Review E* **60**, R2468 (1999).
[58] P. M. Reis, R. A. Ingale, and M. D. Shattuck, "Caging dynamics in a granular fluid," *Physical Review Letters* **98**, 188301 (2007).
[59] **http://www.nsf.gov/crssprgm/reu/**
[60] R. C. Hilborn and R. H. Howes, "Why many undergraduate physics programs are good but few are great," *Physics Today* **56**, 38 (2003).
[61] B. L. Whitten, S. R. Foster, and M. L. Duncombe, "What works for women in undergraduate physics?" *Physics Today* **56**, 38 (2003).

[62] C. A. Manogue and K. S. Krane, "Paradigms in physics: restructuring the upper level," *Physics Today* **56**, 38 (2003).
[63] K. Kohlstedt, A. Snezhko, M.V. Sapoznikov, I. S. Aranson, J. S. Olafsen, and E. Ben-Naim, "Velocity distributions of granular gases with drag and with long-range interactions," *Physical Review Letters* **95**, 068001 (2005).
[64] J. Atwell and J. S. Olafsen, "Anisotropic dynamics in a shaken granular dimer gas experiment," *Physical Review E* **71**, 062301 (2005).
[65] B. Bammes and J. S. Olafsen, "Polymer-like folding of a two-dimensional granular chain in water," *Chaos* **14,** S9 (2004).

11
Structure and patterns in bacterial colonies

NICHOLAS C. DARNTON

11.1 Introduction

Though the movement of a single, isolated bacterium is reasonably well understood, when a large number of interacting bacteria are put together they produce beautiful and often complex phenomena. "Large" here typically means from 10^6 to 10^{12} individual cells: a small population by thermodynamic standards, but certainly unwieldy for any except statistical descriptions. Even restricting ourselves to the simple system of bacteria moving on or in solidified agar plates, the colony structures produced are surprisingly rich. In dilute solutions, swimming cells move independently, interacting with each other through their common consumption of a reservoir of nutrient. A point source of swimming cells expands in concentric rings as successive waves of bacteria chase gradients of nutrients, sometimes condensing into regular geometric patterns by chasing self-generated gradients. More concentrated solutions of swimming cells interact hydrodynamically through the fluid, producing large-scale swirls reminiscent of turbulence. This swirling occurs in two-dimensional surface motility as well, where uncorrelated motion turns into large-scale swirling as surface density increases. At extremely high density, bacteria jam and stop moving, as occurs in colloids. These high densities occur naturally on hard surfaces, where colony expansion is driven by cell growth rather than by motion; as the surface property becomes softer and wetter and bacteria begin to move, the resulting colonies change from fractal-like to radially symmetric and finally to a form dominated by a fast-growing, single-cell-thick outer layer. In this chapter, we will describe these phenomena, introduce some mathematical models that attempt to reproduce these behaviors, and present experimental techniques that are being used to measure colony properties beyond the simple visible structure.

Experimental and Computational Techniques in Soft Condensed Matter Physics, ed. Jeffrey Olafsen.
Published by Cambridge University Press.

Most bacteria can actively move through their environment, at least under some growth conditions. Though the method of locomotion may vary, particularly when cells grow on surfaces [1], the most common and best-studied form of propulsion uses bacterial flagella. Bacterial flagella (not to be confused with eukaryotic flagella) are thin helices, each connected to a rotary motor, whose rotation propels a cell forward. The mechanical characteristics of the motor and of flagella are well known [2–7]. Flagellar propulsion occurs at low Reynolds numbers, so the physics of an isolated, swimming bacterial is understood [8–11]. Bacteria can change the operation of their flagellar motors in response to environmental cues, swimming up (or down) a gradient of chemical attractant (or repellent). This behavior, known as chemotaxis, is best understood in *Escherichia coli*, which modulates the probability of reversing its motors in order to navigate. The phenomenology of chemotaxis, its formal mathematical correspondence to a biased random walk, and the biological control circuitry that produces it, are all well characterized [12–15].

The first experiments describing colony morphology were limited to quantifying, or at least describing evocatively, colony shapes and growth rates that were visible by eye or by microscope. Biophysical models tried to reproduce the observed phenomena using reasonable approximations to the underlying physics and biology. Experiments are now beginning to look beyond simple optical microscopy to probe subtleties of the colony structure and to validate predictions of the models.

We start with a description of bacterial colony growth and of the likely underlying physics. This entire range of population behavior can be accessed in the laboratory by varying the conditions of growth in a simple agar plate. The extremes of pattern formation in low agar plates (dilute populations governed by motility alone) and colony morphology on high agar plates (dense populations governed by growth alone) are the oldest studied and the best understood. The intermediate regime of swarming on medium agar plates (rapid surface colonization depending on both growth and movement) contains the most interesting biology and is currently the most actively researched.

11.2 Phenomenology of bacteria in and on agar

Agar plates are produced by gelling solutions of a seaweed extract in a nutrient broth. They provide flat, featureless surfaces that allow simultaneous access to air (from above) and to water and nutrients (from below). The gelled agar forms a random cross-linked network with a typical pore size that increases as the concentration of agar decreases. When the average pore size is larger than the size of a bacterium (roughly 1 μm for the species of interest here), cells can swim through the agar matrix; when the pore size is smaller than a bacterium, cells

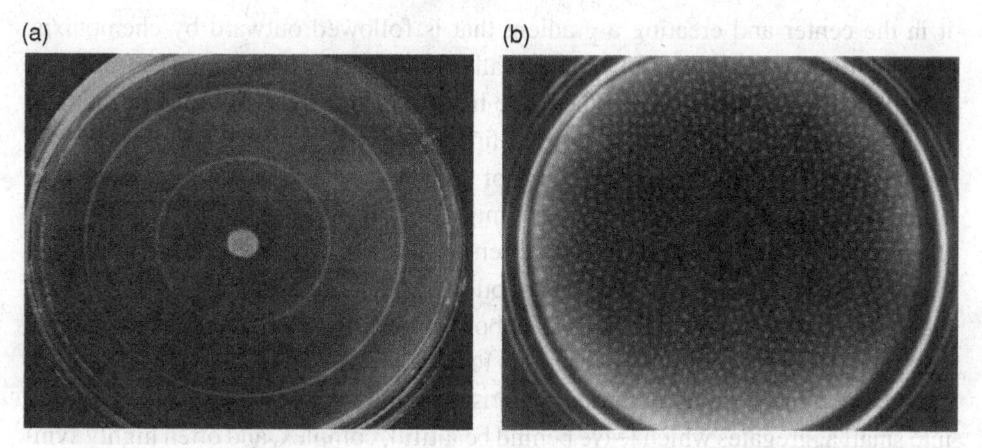

Figure 11.1 (a) Traveling chemotactic waves produced by *E. coli* consuming glucose and galactose [16]. (b) Aggregation pattern produced by chemotaxing *E. coli* [17]. Both patterns are photographed from above the Petri dish with oblique illumination from below. Contrast comes from the slight index of refraction difference between cells and medium; these should be considered only semi-quantitative measures of population density as a function of position. (Reprinted from (a) *Science* **153**, p. 708 (1966) with permission from AAAS; and from (b) *Nature* **349**, p. 630 (1991) with permission from Macmillan Publishers Ltd.)

grow only on the surface of the agar.[1] Low agar concentrations (a few g/l) produce semi-solid plates that resemble soft gelatin; high agar concentrations (10 g/l or higher) act as hard surfaces. Agar is not metabolizable by bacteria; it simply serves as a mechanical support that prevents bulk motion of the broth. From a practical standpoint, this is convenient because it prevents sloshing and mixing of colonies when a plate is moved; from a physics standpoint, this produces large volumes of fluid where advective flow is abolished. When pore sizes are large and bacteria can swim through the agar plate, cells effectively swim freely in three dimensions, except that the agar matrix prevents bulk entrainment of the fluid by the moving bacteria.

11.2.1 Chemotactic patterns

In a low-density agar plate that supports swimming, point inoculation at the center of the plate produces concentric waves of spreading bacteria, as seen in Figure 11.1(a) [16, 18]. Cells consume the most metabolically accessible carbon source, depleting

[1] Unfortunately, historically extremely low agar plates that support free swimming were called 'swarm plates'. This conflicts with the modern usage of the term 'swarming' to refer to surface motility (as in Section 11.2.3) and risks confusing the reader about swimming and swarming phenomena.

it in the center and creating a gradient that is followed outward by chemotaxis. The population that is left behind eventually converts over to metabolizing another, undepleted carbon source, and the cycle begins again; in the experiment pictured in Figure 11.1(a), the waves consume first glucose and then galactose, the two sugars added to the minimal medium of the agar plate. Similar waves occur in a one-dimensional system of bacteria migrating in a sealed capillary tube [19]; the physics of the one-dimensional system is arguably simpler, but the biology is more complicated since oxygen consumption by bacteria causes an additional wave due to aerotaxis, and the dominant metabolic process with and without oxygen are different. The entire process takes hours to days.

In *E. coli*, these expanding rings are unstable under some conditions, collapsing into small aggregates which leave behind beautiful, complex, and often highly symmetrical patterns, as seen in Figure 11.1(b). Microscopic observation of the traveling wave clearly shows the radially expanding front becoming unstable and condensing into islands of cells distributed regularly along the circumference [20, 21]. Rings and points have been observed in many species, including *E. coli* [17, 20, 22], *Salmonella* [23, 24], and *Pseudomonas* [25]. The pattern formation process varies somewhat from species to species; for instance, *Salmonella typhimurium* initially expands in a relatively uniform bacterial lawn, within which form stationary rings which subsequently break up into a symmetrical pattern of point-like aggregates, while *E. coli* produces expanding rings that leave behind point-like aggregates in their wake. In *E. coli* and *S. typhimurium*, aggregation can occur in liquid medium (without agar), though the resulting patterns are less ordered than those produced in agar [17, 24]. Since non-chemotactic (but otherwise normal) bacteria cannot form any of these patterns, the phenomenon must require chemotactic communication of some kind; for instance, the pattern in Figure 11.1(b) results from chemotaxis towards aspartate excreted by the bacteria themselves. A typical pattern requires several days to develop fully.

11.2.2 Colony morphology

At agar concentrations above a few g/l, bacteria cannot fit through the pores in the agar matrix and are restricted to the plate's surface. At the highest agar concentrations (above around 10 g/l) they cannot even move. On these hardest plates, a point inoculant expands into a macroscopic colony by simple cell growth: as bacteria grow and divide, the colony expands in height and girth. As the agar concentration is lowered, bacteria begin to move on the surface, though they are still prevented from entering the plate's interior by the small agar pore size.

Qualitative descriptions of the shapes of growing bacterial colonies date back more than a century, but the Matsushita lab's experiments on *Bacillus subtilis*

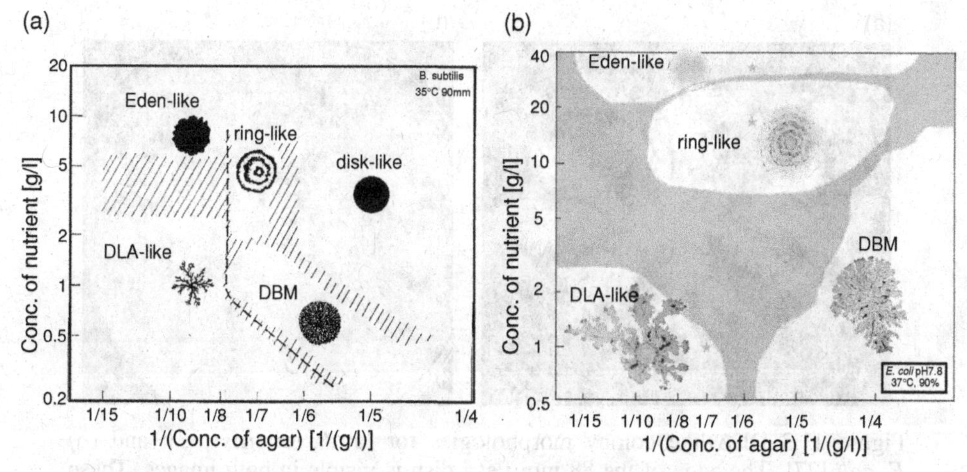

Figure 11.2 Phase diagrams of (a) *B. subtilis* [26] and (b) *E. coli* [27] on agar plates, as a function of the hardness of the agar (abscissa) and of the initial amount of nutrient (ordinate). In the hashed or gray areas between shapes, the observed colony may produce either morphology or switch between morphologies. (Reprinted from (a) *Physica A* **249**, p. 517 (1998) with permission from Elsevier; and from (b) *J. Phys. Soc. Japan* **78**, 074005 (2009) with permission from the Physical Society of Japan.)

prompted modern efforts to map and explain the morphological phase diagram [28–35]. A typical *Bacillus* phase diagram is shown in Figure 11.2(a); it is similar to the morphological phase diagram for other related species, such as *E. coli* (Figure 11.2(b)). Such diagrams generally contain four or five phases: (1) the diffusion-limited-aggregation-like morphology (DLA-like) at high agar and low nutrient concentrations; (2) the dense, rough morphology (Eden-like) at high agar and high nutrient concentration; (3) the dense, branching morphology (DBM) at low agar and low nutrient concentration; (4) the dense, round morphology (disk-like) at low agar and high nutrient concentrations; and (5) the concentric ring morphology (ring-like) between the Eden-like and disk-like morphologies. Cells in the DLA-like and Eden-like colonies are nonmotile. As agar concentration decreases, moving to the right in the phase diagram, cells become free to move on the agar surface. The resulting transitions between colony shapes, namely DLA-like to DBM and Eden-like to ring-like to disk-like, are entirely due to increasing cell motility. The most direct evidence for this is that nonmotile *Bacillus* mutants do not produce the ring-like, disk-like, or DBM morphologies [28]; that is, the nonmotile phase diagram contains only two colony types (DLA-like and Eden-like) and has no dependence on agar concentration. In practice, even at low-agar concentrations only cells on the colony periphery are motile; cells in the interior usually revert to a nonmotile, vegetative state. Though the locations of the boundaries between

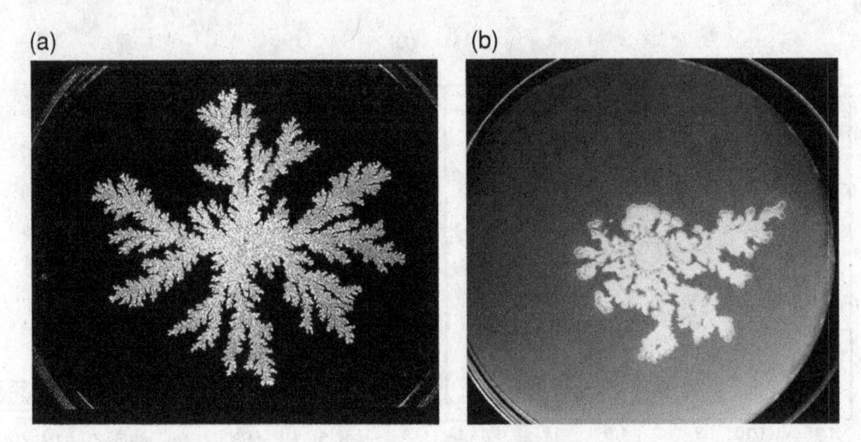

Figure 11.3 DLA-like colony morphologies for (a) *B. subtilis* [33]; and (b) *E. coli* [27]. The edge of the 88 mm Petri dish is visible in both images. These photographs were taken twenty days and one month after inoculation, respectively. (Reprinted by permission from (a) *J. Phys. Soc. Japan* **58**, p. 3875 (1989); and from (b) *J. Phys. Soc. Japan* **78**, 074005 (2009) with permission from the Physical Society of Japan.)

phases may differ, the general features of this phase diagram are similar in close relatives of *B. subtilis* and *E. coli*, such as *Serratia* and *Salmonella*.

The low-nutrient, high-agar portion of the phase diagram is particularly robust. Not just *B. subtilis* (summarized in [34]) but many other bacteria [27, 28, 31–33, 36–42] produce DLA-like colonies. The spreading colony produces a self-similar structure, with the area covered, A, obeying a power law with size l: $A \propto l^{\alpha}$ for distances ranging over several orders of magnitude. The fractal dimension α of a DLA-like colony is typically around 1.7 [34], which is consistent with classical two-dimensional diffusion-limited aggregation. Though many DLA-like colonies have a fractal mass distribution at any particular time, so that the colony is always instantaneously self-similar, the fractal dimension may change with time – usually increasing towards the DLA limit of 1.7 – as the colony expands [43, 44]. DLA-like colonies have the qualitative fine-branching structure associated with diffusion-limited aggregation: an arm grows by intercepting incoming material, which prevents nearby arms from growing, producing characteristic gaps between arms on all distance scales. This is seen qualitatively in Figure 11.3. The occasional slight quantitative discrepancy between colony shape and DLA theory are probably due to the bacterial colony being imperfectly two dimensional: (1) the colony is not flat, as it is normally slightly thicker towards the center of each arm and towards the center of the colony [33]; this is not captured by two-dimensional DLA simulations and is largely ignored by the traditional method of extracting the colony's fractal dimension, which reduces a photograph of the colony to a binary image and assigns areal coverage based on the presence or absence of cells rather than their total surface density, and (2) the diffusion of nutrients occurs in the bulk agar; the agar is typically several mm thick, so nutrient transport in the third, vertical, dimension is not negligible for early, small colonies or for fine arm spacing.

The fractal dimension is only a single measure of morphology and does not capture all important structural details. For instance, some filamentous colonies (which are typical of fungi) have similar fractal dimensions near 1.7, but their structure is clearly different. They lack the screening effect in fractal colonies: filamentous arms cross each other and sometimes intersect, whereas in DLA, arms are mutually repulsive because they competitively deplete the local nutrient supply. Fractal scaling laws have been found in three dimensions in growth [37] and flocculation of bacteria, as analyzed though direct imaging and through small-angle light scattering [45]. Flocculation more closely corresponds to true DLA (cells diffuse and then aggregate), as opposed to DLA-like colony growth, which is more accurately described as diffusion-limited reaction (nutrients diffuse and then 'react' to form bacterial cells).

The Eden-like colony, which occurs at higher nutrient levels, is named for the model of Murray Eden [46]. This two-dimensional model ignores cell motility and differentiation, examining the possible shapes of colonies that grow exclusively by cell division according to simple laws for where and when cells may reproduce. In particular, the model assumes that division rates are determined endogenously, by cell processes, rather than by cells' limited access to nutrients (as occurs for DLA-like morphologies). The Eden model typically produces colonies that are approximately radially symmetric and have significant surface roughness. Extending the fractal measure of colony roughness to the Eden-like colony, Matsushita *et al.* [26] examined the roughness of the Eden-like colony boundary explicitly. They measured the standard deviation W of the boundary width as a function of size L, and found a scaling law $W \propto L^{\alpha}$ with $\alpha = 0.78$. Unlike DLA-like colonies, where many bacteria are found to obey the same scaling law, it is unclear to what extent this scaling holds for other strains under similar conditions.

11.2.3 Swarming

When swimming bacteria are grown on a medium-density agar plate (typically around 0.5% agar) with copious nutrients, they swarm. These conditions provide plentiful access to nutrients (allowing rapid growth), and to water (allowing motility), but the agar pore size is just small enough to prevent bacteria from leaving the surface. After an initial lag period required to turn on swarming genes, the bacteria spread out in a shallow, highly motile layer, covering the surface in a matter of hours. Swarming cells are long (generally multinucleate, resulting from delayed septation), hyperflagellated, and extremely motile.

The swarm consists of a thin, highly motile outer layer with a well-defined edge. The outermost boundary is paused cells: cells that have swum too far onto agar, into an environment that is not sufficiently wet. Inside the boundary is a large monolayer of moving cells that can extend over centimeters radially; eventually, a second layer of moving cells assembles on top of the monolayer, followed by a third, etc. as the swarm thickens. The cells in each layer are in constant motion, and the boundary between successive layers moves slowly outward in concert with overall swarm growth. The boundary of a layer is a dynamically stable structure resulting from continuous, vigorous population exchange between different strata. Toward the center of the swarm, cells return to the shorter, nonmotile vegetative state.

In addition to the direct physical manifestation of the swarm phenotype (increased length and number of flagella), swarm cells often produce a wetting agent, usually a surfactant of some type. For instance, the surfactants in *Serratia* (serrawettins [47]) and *Bacillus* (surfactin [48]) are known. Loss of surfactant dramatically decreases swarming proficiency in *Bacillus* [49, 50] and laboratory strains of *E. coli* do not make surfactant, which is probably why they are such picky swarmers [51]. Surfactants act mainly to maintain surface wetness, which in turn allows cells to swim in the very shallow (often only one cell deep) layer of fluid on the surface of the plate. Increased slime production alone can allow swarming even without the elongation and hypermotility usually present in the swarm phenotype [52]. The onset of swarming is caused by high bacterial density (usually detected by quorum sensing agents [53]), availability of water, nutrients and oxygen, and a viscous environment. In quorum sensing, cells excrete chemical messengers in the medium; if the overall extracellular concentration of these messengers rises above a threshold (signaling a large number of cells in a confined volume), cells respond by activating density-dependent groups of genes. Depending on other environmental factors, these can lead to either swarming or biofilm formation.

The most famous swarmer is *Proteus mirabilis*, which has the above discussed swarm characteristics of elongation, multinucleation, hyperflagellation, and surfactant production, but also synchronizes its differentiation and consolidation along the entire colony front, from vegetative to swarmer cells and back. Old fronts are left behind as terraces in the colony, while successive phases of expansion, occurring at roughly hourlong intervals, expand the colony boundary outward in steps of roughly 2 mm. This time synchronization is more complicated than the monotonic swarm expansion seen in other species, and Proteus swarming has developed into its own sub-discipline: early quantitative efforts include Rauprich *et al.* [54], who examined parameters such as lag time and reconsolidation intervals as a function of agar concentration, temperature, and other environmental variables.

Swarming occurs at a self-maintained high bacterial density, close to 50% of the close-packed surface density. It is possible to artificially confine high of densities bacteria in thin films [55,56]. These systems show some of the properties of swarms, in particular swirling motions on length scales 10 to 100 times larger than the size of a bacterium.

Swarming is technically defined as rapid surface translocation driven by flagella. Though the bodies of most cells are apolar, in that there is no geometrical way to define the head and tail of the cell, this symmetry can be broken by the location of a cell's flagella. There are modes of flagellation that mark one end of a cell as geometrically distinct: polar flagellation (a single flagellum at one end of the cell) and subpolar (a tuft of several flagella at one end of the cell) modes uniquely define the tail of a cell. In contrast, in peritrichous flagellation many flagella are distributed uniformly, and probably randomly, on the cell surface.

Interestingly, all bacteria that swarm are peritrichously flagellated. In fact, some bacteria possess both polar and peritrichous modes of flagellation; they use a single polar flagellum for swimming but switch to (or add) many distributed flagella for swarming. This behavior has been most noticeably reported in several species of *Vibrio*, but it also occurs in many other bacteria: see references in McCarter [57] and in Kirov [58] for further examples. Peritrichous flagellation allows cells to produce more flagella, since additional flagella can be manufactured and inserted anywhere on the cell surface. More flagella presumably increase the swarming cells' speed, and speed is crucial for the colony to quickly colonize a surface, but quantity alone may not explain the correlation of peritrichous flagellation with swarming. When a peritrichously flagellated cell swims, its flagella trail behind it and assemble in a rotating, phase-locked bundle (or in several braid-like bundles if the cell body is longer than the flagella themselves). The location of the bundle dynamically defines the tail of the cell, and this polarity is maintained as long as the cell keeps moving.[2] Because the bacterium itself is fundamentally apolar, if the bundle is disrupted it can reform at the other end of the cell body, effectively flipping the previous roles of head and tail. This occurs when a cell slows or stops, usually because it encounters an obstacle or because too many of its motors reverse [59]. The punctuated polarity associated with motor reversal in bundles of peritrichous flagella appears to be necessary for successful swarming, since mutants whose motors are locked in a single direction of rotation (either clockwise or counterclockwise), and which therefore cannot reverse direction, are unable to swarm [60]. A monoflagellated cell that occasionally reversed its motor would have similar intermittent polarity flips; in practice, however, a monoflagellate's motor slows or stops when it tumbles, but never reverses [61].

In swimming cells, independently operating flagella coalesce into a bundle, so one might also expect co-bundling between the flagella from neighboring cells.

[2] Under normal conditions flagella push a cell, so the point of attachment of a flagellum becomes the trailing end of cell body. There is no physics constraint that prevents a cell from pulling itself forward – it could do so by reversing the direction of rotation of its flagellum – but this almost never occurs.

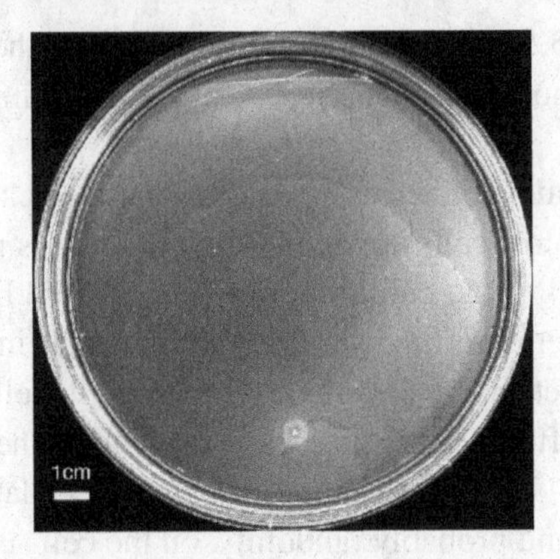

Figure 11.4 A swarming *E. coli* in the process of colonizing an agar plate. The swarm extends about three quarters of the way across the dish, starting from the inoculation site near the lower edge of the Petri dish. This image was taken after overnight incubation. The edge of a swarm of this size typically advances at around 5 μm/s. (Image courtesy of Linda Turner.)

This would only happen if the cells were close together, such as in a swarm, which could explain why swarming flagella assemble into rafts or groups. Co-bundling is suggested by EM images of fixed cells (such as Figure 11.4); however, direct observation of the fluorescently labeled flagella of live swarming cells revealed little co-bundling, perhaps because bundling relies on the twisting up of flagella attached to the same, counter-rotating cell body [62].

11.2.4 Comments

The ability to make precise statements about the spatial and temporal patterns of developing colonies has suffered from the poorly defined nature of agar and of the media used in experiments. This undoubtedly stems from agar's historical role as a substrate for bacterial cultivation rather than as an analytical tool. Agar itself is a poorly defined and understood substance: its primary purpose is to provide a large-pore structure that provides easy diffusive access to the nutrients of the growth medium while preventing bulk motion. In the long run, the use of chemically defined media [63] and synthetic surfaces to replace agar [64] will provide more reproducible and more easily interpretable results; see Julkowska *et al.* [65] for one such example.

11.3 Modeling

The analysis of almost all of these experiments presume that self-generated gradients within the agar matrix or within the developing colony are responsible for the resulting structures. The gradient may be in one or more dissolved nutrients (often a carbon source), oxygen availability, or in a waste product. The role of the gradient is demonstrated indirectly by comparing the observed colony shape to predictions of a reaction–diffusion–chemotaxis theory. Direct observation of the gradient is rare, though the group of Wimpenny has systematically examined spatial variations in pH [66–68], oxygen [69,70], and pyruvate [71] in conjunction with colony growth. Still, experimental validation of the gradients that figure so prominently in models is scarce.

Bacteria adjust their gene expression in response to environmental conditions, which provides an obvious mechanism for generating a physiological pattern. Physical changes in cells are often invisible to the experimenter, however, who may only detect total cell density or, even more often, just the location of the colony edge. When physiological changes are extreme enough, they are considered to correspond to distinct phenotypes; for instance, the differentiation into swarmer cells. Many models introduce such a binary phenotype switch. Some phenotypic variation is abrupt and discontinuous and can be traced to a binary genetic regulatory circuit, especially in the case of extreme phenotypical variations like swarm cell differentiation [72,73], but in many cases a binary switch model of cell activity is an oversimplification of a complex and continuous process of differentiation. Efforts to model bacterial populations are driven by an analogy between the phenotypic pattern in a growing colony and cell differentiation in a developing organism [74,75]. Bacteria colonies are seen as simple versions of multicellular organisms [76], or more fancifully as teams [77], communities [78], or even societies [79].

11.3.1 Chemotactic patterns

The chemotactic bands of Figure 11.1(a) are qualitatively described by a reduced version of the Keller–Segel equations [80–83]:

$$\frac{\partial b}{\partial t} = \vec{\nabla} \cdot (D_b \vec{\nabla} b - \vec{J}_{\mathrm{ch}}) \tag{11.1}$$

$$\frac{\partial n}{\partial t} = D_n \nabla^2 n - c(b, n) \tag{11.2}$$

$$\vec{J}_{\mathrm{ch}} = b\chi(n)\vec{\nabla} n, \tag{11.3}$$

where b is density of bacteria and n is density of a nutrient substrate that functions as an attractant. Bacteria and nutrient diffuse with diffusion coefficients D_b and D_n, respectively. D_b is a macroscopic expression of run-tumble behavior by chemotaxing bacteria; in principle it could be non-constant (for instance, depending on n), but for swimming bacteria the dependence is small and can be neglected.

The last term in Equation (11.2) represents consumption of the nutrient at the rate $c(b, n)$. Since the nutrient consumption rate is proportional the number of bacteria, we expect that $c(b, n) = \tilde{c}(n)b$. Bacteria sense and respond to the nutrient, moving (on average) with a chemotactic speed, also called the drift velocity, that depends on the spatial distribution of n. This drift velocity can be generally written as $\vec{v}_{\text{ch}} = \chi(n)\vec{\nabla}n$, where the entire complicated chemotactic response of bacteria is placed in the chemotactic coefficient $\chi(n)$. This drift acts to advectively transport the local bacterial density, producing the flux $\vec{J}_{\text{ch}} = b\vec{v}_{\text{ch}}$ in Equation (11.3). The sign of the flux term is reversed for a repellent.

The exact form of the chemotactic coefficient $\chi(n)$ depends on assumptions about how the cell detects its environment. From a purely modeling perspective, one can search for the simplest form of $\chi(n)$ that produces reasonable-looking traveling wave solutions to Equations (11.1)–(11.3). This is largely a mathematical question; it may be answered precisely as such, but often at the expense of biological plausibility. For instance, Odell and Keller [81] prove that, if the bacterial population in the wake of the wave is to be zero, $\chi(n)$ must be singular. This is necessary to produce strong enough chemotactic attraction that the pulse of bacteria moves as a group, without leaving any stragglers behind. In reality, there *is* a population left behind after the wave passes; these stragglers are precisely the individuals that are responsible for the second wave, as seen in Figure 11.1(a). This possibility is rejected by Odell and Keller because, without a growth term in Equations (11.1)–(11.3), any residual population left behind would deplete the bacterial pulse, which is not consistent with a traveling wave solution. Despite misgivings about the form of $\chi(n)$, this simple reaction–diffusion–chemotaxis system, with only one reaction term, two diffusion terms, and one chemotaxis term, reproduces the phenomenon of traveling chemotactic bands and matches the observed front speed reasonably well [80, 82], even without including diffusion [18]. Successive waves in Figure 11.1(a) are generated by subpopulations of bacteria consuming different substrates; to model this, the population b is split into separate subpopulations b_i, each consuming with its own substrate n_i. Due to Equation (11.3), this is not a linear system, and we do not expect the principle of superposition to hold. In fact, colliding waves annihilate both in experiment and in simulation [84]; however since it is the concentration gradients that annihilate during the collision, waves chasing different attractants pass through each other unperturbed [85].

Equations (11.1)–(11.3) include diffusion and consumption of a nutrient (or attractant) and chemotactic chasing of the resulting gradient. If cells can *themselves* produce attractant, they can become self-attracting.[3] To describe phenomena such

[3] A self-attraction term is included in the original, full Keller–Segel equations, which sought to describe slime mold aggregation, but is omitted from our reduced version of them in Equations (11.1)–(11.3) [83].

as those in Figure 11.1(b), therefore, an additional production term is added to Equation (11.2):

$$\frac{\partial n}{\partial t} = D_n \nabla^2 n - c(b,n) + p(b,n). \qquad (11.4)$$

The traditional Keller–Segel equations make some assumptions about the form of $c(b,n)$ and $p(b,n)$, namely that $c(b,n) = \tilde{c}(n)n$ and $p(b,n) = \tilde{p}(n)b$. From the modeling perspective, even the simplest functional forms (constant D_b, $\tilde{c}(n)$, $\tilde{p}(n)$, and $\chi(n)$) can produce 'chemotactic collapse': the formation of delta-function spikes in bacterial concentration in two dimensions [86]. This is not a strong statement about the biological relevance, since many other choices of functional forms also lead to spikes in density. These are arguably more physically plausible, since in many cases the resulting density does not diverge; see Horstmann [87] and Hillen and Painter [88] for reviews of modeling efforts.

Alternatively, one can encode as much known biology as possible in the functions that specify substrate consumption and production ($c(b,n)$ and $p(b,n)$) and describe chemotaxis ($\chi(n)$ and $D_b(n)$) and compare the solutions of Equations (11.1), (11.3), and (11.4) to experiments. Based on the binding properties of chemoreceptors, a particular form of the chemotactic coefficient $\chi(n)$ is commonly used. In chemotaxis, a cell measures the time rate of change of attractant; since the cell is in motion during this process, this effectively becomes a measurement of the spatial rate of change of attractant. To first order, molecules of the attractant bind to chemoreceptors independently, so that the fraction of chemoreceptors occupied $f(n)$ obeys a noncooperative Hill function: $f(n) = n/(K+n)$. Assuming the rest of the signal transduction machinery in the cell simply acts to amplify this signal,[4] the eventual chemotactic velocity must be proportional to $\vec{\nabla} f$. That is, $\vec{J}_{\mathrm{ch}} \propto b\vec{\nabla} f \propto b/(K+n)^2 \vec{\nabla} n$; comparing with Equation (11.3), the chemotactic coefficient is (up to a constant of proportionality):

$$\chi(n) = 1/(K+n)^2. \qquad (11.5)$$

In most cases, the process of forming chemotactic aggregates takes several hours to days, so bacterial growth is not negligible. For this reason, the equation describing the evolution of bacterial density acquires a growth term $g(b,n)$:

$$\frac{\partial b}{\partial t} = \vec{\nabla} \cdot (D_b \vec{\nabla} b - \vec{J}_{\mathrm{ch}}) + g(b,n). \qquad (11.6)$$

[4] The signal transduction pathway extends from attractant binding through motor function to flagellar bundling. It is an oversimplification to assert that a single noncooperative binding function captures the entire pathway, including noninstantaneous adaptation responses, plus (because we are trying to model the swimming speed that results) the hydrodynamics of propulsion with several independently controlled flagella. However, this simplification has a kernel of biochemical justification, and should apply over a wider concentration range than, for instance, a constant chemotactic coefficient.

Typically availability of nutrient is not a limiting factor, but overcrowding can be, so the growth term is often set to

$$g(b, n) = \tau^{-1} b(1 - b/b_0), \tag{11.7}$$

where b_0 is the maximum bacterial density (the carrying capacity of the medium) and τ is the bacteria mass doubling time, which is normally the time between bacterial divisions: typically around two hours.

The production and consumption terms in Equation (11.4) are harder to derive rigorously. In an initial model of chemotactic aggregation, Woodward *et al.* [24] used Michaelis–Menton-like production and consumption terms, but were able to reproduce aggregation only by permitting the nonphysical condition that substrate consumption not be proportional to the number of bacteria. Later models were more successful in using biologically plausible production and consumption terms, but only by supplementing the model with additional coupled fields. For instance, Brenner *et al.* [21] and Tyson *et al.* [89] divide the roles of nutrient and attractant in two, positing two separate chemical species, each of which obeys a reaction–diffusion equation similar to Equation (11.4). This has a biological justification, since in reality the carbon source in spot-forming experiments is processed into a separate chemoattractant (for instance, succinate into aspartate in the experiments of Budrene and Berg [20]). It is reassuring that biologically plausible functions for consumption, production, growth, and chemotaxis seem to require both a nutrient and a chemoattractant, which agrees with the experimental observation that chemotactic aggregates form only under conditions where bacteria excrete metabolic products that are chemoattractants. This highlights the crucial difference between stable traveling waves (Figure 11.1(a)) and waves that collapse into point-like aggregates (Figure 11.1(b)): the former occur in a medium containing several nutrients (a different one for each wave) that are themselves chemoattractants; the latter occur in a minimal, chemically defined medium, where the principal nutrient is a negligibly weak attractant compared to the principal excreted by-product.

All these patterns are produced during free swimming, where the bacterial population is relatively dilute and the agar matrix prevents any bulk bio-convection, so there is no appreciable direct interaction between cells. The bacterial population moves coherently only because cells are consuming and detecting the same reservoir of nutrient. Each bacterium is performing normal chemotaxis on its own; in that sense, these experiments are just complex versions of basic chemotaxis. The great attraction of these experiments is that the fundamental chemotactic behavior can be measured accurately in controlled chemotaxis assays, so there is direct data to determine the form of the chemotactic flux in Equation (11.3). See Section 11.4.5 for references that translate from single cell data to population dynamics.

Equations (11.1) and (11.3) incorporate chemotaxis through the parameters D_b (the bacterial diffusion coefficient) and $\chi(n)$ (the chemotactic response). Just as kinetic gas theory predicts macroscopic properties from microscopic statistics, one can calculate D_b and $\chi(n)$ from a complete microscopic knowledge of the swimming chemotactic behavior of cells. Substantial data on the trajectories of swimming cells under different nutrient field conditions is available, and several approaches to deriving macroscopic equations of motion for the bacterial population have been developed. Some approaches start from cell-level information (such as distribution of turn angles and of run lengths) [90–94]. Other approaches start from the sub-cellular level, with the known biochemistry of the signal transduction pathway, simulating the motor's response to external stimuli, inferring the consequences for the cell's swimming motion, and finally calculating the macroscopic consequences [95–97]. The latter approach has the advantage that responses can be simulated under a wider range of stimuli than have been tested in cell trackers, but this is much more computationally expensive and vulnerable to uncertainties in more elements of the model. These practices are usually used to calculate the population-level parameters D_b and $\chi(n)$ in Equations (11.1) and (11.3), which equations are then used to simulate aggregation as outlined above; one can, however, directly simulate individual cells responding to an attractant field, skipping the continuum limit [22].

11.3.2 Colony morphology

Theories of the colony morphology phase diagram generally start from the Keller–Segel equations supplemented with reaction–diffusion equations for nutrients and, in more elaborate models, production of the superficial lubrication layer. By analogy with chemical reactions, growth of cells is represented as conversion of nutrients into new cells. The patterns seen are more complicated than simple reaction–diffusion can produce because (1) cells may perform chemotaxis, which couples cell flux to concentrations of other components, and (2) cell growth (or death) rates depend on components in highly complex way.

11.3.2.1 Diffusion–reaction

The majority of diffusion–reaction equations that describe the various morphologies in the phase diagrams of Figures 11.2(a) and (b) follow a general pattern:

$$\frac{\partial b}{\partial t} = \vec{\nabla} \cdot (D_b(n, s)\vec{\nabla} b) + g(b, n) - d(b) \tag{11.8}$$

$$\frac{\partial n}{\partial t} = D_n \nabla^2 n - \alpha g(b, n) \tag{11.9}$$

$$\frac{\partial s}{\partial t} = d(b). \tag{11.10}$$

This is essentially the diffusion–reaction–chemotaxis equations of the previous section, with chemotaxis removed and another field, the spore density s, coupled to the bacterial density b via a differentiation rate $d(b)$. The principal property of spores is that they are nonmotile – hence the absence of any diffusive term in Equation (11.10). Here we have written $D_b(n, s)$ explicitly as a function of n and s in order to emphasize that a nonconstant bacterial diffusion coefficient D_b is *required* in order to reproduce the morphological phase diagram.[5] While the bacterial diffusion coefficient previously had a good microscopic explanation (as a result of the statistical nature of population movement in the run-tumble behavior of chemotaxis), here D_b represents cell-growth-independent colony spreading, which may occur even without motility, as at high agar concentrations. The physical mechanism of this spreading is unclear when there is no motility to drive it, and in fact models typically choose the bacterial diffusion coefficient to be small (though not zero) on hard agar surfaces where motility is low; increasing D_b as agar concentration drops is a way to encode the fact that bacterial motility increases with lower agar concentration. The growth of bacteria $g(b, n)$ is accomplished through the metabolic conversion of nutrient with an efficiency of α (g nutrient)/(g bacterium); hence, what had been a generic consumption term $c(b, n)$ in Equation (11.4) is now explicitly proportional to growth: $c(b, n) = \alpha g(b, n)$. The exact conversion factor depends slightly on the nature of the nutrient, but it is roughly 3 (g nutrient)/(g bacterium) [98].

As in the previous section, it is possible to reproduce the observed experimental phenomena with fewer terms in the model if one is willing to make each term more complicated and less biologically defensible. For instance, an early model due to Kawasaki *et al.* [99] dispenses with the spore field s completely, but still manages to reproduce the DBM-to-disk transition. Its growth/consumption term is plausibly Michaelis–Menton: $g(b, s) \propto bs/(K + s)$, but bacterial diffusion is increased heuristically by setting $D_b \propto bn$. The latter cannot be justified on physical grounds, but it serves to restrict cell motility to the colony edge (where both b and n are high), as is observed in experiments. Extensions of this model produce a DLA-like morphology by requiring a minimal colony density so that different colony branches remain separated [98]; this can be enforced by using highly nonlinear (step-function) consumption/growth functions [100]: the lower cutoff is meant to reflect that there are no infinitesimally small bacteria.

If cell differentiation is included, as in Equation (11.10), this will abolish motility in the colony interior, which would otherwise gradually fill up the gaps between branches. Still, unless a highly nonlinear growth term is adopted, a

[5] In the previous section, the bacterial diffusion coefficient D_b was generally assumed to be constant, though it was not required by the models.

non-constant bacterial diffusion is necessary to produce fractal growth, and in any case D_b must vary somehow to account for the morphological change with changing agar concentration.

This nonphysical variation of the bacterial diffusion coefficient D_b with agar concentration is critiqued by Pipe and Grimson [101], who instead develop a model that includes height and growth throughout the colony interior. In general, the variable bacterial diffusion coefficient required in many models can be explained as the consequence of locomotion within a lubrication layer, whose depth is controlled by a separate partial differential equation [98, 102]. Since motility depends on surface wetness, the depth of the surface film of water determines the amount of cell movement and, therefore, the size of D_b. The presence and importance of the lubrication layer is clear from experiments, though it is less clear how it affects the bacterial diffusion coefficient. A similar approach is taken by Lega and Passot [103–105] who include reaction–diffusion for cells, water, and nutrients. This model includes cell motility, which contributes an advective term to the cell-density evolution, and prescribes the cell/fluid velocity as resulting from force balance at low Reynolds number (propulsion, drag force, noise, and pressure (collisions)). This produces a complete prediction of colony morphology; much of this information such as the colony height as a function of position ought to be experimentally accessible, but it is rarely reported in the literature.

Equations (11.8) and (11.9) in this section differ from Equations (11.1) and (11.2) in the previous section principally by the inclusion of the additional growth term $g(b, c)$. Simple exponential growth (such as growth limited exclusively by the bacterial doubling time) would have $g(b) \propto b$; in practice, growth is usually capped by setting $g(b, n) \propto b(b_{\max} - b)$ to reflect the fact that the medium can reach a saturation population density (also called its carrying capacity) $b_{\max}$. Variations on this growth-capping scheme are possible: a common method to limit growth in the interior of the colony is to allow cells to transform to a dormant state: for instance, *Bacillus* sporulates, producing a, nongrowing, nonmotile – for purposes of colony evolution, effectively dead – field s, as in Equation (11.10) [106]. In its simplest form, bacterial motility can be represented by an effective bacterial diffusion term that depends on agar concentration [99], which reproduces the transition from DLA-like to DBM to disk-like colonies. Matsushita *et al.* [26, 107] make this conversion from active to inactive populations dependent on nutrient concentration; this modulation reproduces the ring-like colony morphology. These formulations are biologically plausible, but the exact dependence of sporulation rate on environmental conditions is unknown. Note that chemotaxis does not appear explicitly in any of these equations: there is no chemotactic flux term analogous to Equation (11.3). In the extreme high agar and low nutrient part of the morphology phase diagram (the lower left corner of Figures 11.2(a) and (b)), bacteria do not move and the distinction between bacteria and spores in Equations (11.8) and (11.10) vanishes.

In parallel with continuous models, there are models of individual cells that prescribe behavior rules (for growth, death, and locomotion) and simulate colony evolution. For instance, the walker model of Ben-Jacob [108–110], which assumes that discrete spatial portions of the colony (coarse-grained chunks containing many bacteria) transform from active to inactive states, affecting their consumption of a continuously defined nutrient field. Ginovart *et al.* perform a more traditional Monte Carlo model, simulating the growth and division of each individual cell in the colony as well as individual nutrient particles [111]. A more recent model posits two classes of cells, one associated with migration (motility) and proliferation (growth), successfully reproducing the entire morphological phase diagram [112]. The biological relevance of these models depends on the behavior rules of their cells; for instance, the model uses a characteristic time for differentiation to drive periodic waves of colony expansion; this causes terrace formation and reproduces the ringlike colony morphology, the most elusive of the morphologies [112]; however, there is little biological support for a biological clock in *Bacillus* and none in other species.

11.3.2.2 Diffusion-limited aggregation

As its name implies, the DLA-like colony results from growth that is limited by diffusional access to scarce nutrients. This is physically plausible since the DLA-like structure occurs under conditions of poor nutrient availability and extremely slow growth (the DLA-like colonies in Figure 11.3 correspond to several weeks of development). The resulting colony shape is similar to the snowflake pattern that occurs in many nonliving natural processes. DLA was originally designed to model colloidal aggregation [113], but has since been applied to viscous fingering phenomena, electrodeposition of metals, and beyond (reviewed in Ben-Jacob [114]). Growth is limited by diffusion of a species which then aggregates on the growing structure. In the case of the growing bacterial colony, 'aggregation' corresponds to a pseudo-chemical 'reaction' that transforms nutrients into cell mass, which is then visible in the microscope. Some details of the growing structure depend on the geometry (anisotropy) of the interface, which for bacteria would map to the geometrical constraints of cell growth and division, but the general structure is dictated by diffusive access to reactants being the rate-limiting step. To make the growing colony correspond strictly to two-dimensional diffusion-limited aggregation, the surface colony is assumed to have constant surface density, and bacteria are assumed to be nonmotile. The latter assumption is reasonably accurate at high agar concentrations, but the former is only approximately true.

DLA can be modeled discretely as individual particles that perform Brownian motion and aggregate on contact with the colony, or continuously as a density field in contact with a growing, perfectly absorbing colony boundary:

$$\frac{\partial \lambda}{\partial t} = \eta \phi_n \tag{11.11}$$

$$\frac{\partial n}{\partial t} = D_n \nabla^2 n, \tag{11.12}$$

where λ is the linear density of bacteria along the colony edge. Equation (11.11) represents "aggregation." The nutrient flux per colony boundary of length dl is $\phi_n = D_n \vec{\nabla} n \cdot \hat{n}$. By assumption, the colony has a constant surface density b_0, so the accretion in Equation (11.11) produces growth in the colony radius R:

$$b_0 \frac{\partial R}{\partial t} = \eta \phi_n. \tag{11.13}$$

Equation (11.12) is diffusion of the nutrients; since the colony grows slowly, the nutrient field n is in quasi-static equilibrium, so $\partial n / \partial t = 0$ at all times. Anything that reaches the colony edge by diffusion adheres, so $n = 0$ on the colony boundary. This is equivalent to a reaction–diffusion–chemotaxis model with non-motile bacteria ($D_b = 0$), consumption of substrate occurring rigorously only at the colony edge $\vec{R}$ (so $c(b, n) \propto n\delta(\vec{R} - \vec{r})$), and cell growth linked directly to consumption (so $g = \eta c$). Since the bacteria are nonmotile, they cannot perform chemotaxis.

The simplest simulations of diffusion-limited aggregation model the diffusion of individual particles until each particle irreversibly sticks to part of the colony. In two dimensions, these simulations produce a fractal dimension of 1.7, in agreement with the observed fractal dimension for *Bacillus* growth of 1.73. In a bacterial colony, the diffusing particle is a molecule of nutrient, while aggregation is the transformation of the nutrients into cell mass. Since nutrients diffuse through the bulk of the agar plate, rather than on the surface, this corresponds to two-dimensional diffusion-limited aggregation only in the limit that plate thickness (typically a few mm) is negligibly small compared to colony size; in fact, colonies growing in the DLA-like region of the phase diagram are not self-similar when they are very small [33].

The boundary of a diffusion-limited aggregate is extremely unstable to fluctuations. Despite years of effort to model DLA using a partial differential equation approach, it is still standard practice to perform particle-by-particle simulations, precisely because the extreme sensitivity of the equations to perturbation resists continuum analysis. It seems unlikely, therefore, that the DLA-like morphology will be computationally tractable using a continuum description as in Equations (11.8)–(11.10). As a consequence, attempts to explain the entire phase diagram using a single set of principles tend to start from single-cell-level descriptions of behavior, implemented in Monte Carlo simulations.

The DLA-like colony morphology gives way to other morphologies when the controlling factor becomes the bacterial growth rate rather than nutrient availability, which occurs at high nutrient concentration, or when the individual cells become motile, which occurs at high wetness and low agar concentrations. For instance, the transition from DLA-like to Eden-like morphology, along the vertical in Figure 11.2, is caused by increasing nutrient concentration: growth becomes limited by the bacterial doubling time, which is presumably limited by the rate of metabolic process within the cell rather than by diffusion of nutrients. This produces not just a quantitative increase in the speed of colony growth, but a qualitative change in the shape of the colony.

11.3.3 Swarming

Unlike the reaction–diffusion–chemotaxis models of the previous two sections, models of swarming usually start with a set of rules for the motion of individual cells. Most models are variations of Vicsek *et al.* [115], which assumes that cells move at a constant speed v_0, and align their orientation θ with their local neighbors, subject to a certain amount of random noise in the orientation. Mathematically, the orientation, position, and velocity of the ith self-propelled particle follows the rules

$$\theta_i(t+dt) = \langle \theta_j \rangle + \eta(t) \tag{11.14}$$

$$\vec{r}_i(t+dt) = \vec{r}_i(t) + \vec{v}_i(t)dt \tag{11.15}$$

$$\vec{v}_i(t+dt) = v_0 \hat{u}(\theta_i(t+dt)), \tag{11.16}$$

where the average $\langle\rangle$ is over all neighbors j, usually within a certain distance of i, but sometimes over a certain number of nearest neighbors of i. This averaging is not perfect, since it is subject to white noise $\eta(t)$ of a certain strength. The orientation angle θ gives the direction $\hat{u}$ of the velocity vector. Such models generally find, provided the noise is not too strong, that a sufficiently dense, random collection of particles will spontaneously order so that all particles eventually move in the same direction [116, 117]. More complicated swarming models introduce attractive forces (to produce clustering) and/or repulsive forces (to prevent collisions) [118, 119]. Since the physically appropriate inter-particle potential is usually unknown, it is typically chosen for computational convenience, though data on real-world flocking, which may ultimately be used to infer a more reasonable form for the interaction force, are becoming available [120]. These models were inspired by the collective motion of schools of fish and flocks of birds; in these higher organisms, individual fish or birds presumably align themselves with their neighbors deliberately. In contrast, bacteria have no way to sense and respond to nearby cells, so interactions between bacteria are limited to hydrodynamic coupling through the fluid in which they all swim.[6] Though such an interaction is ultimately governed by low-Reynolds-number hydrodynamics, the force exchanged between many geometrically accurate representations of swimming bacteria has not yet been calculated with any certainty. Steric repulsion prevents the cell bodies from occupying the same space at the same time, but flagella are thin and flexible enough that they do not physically exclude other cells. Under most circumstances the repulsion is very short-range; although it presumably has some short distance scale, since it is mediated by the fluid in which the cells swim, for practical purposes it seems equivalent to a hard-core repulsion. There is no direct evidence of an attractive force between swarming bacteria,[7] though this has not been tested rigorously. In many ways, bacterial swarms are probably more similar to nonliving systems ('self-propelled' rods with hard-core interactions [122, 123]) than to birds [120], fish [124, 126], locusts [127, 128], or herd animals [129]; for reviews, see Toner *et al.* [130] and Giardina [131].

The long-range order of two-dimensional swarms is generally predicted to be unstable [132–135]. Although, as mentioned above, cells' velocities spontaneously align locally, the direction of alignment is unstable over long distances. This leads to large-scale swirling patterns and to anomalous diffusion [122, 136]. Even if the underlying hydrodynamics is the same, interactions in two dimensions may differ fundamentally from ones in three dimensions because the presence of a surface changes not just the strength but the scaling of the interaction.[8] To date there is no experimental system that can transition continuously from three dimensions to two dimensions to directly observe the importance of this hydrodynamic screening.

Many swarming models have a self-sustained vortex as a possible solution. Such vortices are found in several bacterial systems, including the eponymous *Bacillus*

[6] Quorum sensing can detect the total concentration of bacteria, and over larger length and time scales bacteria can excrete chemoattractants and perform chemotaxis in response. These behaviors, however, do not act on the sub-second timescale typical for realignment within bacterial swarms. In other contexts, more sophisticated cell–cell interactions are possible; for instance, fimbriae can be used for selective cell–cell adhesion in bacteria, and directional sensing is clearly possible in more slowly evolving situations, such as cell migration during embryonic development.

[7] In contrast, there *are* measurable correlations between pairs of cells swimming in bulk fluid [121], presumably due to an effective interaction potential. Similar effects are likely to occur during locomotion near a surface, as occurs in swarming.

[8] For simple geometries, the interaction between an object and a surface (and its attendant no-slip boundary condition) is mathematically equivalent to the interaction with an image object located on the other side of the surface. Just as in electrostatics, where this image charge steepens the power law dependence of the electric field from a point charge, this causes flow fields from the object to die off more rapidly with distance.

circulans [137] and *Paenibacillus vortex* [138], as well as in fish keratocytes [139] and in three dimensions in the zooplankton *Daphnia*. One can see a macroscopic example of this phenomenon from the viewing areas above the large tank in any major aquarium, where schools of fish continuously circle the inner wall. In order to create a cohesive swarm with a well-defined edge, the constituent particles must be prevented from wandering off. As mentioned above, this can be accomplished mathematically by adding an attractive force between particles, but in real bacterial swarms the system is more likely to be confined by geometry. Swarming bacteria move in the thin layer of fluid above the agar; the boundary of the swarm is defined by the edge of the wetting layer. Cells that attempt to swim onto insufficiently wet agar stall, and usually remain stranded until the expanding swarm rescues them. This is a less extreme example of the phenomenon of bacterial rafts in *Proteus* swarming: isolated swarming cells are nonmotile, but pairs of adjacent cells can move and large rafts of cells are the fastest. The local swirling structure predicted by self-propelled particle models of swarming should be relatively insensitive to either a large-scale attractive potential or to distant boundaries, but a full description of colony evolution would need to include the boundary conditions. This likely involves modeling the lubrication layer and the convective effects of motile swarmer cells, patterns of surfactant production, and the interaction between wetting agents and the agar undersurface on the spread of the lubrication layer.

Proteus swarming has the unique added feature of synchronized differentiation. The times and distances associated with the formation of successive terraces have been measured experimentally [140]. *Proteus*-specific growth models attempt to reproduce this behavior by, for example, adding an inherent time constant to the equations relating diffusion, growth, and interconversion of the two *Proteus* phenotypes [140–143]. The ring-like morphology of *Bacillus* colonies involves coordinated swarming and consolidation phases, a less synchronized version of the phenomenon in *Proteus*. In both *Bacillus* and *Proteus* the cycle time is independent of nutrient concentration [54, 144, 145] (as expected for growth on rich media where dynamics are limited by essential cell processes rather than by uptake of nutrients) and the cycle phase cannot be transported by individual cells (indicating that there is no fundamental biological clock involved), leading to models that waves of differentiation are triggered by local cell density thresholds, and converted to characteristic times via the cell growth rate.

11.3.4 Comments

The Keller–Segel Equations ((11.1)–(11.3) and variants) are one of the most extensively studied sets of equations in mathematical biology. The preceding sections give a few examples of how they have been applied to understanding colony growth and shape in the particular case of bacteria growing on agar surfaces. This is meant to give a sense of the types of approaches that have been used to model these phenomena, with more or less success and more or less contact with the underlying biology. The literature in this area is vast and continually increasing, so the examples chosen here are often the earliest, simplest, or most closely linked to real experimental conditions, rather than the latest or most sophisticated. Currently, the theoretical effort seems to have outpaced experimental confirmation; in the following section we present experimental techniques that are, can, or should be used to produce more and different data to confront models.

11.4 Detection

As illustrated in the previous sections, bacteria on agar plates produce highly diverse structures in response to simple variations in agar concentration and nutrient level. This range of behavior can only be explained by the bacteria responding to external conditions in nontrivial ways. That is, cells develop 'internal states' that depend on their history of environmental exposures, not merely on the instantaneous values of local nutrients, temperatures, etc. [79]. This bacterial memory is encoded as stable, altered patterns of gene expression that ultimately produce a spatial or temporal pattern of different phenotypes in different parts of the population. The most extreme examples of this phenomenon are sporulation in the interior of *Bacillus* colonies and differentiation into swarmer cells at the edge of swarming colonies of *Bacillus* and related species. A small number of authors have directly measuring gene expression in concert with colony development. For instance, Wang *et al.* [147] have looked at differential gene expression in swarming *Salmonella* by mechanically removing edge cells and assaying them; Kim and Surette [73] have shown that differentiation persists even once swarm-promoting conditions are removed; while Mendelson and Salhi [48] have indirectly looked at gene expression near the DBM morphology phase boundary in *B. subtilis*. However, many sophisticated techniques for mapping physiological states are available, often in the biofilm community [148], which are only starting to be applied to physiological patterns in colony growth; these will be addressed in Section 11.4.2 below.

11.4.1 Optical microscopy

Historically, optical microscopy has most often been used to simply report the physical extent of colony growth. This ignores subtleties in colony structure, such as varying patterns of density or motility. Many colony models predict colony structure, either implicitly through cell surface density or explicitly as colony height. The integrated surface density can be inferred from absorbance measurements using transmission microscopy [149], and confocal microscopy can measure the colony height directly [150, 151]. The global colony shape affects the small-angle light scattering intensity; this was used to rapidly distinguish *Listeria* colony shapes [152, 153]. While many colony models assume two-dimensional colony growth, in practice variations in colony height are not always negligible; see Pipe and Grimson [101] for a review of scaling laws for radial growth and height profiles of compact colonies. In particular, oxygen can be scarce in the interior of a thick colony. In organisms other than bacteria, growth in mounds is limited by diffusive access to nutrients and oxygen, which dictates the height profile as a function of radius; for example, Kamath and Bungay [154] have demonstrated this in yeast.

The motion of individual bacteria in dilute solutions can be tracked optically on a cell-by-cell level. The run-tumble description of chemotaxis was originally developed from this type of data [155]. More recently, higher-throughput methods have been developed that can track many chemotaxing bacteria simultaneously: the two-dimensional locations of a large number of cells are recorded by video camera, and the third dimension (vertical distance from the focal plane) is reconstructed from the defocusing of each cell; this method can follow around 100 cells simultaneously [156]. Cells are tracked using a particle tracking velocimetry algorithm, so successive position measurements can be reliably assigned to the proper cells only if the distance moved is small compared to the inter-particle distance. This process breaks down at high cell density and high cell speeds: tracking individual cells in a dense, highly motile colony (like a swarm) is very challenging. If one is willing to give up single-cell statistics, the overall motion can be measured at lower position resolution using particle imaging velocimetry (Chapter 7), as illustrated in Figure 11.5. This has been employed in swarming *Serratia marcescens* [157] or *P. vortex* [138] and in self-concentrated 3D motion of *B. subtilis* [158]. In these experiments, the 'particles' imaged were actually cells, either unlabeled or fluorescently labeled. In traditional tracking and imaging velocimetry, the fluid flow is marked with tracer particles, which are usually fluorescent microspheres; in biological contexts, this kind of tracer technique can be used in larger systems [159] and was originally used to track the swirling motion of swarming *B. subtilis* by hand [160]. Because of the widespread use of beads as tracers, many tracking algorithms implicitly assume that the particles being tracked are radially symmetric, so images are usually querying against a library of radially symmetric patterns. Although most bacteria have length:width ratios of 4:1 or higher, in practice standard tracking algorithms generally perform relatively well. High-density tracking could be improved by incorporating information about the shape of the tracked object into the tracking algorithm, something which is not currently standard practice.

At high bacterial densities or low magnifications, when it is impossible or impractical to resolve single cells, a labeled subpopulation can be introduced. This subpopulation can be tracked as in a low-density experiment. For instance,

Figure 11.5 Velocity field of swarming *E. coli*, measured by particle image velocimetry, showing the characteristic swirling pattern. The velocity vector was computed for spatial bins that contained several bacterial cells. The field of view was $90 \times 120\ \mu\text{m}$; the maximum swarm speed was around $100\ \mu\text{m/s}$. (Image courtesy of Rongjing Zhang.)

Mittal *et al.* labeled a fraction of chemotaxing *E. coli* using a green fluorescent protein marker [22]. Provided the expression level does not affect cell behavior, this gives a statistically valid description of cell motility of the colony, but at the expense of knowledge of the immediate environment of a tracked cell. Individual cells can be tracked by hand, though the process is very labor-intensive, by using cell size, shape, and continuity of direction as additional clues to cell identity. Automating this process is difficult without very good picture quality, which is compromised by imaging through a turbid agar plate. A small data set of hand-tracked cells gave correlations between cells' velocities; these correlations extended for a shorter distance ahead of a cell (where 'front' is defined as the direction of motion) than behind it, presumably because the cell has had more opportunities to interact, and align, with cells in the trailing direction (Figure 11.6(a)). This correlation length is consistent with the $9\ \mu\text{m}$ correlation length, averaged over all orientation, measured by Steager *et al.* [157]. In a swarm-like thin film whose bacterial density they can

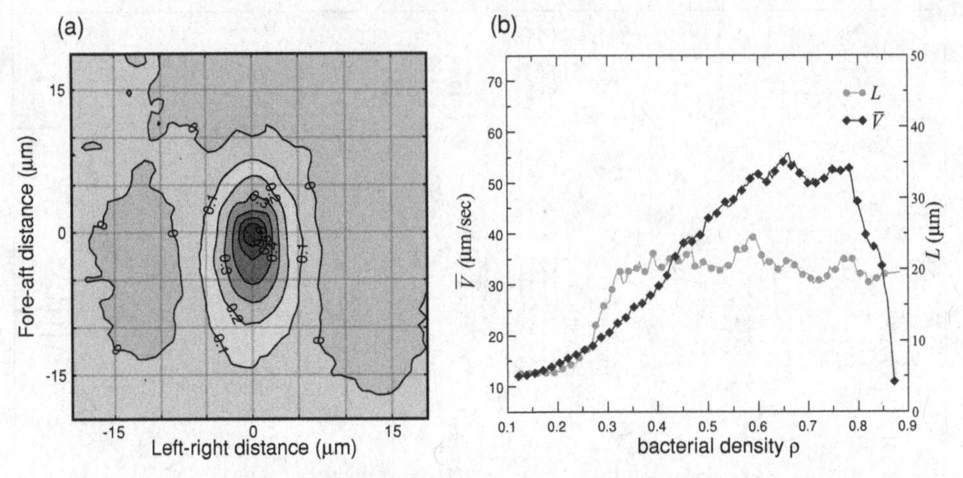

Figure 11.6 (a) Velocity–velocity correlation (the average of the cosine of the angle between two cells' velocities) as a function of the separation between two cells, for swarming *E. coli*. The axes are aligned locally with the major body axis of one of the two cells. (b) Mean swimming speed and velocity–velocity correlation length (averaged over all orientations) for *B. subtilis* swimming in a thin film, as a function of concentration fraction ρ [55]. $\rho = 1$ is the close-packed limit. ((b) reprinted from *Phys. Rev. Lett.* **98**, 158102 (2007) with permission from the American Physical Society.)

control, Sokolov *et al.* map correlation as a function of density; at high density they report a 20 μm correlation length, which is not inconsistent with the previously quoted measurements considering that it is likely that less hydrodynamic screening occurs in a thin film than on a swarm plate [55]. At high packing fractions, they observe an abrupt drop in mean cell speed caused by jamming (see Chapter 2).

11.4.2 Fluorescent reporters

Fluorescent proteins (FPs) are versatile, convenient, endogenously manufactured *in vivo* cellular labels. They can be expressed constitutively, so that they simply make the cell easier to see, or they can be expressed in tandem with a gene of interest, so that their presence indicates the level of the gene's expression. Fluorescing markers in general have much higher signal to noise ratios than simple pigments, and are more convenient that other expressible fluorescing systems (such as luciferase) because they require no substrate or cofactors. Variants of FPs are now available that span the entire optical spectrum, with a variety of Stokes shifts, fast folding times, and high quantum yields. It is possible to choose fluorescent proteins whose emission and excitation spectra overlap, so that they function as resonant energy transfer (FRET) pairs, as well as more exotic versions such as photo-activatable FPs [162]. At its simplest, the gene for a fluorescent protein can be introduced into a bacterial cell on a plasmid. The promoter that governs the gene's expression can be constitutively active (always on) or inducible (tunable by an externally supplied chemical). If the molecular biology is amenable, adding an FP on a plasmid is a straightforward and benign way to make a non-fluorescent cell visible. The only requirements are that it be possible to transfect the target cell, that the plasmid be stably maintained once it is introduced,[9] and that the FP not adversely affect the cell at expression levels that makes the fluorescent signal large compared to background. The FP tagging technique is often applied in the study of biofilms, communities of bacteria enmeshed in an excreted extracellular matrix. Biofilms are medically important because they can shield bacteria from antibiotics. They frequently incorporate more than one bacterial strain, and the interaction and communication between strains plays an important role in biofilm development, so the temporal and spatial distribution of species in a biofilm is of interest. Each bacterial strain is labeled with its own FP color before the biofilm is seeded; once the biofilm has formed, the spatial arrangement of FPs reveals where each population has flourished [163–166]. Many variations on this technique are possible; for instance, Rani *et al.* (*Staphylococcus* [148]) and Werner *et al.* (*Pseudomonas aeruginosa* [167]) use inducible FPs in conjunction with a pulse-labeling protocol. FP expression is induced after several days of biofilm growth; since only active cells take up the inducer and produce FP, a visual map of fluorescence distinguishes between active and dormant or dead cells.

Even within a single bacterial species, we expect that different regions of a bacterial colony should correspond to different patterns of gene expression. If the effect of a gene of interest is not directly observable, it may be possible to introduce an FP that is linked to the target gene; the level of FP fluorescence then acts as a visual signature of the expression of the target gene. For instance, Figure 11.7 shows a population of *B. subtilis* containing both vegetative (small) and swarmer (long) cells. The cells contain a red fluorescent protein that is regulated by the same promoter that regulates the main protein that comprises flagella; since flagellar synthesis is up-regulated in swarmer cells, the swarmer cells glow red, distinguishing the swarm phenotype among genetically identical cells.[10] Swarm differentiation is an abrupt, all-or-none phenomenon, but FP tagging can be also be used to monitor continuous variations in expression levels. Since colony growth

[9] This is traditionally accomplished by including antibiotic resistance on the plasmid and applying selection pressure by growing cells in the presence of antibiotic. Such a process is not always feasible under the desired experimental conditions; for instance, if the labeled cell is to be grown in a background of unlabeled, non-antibiotic-resistance cells.

[10] In this case, of course, swarmer and vegetative cells can be distinguished by the difference in their lengths, but the tagging principle is generally applicable, even to genes that have no effect that is visible under the microscope.

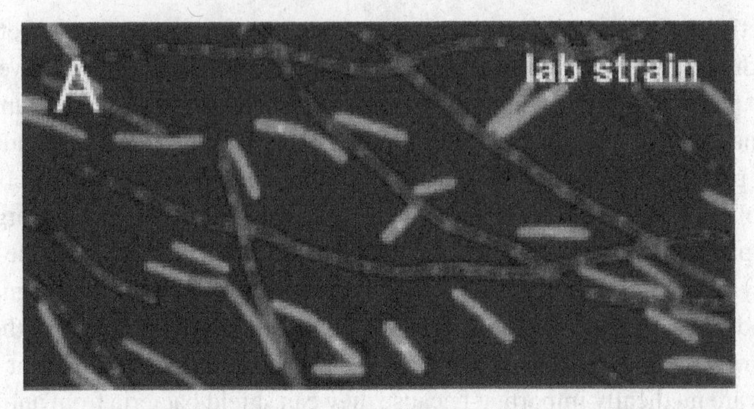

Figure 11.7 Vegetative cells (short) and swarmer cells (long) of *B. subtilis* [161]. The swarmer cells are labeled with a red fluorescent protein fused to the promoter for flagellin; since flagellin production is increased in swarmer cells, those cells produce red fluorescent protein which appears dim in this image, compared to the brighter green autofluorescence of vegetative cells. (Reprinted from *Genes & Dev.* **19**, p. 3083 (2005) with permission from Cold Spring Harbor Press.)

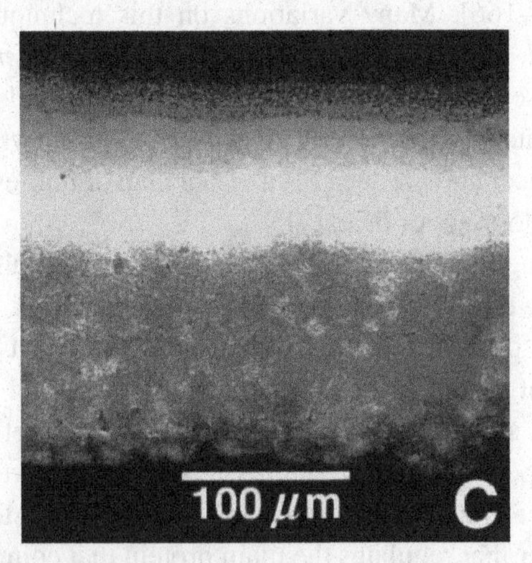

Figure 11.8 Stratified expression of rrnBp$_1$ (a growth-correlated gene) in a *P. aeruginosa* biofilm, as visualized by GFP fused to the rrnBp$_1$ promoter [167]. Reprinted from *Appl. Env. Microbiol.* **70**, p. 6188 (2004) with permission from the American Society for Microbiology.

is generally nonuniform, with more growth at the outside of the colony where resources such as nutrients and oxygen are more abundant, we expect metabolic activity to be higher near the edge of the colony. This is a relatively minor effect in the thin colonies discussed in Sections 11.3.2 and 11.3.3, but not so in thicker structures like biofilms. A typical experiment is shown in Figure 11.8, which reports

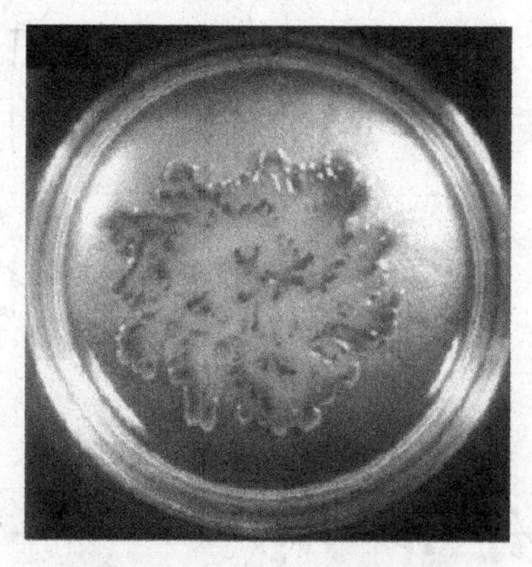

Figure 11.9 Expression pattern of motility-related genes in a colony of *B. subtilis* growing on high nutrient and moderately high agar plates [48]. The dark blue color, which is generally located near the edge of the colony, indicates high activity of *lacZ*, which has been transformed into the *Bacillus* chromosome. (Reprinted from *J. Bacteriol.* **178**, p. 1980 (1996) with permission from the American Society for Microbiology.)

the activity of a growth-correlated gene in a cross section of a *P. aeruginosa* biofilm. The gene's promoter was fused to an FP tag[11] so the FP level is proportional to the activation of that particular promoter, and therefore proportional to the expression level of the gene of interest [167].

A more traditional method of revealing expression patterns uses the lactase gene *lacZ* as a reporter. When *lacZ* is active, lactase is produced and excreted by cells, cleaving a colorless galactose analog in the medium and turning it blue. The color change reveals the pattern of expression of genes adjacent to *lacZ*, which can be inserted at a location of interest. Figure 11.9 shows a reporter that is expressed mainly at the tips of a growing colony of *B. subtilis*. The pattern of expression is relatively stable over time, indicating that there is little population exchange between the colony edge (dark) and interior (white).

11.4.3 Infrared and Raman spectroscopy

Nongenetic techniques can also be used to map spatial patterns of cell phenotype in a colony. Keirsse *et al.* use an infrared probe to map the swarmer cell distribution in *P. mirabilis* (Figure 11.10), showing that the swarmer phenotype is restricted to a narrow band on the outside of the colony [168]. They distinguish between swarmer and vegetative cells based on the differences in infrared absorbance [169].

[11] The FP gene was incorporated into the bacterial chromosome in order to avoid the possibility of losing the gene, which is always a risk if the gene is carried on a plasmid.

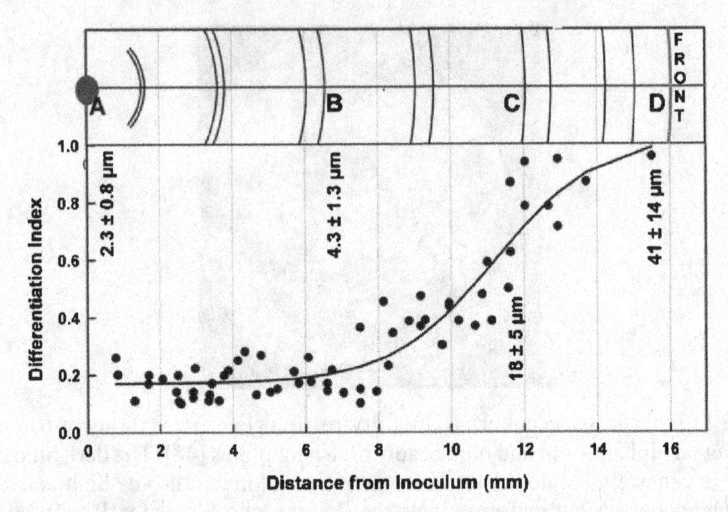

Figure 11.10 Degree of *P. mirabilis* cell differentiation as a function of radial position in a swarm, as determined by real-time infrared spectroscopy of the colony surface [168]. The ordinate is an estimate of the population fraction of swarmer cells, based on a decomposition of the observed infrared spectra into the characteristic spectra of swarming and stationary cells. Above the plot is a schematic depiction of the colony structure, showing six terraces and the swarm front, which has expanded 16 mm from the point of inoculation. The vertical numbers show the increase in average cell lengths, at four locations in the colony, as the fraction of swarmer cells increases. (Reprinted from *Appl. Spectros.* **60**, p. 584 (2006) with permission from the Society for Applied Spectroscopy.)

Spectral properties fortunately show large differences between swarmer and vegetative cells, an identifiable signature due to flagella, and a signal due to excreted 'slime' (lipopolysaccharide and exopolysaccharide); the respective spectra are determined empirically, by harvesting and assaying the appropriate cells or substances from a developing swarm. These data show that elongation (associated with changes in the cell membrane) occurs while slime production is increasing, but before the surge in flagellar synthesis. This type of information is available through genetic screens, but infrared spectroscopy has the advantage that, provided a useful signature is available, it does not require modification (chemical or genetic) of the organism or the addition of any labeling agents.

Similar types of information about colony structure can be obtained using Raman microspectroscopy. For instance, spatially resolved measurements (μm-size volume elements, defined confocally) showed measurable differences between the interior and surface of a colony after a few days' growth [170]; as with infrared spectroscopy, the characteristic spectra associated with each phenotype are determined empirically, in this case by using principal component analysis to construct maximally independent signatures. The spatial resolution is slightly larger than single cells, though confocal Raman spectroscopy has high enough sensitivity to classify individual cells [171]. This technique is typically used to identify medically important pathogenic bacteria, as reviewed in Maquelin *et al.* [172] and Harz *et al.* [173]. Since position resolution is generally coarser than the cell size, single-cell measurements are used on isolated cells rather than on cells in the middle of a dense colony [174].

Spectrally resolved confocal microscopy can be performed in the visual range as well. As with confocal Raman microscopy, the probed volume is several μm on a side, so in a close-packed bacterial colony single cells are not resolved. The spectral information from a fluorescently labeled subpopulation at each position is used to identify the target. The signatures used are similar to those employed in traditional flow cytometry. Of course, this technique requires either an external fluorescent label, an engineered endogenously produced fluorescent protein, or some easily identifiable autofluorescence.

A powerful, though indirect, tool for probing the important parameters that govern colony growth is to change the geometry of the agar plate and observe the consequences for the population pattern. Shimada *et al.* [145] investigated the mechanism of concentric ring formation in the ring-like colony morphology of *B. subtilis*. By transferring portions of swarms to new plates (either by cutting and pasting portions of the colony or by blotting cells from one surface to another), they show that bacterial density is the phase-determining factor in colony cycling; in particular, that there is no biological clock, consistent with the conclusions about *Proteus* swarming referenced in Section 11.4.4. Bacteria can be excluded from a geometrically defined section of the plate by applying a pattern of ultraviolet light. Depending on the intensity of the light, this can either kill or inhibit growth, and the light pattern can be moved to measure the colony's dynamic response to changing conditions [175, 176]. A nutrient gradient can be added to the agar plate by, for instance, inserting a small plug of high-nutrient agar into a low-nutrient plate [177].

11.4.4 Other techniques

Debois *et al.* used mass spectroscopy to map the concentration of surfactant produced by a developing *B. subtilis* colony [178]. The surface of the developing colony was blotted onto a silicon wafer, and its mass-to-charge ratio was measured using time-of-flight mass spectroscopy. Scanning the ion beam yielded a position resolution as small as two μm – essentially single-cell size. Figure 11.11 shows the total surfactant density; the chemical structure of the surfactins present in different parts of the colony can be inferred from the time of flight spectra (not shown). The highest surfactant concentration occurs at the center of the colony; apparently the

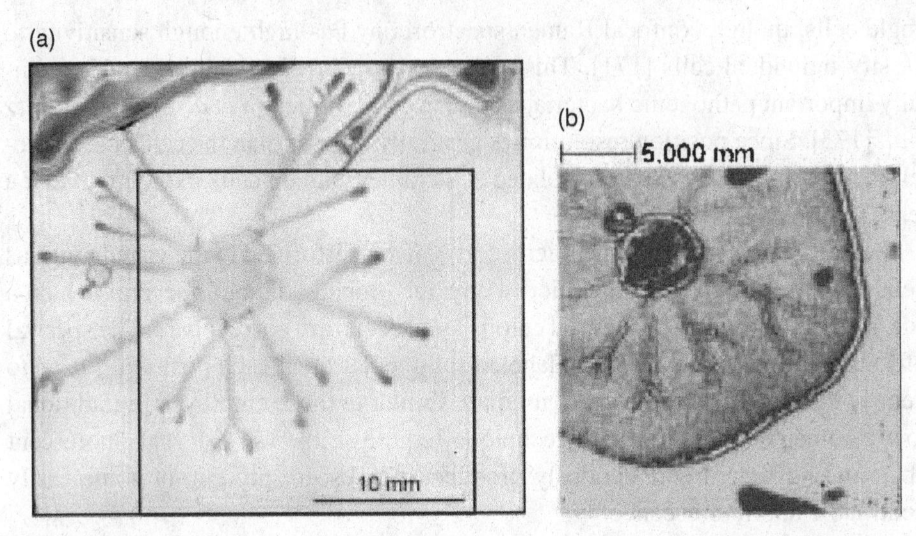

Figure 11.11 A section of a *B. subtilis* swarm colony (a) and its surfactin distribution (b) as mapped by spatially resolved mass spectrometry [178]. (Reprinted from *Proteomics* **8**, p. 3682 (2008) with permission from Wiley-VCH Verlag GmbH & Co.)

large number of cells in the thick center compensates for the fact that a substantial fraction of those cells are dormant.

Lahaye *et al.* [151] measured the microrheological properties of an intact *Proteus* swarm, finding significant differences between the leading edge of the swarm and its interior (at a given time), and periodic variations in the swarm over a cycle of expansion and consolidation (at a given location). They measured the swarm surface's strain in response to local stress produced by a 50-μm-sized puff of air. Based on microbiology, viscosity is clearly involved in triggering swarm differentiation [179, 180], but this conflicts with the molecular understanding of the flagellar motor, which has no known mechanism to sense load and therefore no way to sense the viscosity of the external medium. Based on the microrheological data, Lahaye *et al.* proposed that changes in the viscoelastic properties of the *Proteus* swarm front, caused by the production of exopolysaccharides and the absorption of water from agar, serve to coordinate differentiation and consolidation in the cell population. This explanation ascribes *Proteus* cycling to cells' response to varying viscoelasticity (which has been observed in many species, though its molecular mechanism is unknown) instead of to a synchronized biological clock (which would be specific to *Proteus*, but for which there is no direct evidence).

A typical bacterial cell is roughly the same size as the wavelength of light, so none of the previously mentioned techniques can resolve subcellular parts. Scanning electron microscopy (SEM), on the other hand, can image cell organelles

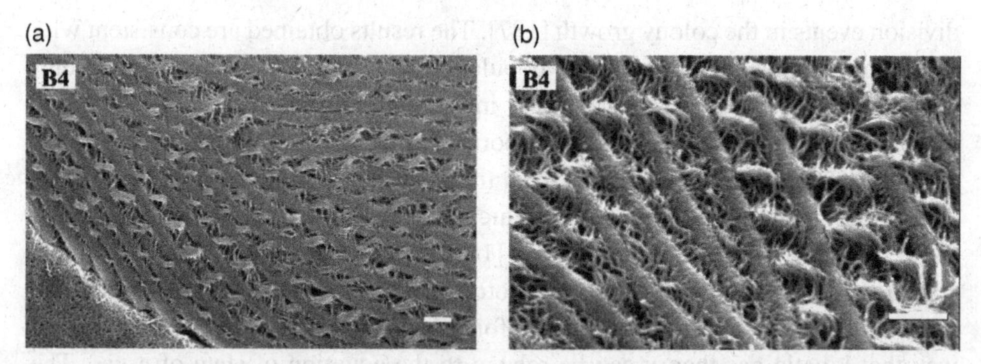

Figure 11.12 Scanning electron micrograph of fixed *P. mirabilis* obtained from a swarm [181]. Individual flagella are visible between the long cell bodies. most of the flagella have assembled in large bundles that appear to include flagella from neighboring cells. Scale bars are approximately 1 μm long. (Reprinted from *Inf. Immun.* **72**, p. 3941 (2004) with permission from the American Society for Microbiology.)

such as flagella and pili. Since it must be done in a vacuum, SEM requires the cell sample first be fixed and dried. This preparation process risks disrupting normal interactions between cells and introducing spurious ones. Nevertheless, it produces evocative images such as the massive flagellar bundles between *Proteus* cells shown in Figure 11.12.[12]

11.4.5 Single-cell measurements

Knowledge of bacterial behavior at the single-cell level will provide a rigorous basis for the constructing the rules of cellular automaton models and a way to calculate parameters in continuum models of population dynamics. Unfortunately, it is difficult to resolve individual cells while maintaining the crowded environment typical of colony growth, so many of the experimental characterizations of individual cell behavior must be obtained at lower population densities.

The statistics of cell division in *E. coli* and *Listeria innocua* have been recorded inside a flow cytometer that removes daughter cells [182], and these data have been used to model statistics of individual growth [183–185]. Under conditions more relevant for colony growth, individual divisions on agar were resolved through image analysis (though without single-cell precision imaging) to give lag time distributions in *E. coli*, *Listeria monocytogenes* and *P. aeruginosa* [186] division, and the distribution of *Listeria* colony sizes was used to infer lag times in the early

[12] When this technique was applied to *P. vortex*, it revealed 'entangled' flagella. Since these bacteria were motile before fixation, and since no such entanglement has been reported in other contexts, even at extremely high cell density, this apparent entanglement is probably an artifact of fixation [138].

division events in the colony growth [187]. The results obtained are consistent with those obtained for growth of dilute populations, but all of these are nonnutrient-limited situations that are relevant only in the high nutrient region of the colony morphology phase diagram (the upper portion of Figures 11.2(a) and (b)).

Most data for chemotaxis come from tracking the behavior of individual cells, either freely swimming [156, 188] (in which case the run-tumble statistics are measured directly), or attached to surfaces [189, 190] (in which case the run-tumble statistics must be inferred from the motor reversal probabilities). This detailed knowledge can be used to derive rules for population-wide behavior in the same way that kinetic gas theory can be related to the equation of state of a gas. The connection of individual cell behavior to a population-level chemotaxis equation has been made with increasing generality, starting from either particular single-cell-level distributions (for instance, experimentally motivated distributions are used in Patlak [191] and Alt [90]), or more general single-cell-level distributions (for instance, in Chubin *et al.* [192] and Erban and Othmer [96]). These papers show that the Keller–Segel equations (our Equations (11.1)–(11.3)) govern population dynamics in the long-time limit, and give explicit rules for calculating the bacterial diffusion coefficient D_b and the chemotactic coefficient $\chi(n)$ from microscopic data. Lewus and Ford experimentally verified the translation from single-cell statistics to population-scale chemotaxis in *E. coli* [193].

11.5 Comments

Invocation of chemotaxis in surface growth systems is appealing but dangerous [98]. Its appeal lies in the elegant mathematical descriptions of chemotaxis – it maps to a biased random walk with step lengths that depend on the chemoattractant concentration – but these results apply *only* to chemotaxis by swimming cells in bulk fluid. Microscopic observation of cells on surfaces shows that the run-tumble behavior underlying the theory of chemotaxis does not apply to swarming motility. It is unlikely to apply to *any* surface motility because viscous coupling to the surface will significantly impede active reorientation by cells, a necessary condition for conventional chemotaxis. Chemotaxis produces diffusion as well as transport, and diffusion is a major component of most of the processes described in this chapter, but the correspondence between diffusion coefficients in different contexts is unclear. In particular, cell motility becomes more and more constrained as locomotion goes from three-dimensional bulk swimming (Section 11.2.1) to two-dimensional surface swimming (Section 11.2.3) to two-dimensional growth-driven spreading (Section 11.2.2); the consequences for the effective bacterial diffusion coefficient are not well understood. Though the traditional features of swimming chemotaxis (random-walk trajectories with active modulation of step length in response to environment) are not observed in dense, two-dimensional populations of bacteria (as occur on all but the softest agar plates), nonetheless colonies can perform 'effective' chemotaxis towards nutrient sources, expanding more quickly towards a nutrient-rich region of an agar plate [29, 33]. This occurs even in systems where individual cells are unable to do chemotaxis [177].

Chemotactic-like response of a whole colony would occur if cell growth rates or swimming speeds changed in response to nutrient levels – a plausible though unproven effect – but neither of these mechanisms would legitimize the application the traditional biased random walk model of chemotaxis. Run-tumble chemotaxis produces a statistical drift along a chemical gradient; mathematically, this corresponds to both a bacterial diffusion term[13] D_b and a flux term $\vec{J}_{\text{ch}}$ in Equations (11.1) and (11.3). Some formulations conspicuously omit the flux term but retain the diffusion term, as in Equation (11.8): this can represent an implicit dependence on only a portion of chemotactic behavior. Indeed, successful swarming seems to require the random part of chemotaxis but not the directed part – nonchemotactic cells that do not reorient do not swarm, while similar cells that randomly change direction do swarm. Though chemotaxis is critical to the phenomenon of pattern formation, it is surprisingly unimportant in colony development: it is not necessary for successful swarming [194], or for the formation of ring-like colonies of *Bacillus* [145]. In simulations, nonchemotactic cells produced colonies with the same fractal structure as chemotactic cells, though they grew more slowly [195]. That chemotaxis is inhibited in dense colonies was noted some time ago; for example, Golding *et al.* [98] commented that chemotaxis is reduced in colonies, but that random-walk-like motility of cells is visible under the microscope. This observation implies that, although diffusion occurs, there is no net population transport, as mentioned above. In contrast with the long experimental history for chemotaxis in swimming bacteria [188, 196], experimental attention to the individual cell statistics of surface locomotion in bacterial colonies is just beginning.

From a physicist's point of view, differentiation of cells within a colony – and in particular, differentiation into a nonmotile form, as in the coupling between Equations (11.8) and (11.10) – is a mathematical device required to model colony structures. For a biologist, however, this differentiation foreshadows the complex body structures in higher organisms, and is therefore a central reason to study colony development. Empirically, differentiated forms (either dormant cells, spores, or swarmers) are important in colony development, but colony models' rules for differentiation are currently rather arbitrary; the differentiation model

[13] The bacterial diffusion coefficient, which we have called D_b, is often termed 'bacterial random motility' and designated $\mu(s)$. We have renamed it to emphasize the parallels between modeling of pattern formation (Section 11.2.1) and colony morphology (Section 11.2.2).

ought to adhere to known microbiology, though a complete description will require more biological information than is currently available. If the appearance, disappearance, and distribution of phenotypes are measured more accurately, the forces that drive these dramatic transformations in cell appearance, metabolism, and gene expression may become clear. Ultimately, this may lead to an understanding of the forces that drive epigenetic differentiation in eukaryotes as well.

Though the physics of low Reynolds number hydrodynamics is well understood,[14] its application to the complicated geometry of bacterial cells, and particularly to cells near surfaces, is still being developed. Most of the work is numerical, though there are some experimental results for cells in defined geometries [121] and for scaled-up mechanical models [200]. The locomotion of many interacting bacteria is clearly more complicated than the locomotion of a single, isolated cell; experimental observations of the two-body problem have revealed nonisotropic hydrodynamic coupling [121, 201]. Adding a boundary complicates the problem further. Some of the consequences are known: for instance, individual cells are attracted to surfaces [202] and swim in circles near a surface [203]. This can have macroscopic consequences for colony growth, leading to a chiral pattern of the colony itself [204]. A complete understanding of the low-Reynolds-number many-body problem, particularly near a surface, will allow us to better understand the properties of self-propelled bacteria moving on an agar plate.

Reaction-diffusion-chemotaxis models have increased in complexity as they attempt to explain more phenomena, from bacterial waves to chemotactic aggregation to swarming to variations in colony morphology. The theory has incorporated structures observed qualitatively on a plate (such as lubrication layers, zones of activity, and patterns of cell differentiation) or quantitatively in simplified environments (such as chemotactic responses and cell growth rates measured *in vitro*). Experimental techniques from molecular biology and microbiology can probe some of these features on an intact colony, delivering information beyond the early, basic measurements of population distributions in space and time. There are currently several models of colony evolution. As advanced techniques penetrate the experimental community and knowledge of patterns of differentiation and metabolic activity improves, including more accurate microscopic measurements of population structures and individual cell characteristics, it will become possible to distinguish between competing models.

[14] Viscoelastic fluids are considerably more complicated than Newtonian fluids, which have linear stress-strain relations. Most of the environments discussed in this chapter are reasonably Newtonian, with the possible exception of swarms that produce particularly large amounts of extracellular slime [197]. Viscoelastic fluids are known to change bacterial swimming behavior *in vitro* [198, 199], but the importance of this phenomenon on agar plates – or, indeed, more generally in biologically relevant environments – is unclear.

Acknowledgments

Swarming work included in the chapter was performed in the laboratory of Howard C. Berg at the Rowland Institute at Harvard, in collaboration with Howard Berg, Linda Turner and Svetlana Rojevskaya. Figures 11.4 and 11.5 were kindly supplied by Linda Turner and Rongjing Zhang, respectively.

References

[1] J. Henrichsen, "Bacterial surface translocation: a survey and a classification," *Bacteriol. Rev.* **36**, 478–503 (1972).
[2] M. Baker and R. M. Berry, "An introduction to the physics of the bacterial flagellar motor: a nanoscale rotary electric motor," *Contemp. Phys.* **50**, 617–32 (2009).
[3] Y. Sowa and R. M. Berry, "Bacterial flagellar motor," *Q. Rev. Biophys.* **41**, 103–32 (2008).
[4] H. Terashima, S. Kojima, and M. Homma, "Flagellar motility in bacteria structure and function of flagellar motor," *Int. Rev. Cell. Mol. Biol.* **270**, 39–85 (2008).
[5] A. Kitao, "Switch interactions control energy frustration and multiple flagellar filament structures," *Proc. Nat. Acad. Sci. USA* **103**, 4894–9 (2006).
[6] K. Yonekura, S. Maki-Yonekura, and K. Namba, "Complete atomic model of the bacterial flagellar filament by electron cryomicroscopy," *Nature* **424**, 643–50 (2003).
[7] K. Namba and F. Vonderviszt, "Molecular architecture of bacterial flagellum," *Q. Rev. Biophys.* **30**, 1–65 (1997).
[8] E. Lauga and T. R. Powers, "The hydrodynamics of swimming microorganisms," *Rep. Prog. Phys.* **72**, 096601 (2009).
[9] S. Chattopadhyay and X.-L. Wu, "The effect of long-range hydrodynamic interaction on the swimming of a single bacterium," *Biophys. J.* **96**, 2023–8 (2009).
[10] N. C. Darnton, L. Turner, S. Rojevsky, and H. C. Berg, "On torque and tumbling in swimming *Escherichia coli*," *J. Bacteriol.* **189**, 1756–64 (2007).
[11] S. Chattopadhyay, R. Moldovan, C. Yeung, and X. L. Wu, "Swimming efficiency of bacterium *Escherichia coli*," *Proc. Nat. Acad. Sci. USA* **103**, 13712–17 (2006).
[12] G. Hazelbauer, J. Falke, and J. Parkinson, "Bacterial chemoreceptors: high-performance signaling in networked arrays," *Trends Biochem. Sci.* **33**, 9–19 (2008).
[13] A. Vaknin and H. C. Berg, "Physical responses of bacterial chemoreceptors," *J. Mol. Biol.* **366**, 1416–23 (2007).
[14] G. H. Wadhams and J. P. Armitage, "Making sense of it all: bacterial chemotaxis," *Nat. Rev. Mol. Cell Biol.* **5**, 1024–37 (2004).
[15] V. Sourjik, "Receptor clustering and signal processing in *E. coli* chemotaxis," *Trends Microbiol.* (2004).
[16] J. Adler, "Chemotaxis in bacteria," *Science* **153**, 708–16 (1966).
[17] E. O. Budrene and H. C. Berg, "Complex patterns formed by motile cells of *Escherichia coli*," *Nature* **349**, 630–3 (1991).
[18] R. Nossal, "Growth and movement of rings of chemotactic bacteria," *Exp. Cell Res.* **75**, 138–42 (1972).
[19] J. Adler, "Effect of amino acids and oxygen on chemotaxis in *Escherichia coli*," *J. Bacteriol.* **92**, 121–9 (1966).

[20] E. O. Budrene and H. C. Berg, "Dynamics of formation of symmetrical patterns by chemotactic bacteria," *Nature* **376**, 49–53 (1995).
[21] M. P. Brenner, L. S. Levitov, and E. O. Budrene, "Physical mechanisms for chemotactic pattern formation by bacteria," *Biophys. J.* **74**, 1677–93 (1998).
[22] N. Mittal, E. O. Budrene, M. P. Brenner, and A. V. Oudenaarden, "Motility of *Escherichia coli* cells in clusters formed by chemotactic aggregation," *Proc. Nat. Acad. Sci. USA* **100**, 13259–63 (2003).
[23] Y. Blat and M. Eisenbach, "Tar-dependent and -independent pattern formation by *Salmonella typhimurium*," *J. Bacteriol.* **177**, 1683–91 (1995).
[24] D. E. Woodward, R. Tyson, M. R. Myerscough, J. D. Murray, E. O. Budrene, and H. C. Berg, "Spatio-temporal patterns generated by *Salmonella typhimurium*," *Biophys. J.* **68**, 2181–9 (1995).
[25] D. Emerson, "Complex pattern formation by *Pseudomonas* strain kc in response to nitrate and nitrite," *Microbiology* **145**, 633–41 (1999).
[26] M. Matsushita, J. Wakita, H. Itoh, I. Rafols, T. Matsuyama, H. Sakaguchi, and M. Mimura, "Interface growth and pattern formation in bacterial colonies," *Physica A* **249**, 517–24 (1998).
[27] R. Tokita, T. Katoh, Y. Maeda, J.-I. Wakita, M. Sano, T. Matsuyama, and M. Matsushita, "Pattern formation of bacterial colonies by *Escherichia coli*," *J. Phys. Soc. Jpn* **78**, 074005 (2009).
[28] M. Ohgiwari, M. Matsushita, and T. Matsuyama, "Morphological-changes in growth phenomena of bacterial colony patterns," *J. Phys. Soc. Jpn* **61**, 816–22 (1992).
[29] M. Matsushita and H. Fujikawa, "Diffusion-limited growth in bacterial colony formation," *Physica A* **168**, 498–506 (1990).
[30] T. Vicsek, M. Cserzo, and V. Horvath, "Self-affine growth of bacterial colonies," *Physica A* **167**, 315–321.
[31] H. Fujikawa, "Periodic growth of *Bacillus subtilis* colonies on agar plates," *Physica A* **189**, 15–21 (1992).
[32] H. Fujikawa and M. Matsushita, "Bacterial fractal growth in the concentration field of nutrient," *J. Phys. Soc. Jpn* **60**, 88–94 (1991).
[33] H. Fujikawa and M. Matsushita, "Fractal growth of *Bacillus subtilis* on agar plates," *J. Phys. Soc. Jpn* **58**, 3875–8 (1989).
[34] H. Fujikawa, "Diversity of the growth patterns of *Bacillus subtilis* colonies on agar plates," *FEMS Microbiol. Ecol.* **13**, 159–67 (1994).
[35] I. Rafols, "Formation of concentric rings in bacterial colonies," Chuo University (1998).
[36] T. Matsuyama, M. Sogawa, and Y. Nakagawa, "Fractal spreading growth of *Serratia marcescens* which produces surface active exolipids," *FEMS Microbiol. Lett.* **61**, 243–6 (1989).
[37] T. Matsuyama and M. Matsushita, "Self-similar colony morphogenesis by gram-negative rods as the experimental model of fractal growth by a cell population," *Appl. Environ. Microbiol.* **58**, 1227–32 (1992).
[38] T. Matsuyama and M. Matsushita, "Fractal morphogenesis by a bacterial cell population," *Crit. Rev. Microbiol.* **19**, 117–35 (1993).
[39] A. Nakahara, Y. Shimada, J. Wakita, M. Matsushita, and T. Matsuyama, "Morphological diversity of the colony produced by bacteria *Proteus mirabilis*," *J. Phys. Soc. Jpn* **65**, 2700–6 (1996).
[40] I. Das, A. Kumar, and U. Singh, "Nonequilibrium growth of *Klebsiella ozaenae* on agar plates," *Indian J. Chem. A* **36**, 197–200 (1997).

[41] I. Das, A. Kumar, and U. Singh, "Dynamic instability and non-equilibrium patterns during the growth of *E. coli*," *Indian J. Chem. A* **36**, 1018–22 (1997).

[42] M. Tcherpakov, E. Ben-Jacob, and D. L. Gutnick, "*Paenibacillus dendritiformis* sp. nov., proposal for a new pattern-forming species and its localization within a phylogenetic cluster," *Int. J. Syst. Bacteriol.* **49**, 239–46 (1999).

[43] M. Ruzicka, M. Fridrich, and M. Burkhard, "A bacterial colony is not self-similar," *Physica A* **216**, 382–5 (1995).

[44] M. Obert, P. Pfeifer, and M. Sernetz, "Microbial growth patterns described by fractal geometry," *J. Bacteriol.* **172**, 1180–5 (1990).

[45] S. Tang, Y. Ma, and I. Sebastine, "The fractal nature of *Escherichia coli* biological flocs," *Colloids and Surfaces B: Biointerfaces* **20**, 211–18 (2001).

[46] M. Eden, "A two-dimensional growth process," *Proc. Berkeley Symp. Math. Stat. Prob.* **4**, 233 (1961).

[47] P. W. Lindum, U. Anthoni, C. Christophersen, L. Eberl, S. Molin, and M. Givskov, "N-acyl-l-homoserine lactone autoinducers control production of an extracellular lipopeptide biosurfactant required for swarming motility of *Serratia liquefaciens* mg1," *J. Bacteriol.* **180**, 6384–8 (1998).

[48] N. H. Mendelson and B. Salhi, "Patterns of reporter gene expression in the phase diagram of *Bacillus subtilis* colony forms," *J. Bacteriol.* **178**, 1980–9 (1996).

[49] D. B. Kearns and R. Losick, "Swarming motility in undomesticated *Bacillus subtilis*," *Mol. Microbiol.* **49**, 581–90 (2003).

[50] D. Julkowska, M. Obuchowski, I. B. Holland, and S. J. Séror, "Comparative analysis of the development of swarming communities of *Bacillus subtilis* 168 and a natural wild type: critical effects of surfactin and the composition of the medium," *J. Bacteriol.* **187**, 65–76 (2005).

[51] R. M. Harshey and T. Matsuyama, "Dimorphic transition in *Escherichia coli* and *Salmonella typhimurium*: surface-induced differentiation into hyperflagellate swarmer cells," *Proc. Nat. Acad. Sci. USA* **91**, 8631–5 (1994).

[52] M.-P. Zorzano, D. Hochberg, M.-T. Cuevas, and J.-M. Gómez-Gómez, "Reaction-diffusion model for pattern formation in *E. coli* swarming colonies with slime," *Phys. Rev. E* **71**, 031908 (2005).

[53] R. Daniels, J. Vanderleyden, and J. Michiels, "Quorum sensing and swarming migration in bacteria," *FEMS Microbiol. Rev.* **28**, 261–89 (2004).

[54] O. Rauprich, M. Matsushita, C. J. Weijer, F. Siegert, S. E. Esipov, and J. A. Shapiro, "Periodic phenomena in *Proteus mirabilis* swarm colony development," *J. Bacteriol.* **178**, 6525–38 (1996).

[55] A. Sokolov, I. S. Aranson, J. O. Kessler, and R. E. Goldstein, "Concentration dependence of the collective dynamics of swimming bacteria," *Phys. Rev. Lett.* **98**, 158102 (2007).

[56] X. L. Wu and A. Libchaber, "Particle diffusion in a quasi-two-dimensional bacterial bath," *Phys. Rev. Lett.* **84**, 3017–20 (2000).

[57] L. L. McCarter, "Polar flagellar motility of the *Vibrionaceae*," *Microbiol. Mol. Biol. Rev.* **65**, 445–62 (2001).

[58] S. M. Kirov, "Bacteria that express lateral flagella enable dissection of the multifunctional roles of flagella in pathogenesis," *FEMS Microbiol. Lett.* **224**, 151–9 (2003).

[59] L. Cisneros, C. Dombrowski, R. E. Goldstein, and J. O. Kessler, "Reversal of bacterial locomotion at an obstacle," *Phys. Rev. E* **73**, 030901 (2006).

[60] S. Mariconda, Q. Wang, and R. M. Harshey, "A mechanical role for the chemotaxis system in swarming motility," *Mol. Microbiol.* **60**, 1590–602 (2006).

[61] J. P. Armitage, T. P. Pitta, M. A.-S. Vigeant, H. L. Packer, and R. M. Ford, "Transformations in flagellar structure of *Rhodobacter sphaeroides* and possible relationship to changes in swimming speed," *J. Bacteriol.* **181**, 4825–33 (1999).
[62] T. R. Powers, "Role of body rotation in bacterial flagellar bundling," *Phys. Rev. E* **65**, 040903 (2002).
[63] C. Takahashi, T. Nozawa, T. Tanikawa, Y. Nakagawa, J. Wakita, M. Matsushita, and T. Matsuyama, "Swarming of *Pseudomonas aeruginosa* pao1 without differentiation into elongated hyperflagellates on hard agar minimal medium," *FEMS Microbiol. Lett.* **280**, 169–75 (2008).
[64] T. Sams, K. Sneppen, M. Jensen, C. Ellegaard, B. Christensen, and U. Thrane, "Morphological instabilities in a growing yeast colony: experiment and theory," *Phys. Rev. Lett.* **79**, 313–16 (1997).
[65] D. Julkowska, M. Obuchowski, I. B. Holland, and S. J. Séror, "Branched swarming patterns on a synthetic medium formed by wild-type *Bacillus subtilis* strain 3610: detection of different cellular morphologies and constellations of cells as the complex architecture develops," *Microbiology* **150**, 1839–49 (2004).
[66] T. P. Robinson, J. W. Wimpenny, and R. G. Earnshaw, "pH gradients through colonies of *Bacillus cereus* and the surrounding agar," *J. Gen. Microbiol.* **137**, 2885–9 (1991).
[67] A. Ouvry, R. Cachon, and C. Divies, "Application of microelectrode technique to measure pH and oxidoreduction potential gradients in gelled systems as model food," *Biotechnol. Lett.* **23**, 1373–7 (2001).
[68] S. Walker, T. Brocklehurst, and J. Wimpenny, "The effects of growth dynamics upon pH gradient formation within and around subsurface colonies of *Salmonella typhimurium*," *J. Appl. Microbiol.* **82**, 610–14 (1997).
[69] J. W. Wimpenny and J. P. Coombs, "Penetration of oxygen into bacterial colonies," *J. Gen. Microbiol.* **129**, 1239–42 (1983).
[70] A. C. Peters, J. W. Wimpenny, and J. P. Coombs, "Oxygen profiles in, and in the agar beneath, colonies of *Bacillus cereus*, *Staphylococcus albus* and *Escherichia coli*," *J. Gen. Microbiol.* **133**, 1257–63 (1987).
[71] S. Belova, A. Dorofeev, and N. Panikov, "Growth and substrate utilization by bacterial lawn on the agar surface: experiment and one-dimensional distributed model," *Microbiology* **65**, 690–4 (1996).
[72] D. B. Kearns, F. Chu, R. Rudner, and R. Losick, "Genes governing swarming in *Bacillus subtilis* and evidence for a phase variation mechanism controlling surface motility," *Mol. Microbiol.* **52**, 357–69 (2004).
[73] W. Kim and M. G. Surette, "Metabolic differentiation in actively swarming *Salmonella*," *Mol. Microbiol.* **54**, 702–14 (2004).
[74] M. Saier, "Bacterial diversity and the evolution of differentiation," *ASM News* **66**, 337–43 (2000).
[75] J. A. Shapiro, "Thinking about bacterial populations as multicellular organisms," *Annu. Rev. Microbiol.* **52**, 81–104 (1998).
[76] C. Aguilar, H. Vlamakis, R. Losick, and R. Kolter, "Thinking about *Bacillus subtilis* as a multicellular organism," *Curr. Opin. Microbiol.* **10**, 638–43 (2007).
[77] E. P. Greenberg, "Bacterial communication: tiny teamwork," *Nature* **424**, 134 (2003).
[78] P. Stoodley, K. Sauer, D. G. Davies, and J. W. Costerton, "Biofilms as complex differentiated communities," *Annu. Rev. Microbiol.* **56**, 187–209 (2002).
[79] E. Ben-Jacob, "Social behavior of bacteria: from physics to complex organization," *Eur. Phys. J. B* **65**, 315–22 (2008).

[80] G. M. Odell and E. F. Keller, "Letter: traveling bands of chemotactic bacteria revisited," *J. Theor. Biol.* **56**, 243–7 (1976).
[81] G. M. Odell and E. F. Keller, "Necessary and sufficient conditions for chemotactic bands," *Math. Biosci.* **27**, 309–17 (1975).
[82] E. F. Keller and L. A. Segel, "Traveling bands of chemotactic bacteria: a theoretical analysis," *J. Theor. Biol.* **30**, 235–48 (1971).
[83] E. F. Keller and L. A. Segel, "Initiation of slime mold aggregation viewed as an instability," *J. Theor. Biol.* **26**, 399–415 (1970).
[84] K. Agladze, L. Budriene, G. Ivanitsky, V. Krinsky, V. Shakhbazyan, and M. Tsyganov, "Wave mechanisms of pattern formation in microbial populations," *Proc. Biol. Sci.* **253**, 131–5 (1993).
[85] J. Adler, "The sensing of chemicals by bacteria," *Sci. Am.* **234**, 40–7 (1976).
[86] S. Childress and J. K. Percus, "Nonlinear aspects of chemotaxis," *Math. Biosci.* **56**, 217–37 (1981).
[87] D. Horstmann, "From 1970 until present: the Keller-Segel model in chemotaxis and its consequences," *Jahresbericht der DMV* **105**, 103–65 (2003).
[88] T. Hillen and K. J. Painter, "A user's guide to PDE models for chemotaxis," *J. Math. Biol.* **58**, 183–217 (2009).
[89] R. Tyson, S. Lubkin, and J. Murray, "A minimal mechanism for bacterial pattern formation," *Proc. Nat. Acad. Sci. USA* **266**, 299–304 (1999).
[90] W. Alt, "Biased random walk models for chemotaxis and related diffusion approximations," *J. Math. Biol.* **9**, 147–77 (1980).
[91] A. Stevens, "The derivation of chemotaxis equations as limit dynamics of moderately interacting stochastic many-particle systems," *SIAM J. Appl. Math.* **61**, 183–212 (2000).
[92] M. Schnitzer, "Theory of continuum random walks and application to chemotaxis," *Phys. Rev. E* **48**, 2553–68 (1993).
[93] K. C. Chen, R. M. Ford, and P. T. Cummings, "The global turning probability density function for motile bacteria and its applications," *J. Theor. Biol.* **195**, 139–55 (1998).
[94] K. C. Chen, R. M. Ford, and P. T. Cummings, "Cell balance equation for chemotactic bacteria with a biphasic tumbling frequency," *J. Math. Biol.* **47**, 518–46 (2003).
[95] R. Erban and H. Othmer, "From signal transduction to spatial pattern formation in *E. coli*: A paradigm for multiscale modeling in biology," *Multiscale Modeling and Simulation* **3**, 362–94 (2005).
[96] R. Erban and H. Othmer, "From individual to collective behavior in bacterial chemotaxis," *SIAM J. Appl. Math.* **65**, 361–91 (2004).
[97] T. Emonet, C. M. Macal, M. J. North, C. E. Wickersham, and P. Cluzel, "Agentcell: a digital single-cell assay for bacterial chemotaxis," *Bioinformatics* **21**, 2714–21 (2005).
[98] I. Golding, Y. Kozlovsky, I. Cohen, and E. Ben-Jacob, "Studies of bacterial branching growth using reaction-diffusion models for colonial development," *Physica A* **260**, 510–54 (1998).
[99] K. Kawasaki, A. Mochizuki, M. Matsushita, T. Umeda, and N. Shigesada, "Modeling spatio-temporal patterns generated by *Bacillus subtilis*," *J. Theor. Biol.* **188**, 177–85 (1997).
[100] D. A. Kessler and H. Levine, "Fluctuation-induced diffusive instabilities," *Nature* **394**, 556–8 (1998).

[101] L. Z. Pipe and M. J. Grimson, "Spatial-temporal modelling of bacterial colony growth on solid media," *Molecular BioSystems* **4**, 192 (2008).
[102] Y. Kozlovsky, I. Cohen, I. Golding, and E. Ben-Jacob, "Lubricating bacteria model for branching growth of bacterial colonies," *Phys. Rev. E* **59**, 7025–35 (1999).
[103] J. Lega and T. Passot, "Hydrodynamics of bacterial colonies: phase diagrams," *Chaos* **14**, 562–70 (2004).
[104] J. Lega and T. Passot, "Hydrodynamics of bacterial colonies: a model," *Phys. Rev. E* **67**, 031906 (2003).
[105] J. Lega and T. Passot, "Hydrodynamics of bacterial colonies," *Nonlinearity* **20**, C1–C16 (2006).
[106] J. Y. Wakano, A. Komoto, and Y. Yamaguchi, "Phase transition of traveling waves in bacterial colony pattern," *Phys. Rev. E* **69**, 051904 (2004).
[107] M. Mimura, H. Sakaguchi, and M. Matsushita, "Reaction-diffusion modelling of bacterial colony patterns," *Physica A* **282**, 283–303 (2000).
[108] E. Ben-Jacob, H. Shmueli, O. Shochet, and A. Tenenbaum, "Adaptive self-organization during growth of bacterial colonies," *Physica A* **187**, 378–424 (1992).
[109] E. Ben-Jacob, O. Schochet, A. Tenenbaum, I. Cohen, A. Czirók, and T. Vicsek, "Generic modelling of cooperative growth patterns in bacterial colonies," *Nature* **368**, 46–9 (1994).
[110] E. Ben-Jacob, "From snowflake formation to growth of bacterial colonies. 2. cooperative formation of complex colonial patterns," *Contemp. Phys.* **38**, 205–41 (1997).
[111] M. Ginovart, D. Lopez, J. Valls, and M. Silbert, "Individual based simulations of bacterial growth on agar plates," *Physica A* **305**, 604–18 (2002).
[112] M. Badoual, P. Derbez, M. Aubert, and B. Grammaticos, "Simulating the migration and growth patterns of *Bacillus subtilis*," *Physica A* **388**, 549–59 (2009).
[113] T. A. Witten and L. M. Sander, "Diffusion-limited aggregation, a kinetic critical phenomenon," *Phys. Rev. Lett.* **47**, 1400–3 (1981).
[114] E. Ben-Jacob, "From snowflake formation to growth of bacterial colonies. 1. diffusive patterning in azoic systems," *Contemp. Phys.* **34**, 247–73 (1993).
[115] T. Vicsek, A. Czirók, E. Ben-Jacob, I. Cohen, and O. Shochet, "Novel type of phase transition in a system of self-driven particles," *Phys. Rev. Lett.* **75**, 1226–9 (1995).
[116] H. Chate, F. Ginelli, G. Gregoire, and F. Raynaud, "Collective motion of self-propelled particles interacting without cohesion," *Phys. Rev. E* **77**, 046113 (2008).
[117] F. Peruani, A. Deutsch, and M. Bär, "A mean-field theory for self-propelled particles interacting by velocity alignment mechanisms," *Eur. Phys. J. Spec. Top.* **74**, 030904 (2008).
[118] G. Gregoire, H. Chaté, and Y. Tu, "Moving and staying together without a leader," *Physica A* **181**, 157–70 (2003).
[119] H. Levine, W. Rappel, and I. Cohen, "Self-organization in systems of self-propelled particles," *Phys. Rev. E* **63**, 017101 (2000).
[120] M. Ballerini, N. Cabibbo, R. Candelier, A. Cavagna, E. Cisbani, I. Giardina, A. Orlandi, G. Parisi, A. Procaccini, M. Viale, and V. Zdravkovic, "Empirical investigation of starling flocks: a benchmark study in collective animal behaviour," *Anim. Behav.* **76**, 201–15 (2008).
[121] Q. Liao, G. Subramanian, M. P. Delisa, D. L. Koch, and M. Wu, "Pair velocity correlations among swimming *Escherichia coli* bacteria are determined by force-quadrupole hydrodynamic interactions," *Phys. Fluids* **19**, 061701 (2007).
[122] L. J. Daniels, Y. Park, T. Lubensky, and D. J. Durian, "Dynamics of gas-fluidized granular rods," *Phys. Rev. E* **79**, 041301 (2009).

[123] V. Narayan, S. Ramaswamy, and N. Menon, "Long-lived giant number fluctuations in a swarming granular nematic," *Science* **317**, 105–8 (2007).
[124] N. C. Makris, P. Ratilal, S. Jagannathan, and Z. Gong, "Critical population density triggers rapid formation of vast oceanic fish shoals," *Science* **323**, 1734–1737 (2009).
[125] B. Birnir, "An ODE model of the motion of pelagic fish," *J. Stat. Phys.* **128**, 535–568 (2007).
[126] C. Becco, N. Vandewalle, J. Delcourt, and P. Poncin, "Experimental evidences of a structural and dynamical transition in fish school," *Physica A* **367**, 487–93 (2006).
[127] J. Buhl, D. Sumpter, I. D. Couzin, J. J. Hale, E. Despland, E. R. Miller, and S. J. Simpson, "From disorder to order in marching locusts," *Science* **312**, 1402–6 (2006).
[128] L. Edelstein-Keshet, J. Watmough, and D. Grunbaum, "Do travelling band solutions describe cohesive swarms? an investigation for migratory locusts," *J. Math. Biol.* **36**, 515–49 (1998).
[129] S. Gueron, "The dynamics of herds: from individuals to aggregations," *J. Theor. Biol.* **182**, 85–98 (1996).
[130] J. Toner, Y. Tu, and S. Ramaswamy, "Hydrodynamics and phases of flocks," *Ann. Phys.* **318**, 170–244 (2005).
[131] I. Giardina, "Collective behavior in animal groups: theoretical models and empirical studies," *HFSP J.* **2**, 205–19 (2008).
[132] N. Sambelashvili, A. W. C. Lau, and D. Cai, "Dynamics of bacterial flow: emergence of spatiotemporal coherent structures," *Phys. Lett. A* **360**, 507–11 (2007).
[133] S. Sankararaman and S. Ramaswamy, "Instabilities and waves in thin films of living fluids," *Phys. Rev. Lett.* **102**, 118107 (2009).
[134] R. Simha and S. Ramaswamy, "Hydrodynamic fluctuations and instabilities in ordered suspensions of self-propelled particles," *Phys. Rev. Lett.* **89**, 058101 (2002).
[135] D. Saintillan and M. J. Shelley, "Instabilities and pattern formation in active particle suspensions: kinetic theory and continuum simulations," *Phys. Rev. Lett.* **100**, 178103 (2008).
[136] D. Saintillan and M. J. Shelley, "Orientational order and instabilities in suspensions of self-locomoting rods," *Phys. Rev. Lett.* **99**, 058102 (2007).
[137] A. Komoto, K. ichi Hanaki, S. Maenosono, J. Y. Wakano, Y. Yamaguchi, and K. Yamamoto, "Growth dynamics of *Bacillus circulans* colony," *J. Theor. Biol.* **225**, 91–7 (2003).
[138] C. J. Ingham and E. B. Jacob, "Swarming and complex pattern formation in *Paenibacillus vortex* studied by imaging and tracking cells," *BMC Microbiol.* **8**, 36 (2008).
[139] B. Szabó, G. J. Szöllösi, B. Gönci, Z. Jurányi, D. Selmeczi, and T. Vicsek, "Phase transition in the collective migration of tissue cells: Experiment and model," *Phys. Rev. E* **74** (2006).
[140] A. Czirok, M. Matsushita, and T. Vicsek, "Theory of periodic swarming of bacteria: application to *Proteus mirabilis*," *Phys Rev E* **63**, 031915 (2001).
[141] S. Esipov and J. Shapiro, "Kinetic model of *Proteus mirabilis* swarm colony development," *J. Math. Biol.* **36**, 249–68 (1998).
[142] B. P. Ayati, "A structured-population model of *Proteus mirabilis* swarm-colony development," *J. Math. Biol.* **52**, 93–114 (2006).
[143] S. Arouh, "Analytic model for ring pattern formation by bacterial swarmers," *Phys. Rev. E* **63**, 031908 (2001).

[144] J. Wakita, H. Shimada, H. Itoh, T. Matsuyama, and M. Matsushita, "Periodic colony formation by bacterial species *Bacillus subtilis*," *J. Phys. Soc. Jpn* **70**, 911 (2001).

[145] H. Shimada, T. Ikeda, J. Wakita, H. Itoh, S. Kurosu, F. Hiramatsu, M. Nakatsuchi, Y. Yamazaki, T. Matsuyama, and M. Matsushita, "Dependence of local cell density on concentric ring colony formation by bacterial species *Bacillus subtilis*," *J. Phys. Soc. Jpn* **73**, 1082–9 (2004).

[146] Itoh, J. Wakita, T. Matsuyama, and M. Matsushita, "Periodic pattern formation of bacterial colonies," *J. Phys. Soc. Jpn* **68**, 1436 (1999).

[147] Q. Wang, J. G. Frye, M. McClelland, and R. M. Harshey, "Gene expression patterns during swarming in *Salmonella typhimurium*: genes specific to surface growth and putative new motility and pathogenicity genes," *Mol. Microbiol.* **52**, 169–87 (2004).

[148] S. A. Rani, B. Pitts, H. Beyenal, R. A. Veluchamy, Z. Lewandowski, W. M. Davison, K. Buckingham-Meyer, and P. S. Stewart, "Spatial patterns of DNA replication, protein synthesis, and oxygen concentration within bacterial biofilms reveal diverse physiological states," *J. Bacteriol.* **189**, 4223–33 (2007).

[149] A. McKay, A. C. Peters, and J. Wimpenny, "Determining specific growth rates in different regions of *Salmonella typhimurium* colonies," *Lett. Appl. Microbiol.* **24**, 74–6 (1997).

[150] M. S. Mary, J. Gopal, B. V. R. Tata, T. S. Rao, and S. Vincent, "A confocal microscopic study on colony morphology and sporulation of *Bacillus* sp," *World. J. Microbiol. Biotechnol.* **24**, 2435–2 (2008).

[151] E. Lahaye, T. Aubry, V. Fleury, and O. Sire, "Does water activity rule *P. mirabilis* periodic swarming? ii. viscoelasticity and water balance during swarming," *Biomacromolecules* **8**, 1228–35 (2007).

[152] E. Bae, P. P. Banada, K. Huff, A. K. Bhunia, J. P. Robinson, and E. D. Hirleman, "Analysis of time-resolved scattering from macroscale bacterial colonies," *J. Biomed. Optic.* **13**, 014010 (2008).

[153] P. P. Banada, S. Guo, B. Bayraktar, E. Bae, B. Rajwa, J. P. Robinson, E. D. Hirleman, and A. K. Bhunia, "Optical forward-scattering for detection of *Listeria monocytogenes* and other *Listeria* species," *Biosens. Bioelectron.* **22**, 1664–71 (2007).

[154] R. S. Kamath and H. R. Bungay, "Growth of yeast colonies on solid media," *J. Gen. Microbiol.* **134**, 3061–9 (1988).

[155] H. C. Berg and D. Brown, "Chemotaxis in *Escherichia coli* analyzed by three-dimensional tracking. addendum," *Antibiot. Chemother.* **19**, 55–78 (1974).

[156] M. Wu, J. W. Roberts, S. Kim, D. L. Koch, and M. P. Delisa, "Collective bacterial dynamics revealed using a three-dimensional population-scale defocused particle tracking technique," *Appl. Environ. Microbiol.* **72**, 4987–94 (2006).

[157] E. B. Steager, C.-B. Kim, and M. J. Kim, "Dynamics of pattern formation in bacterial swarms," *Phys. Fluids* **20**, 073601 (2008).

[158] C. Dombrowski, L. Cisneros, S. Chatkaew, R. E. Goldstein, and J. O. Kessler, "Self-concentration and large-scale coherence in bacterial dynamics," *Phys. Rev. Lett.* **93**, 098103 (2004).

[159] M. B. Short, C. A. Solari, S. Ganguly, T. R. Powers, J. O. Kessler, and R. E. Goldstein, "Flows driven by flagella of multicellular organisms enhance long-range molecular transport," *Proc. Nat. Acad. Sci. USA* **103**, 8315–19 (2006).

[160] N. H. Mendelson, A. Bourque, K. Wilkening, K. R. Anderson, and J. C. Watkins, "Organized çell swimming motions in *Bacillus subtilis* colonies: patterns of short-lived whirls and jets," *J. Bacteriol.* **181**, 600–9 (1999).

[161] D. B. Kearns and R. Losick, "Cell population heterogeneity during growth of *Bacillus subtilis*," *Gene. Dev.* **19**, 3083–94 (2005).

[162] N. Shaner, P. Steinbach, and R. Tsien, "A guide to choosing fluorescent proteins," *Nat. Meth.* **2**, 905–9 (2005).

[163] H. Daims and M. Wagner, "Quantification of uncultured microorganisms by fluorescence microscopy and digital image analysis," *Appl. Microbiol. Biotechnol.* **75**, 237–48 (2007).

[164] G. V. Bloemberg, A. H. Wijfjes, G. E. Lamers, N. Stuurman, and B. J. Lugtenberg, "Simultaneous imaging of *Pseudomonas fluorescens* wcs365 populations expressing three different autofluorescent proteins in the rhizosphere: new perspectives for studying microbial communities," *Mol. Plant Microbe Interact.* **13**, 1170–6 (2000).

[165] C. Ramos, L. Mølbak, and S. Molin, "Bacterial activity in the rhizosphere analyzed at the single-cell level by monitoring ribosome contents and synthesis rates," *Appl. Environ. Microbiol.* **66**, 801–9 (2000).

[166] S. Møller, C. Sternberg, J. B. Andersen, B. B. Christensen, J. L. Ramos, M. Givskov, and S. Molin, "In situ gene expression in mixed-culture biofilms: evidence of metabolic interactions between community members," *Appl. Environ. Microbiol.* **64**, 721–32 (1998).

[167] E. Werner, F. Roe, A. Bugnicourt, M. J. Franklin, A. Heydorn, S. Molin, B. Pitts, and P. S. Stewart, "Stratified growth in *Pseudomonas aeruginosa* biofilms," *Appl. Environ. Microbiol.* **70**, 6188–96 (2004).

[168] J. Keirsse, E. Lahaye, A. Bouter, V. Dupont, C. Boussard-Plédel, B. Bureau, J.-L. Adam, V. Monbet, and O. Sire, "Mapping bacterial surface population physiology in real-time: infrared spectroscopy of *Proteus mirabilis* swarm colonies," *Appl. Spectros.* **60**, 584–91 (2006).

[169] M. Gué, V. Dupont, A. Dufour, and O. Sire, "Bacterial swarming: a biochemical time-resolved FTIR-ATR study of *Proteus mirabilis* swarm-cell differentiation," *Biochemistry* **40**, 11938–45 (2001).

[170] L. P. Choo-Smith, K. Maquelin, T. van Vreeswijk, H. A. Bruining, G. J. Puppels, N. A. N. Thi, C. Kirschner, D. Naumann, D. Ami, A. M. Villa, F. Orsini, S. M. Doglia, H. Lamfarraj, G. D. Sockalingum, M. Manfait, P. Allouch, and H. P. Endtz, "Investigating microbial (micro)colony heterogeneity by vibrational spectroscopy," *Appl. Environ. Microbiol.* **67**, 1461–9 (2001).

[171] K. C. Schuster, E. Urlaub, and J. R. Gapes, "Single-cell analysis of bacteria by Raman microscopy: spectral information on the chemical composition of cells and on the heterogeneity in a culture," *J. Microbiol. Meth.* **42**, 29–38 (2000).

[172] K. Maquelin, C. Kirschner, L.-P. Choo-Smith, N. van den Braak, H. P. Endtz, D. Naumann, and G. J. Puppels, "Identification of medically relevant microorganisms by vibrational spectroscopy," *J. Microbiol. Meth.* **51**, 255–71 (2002).

[173] M. Harz, P. Rösch, and J. Popp, "Vibrational spectroscopy – a powerful tool for the rapid identification of microbial cells at the single-cell level," *Cytometry A* **75**, 104–13 (2009).

[174] P. Rösch, M. Harz, M. Schmitt, K.-D. Peschke, O. Ronneberger, H. Burkhardt, H.-W. Motzkus, M. Lankers, S. Hofer, H. Thiele, and J. Popp, "Chemotaxonomic identification of single bacteria by micro-Raman spectroscopy: application to clean-room-relevant biological contaminations," *Appl. Environ. Microbiol.* **71**, 1626–37 (2005).

[175] A. M. Delprato, A. Samadani, A. Kudrolli, and L. S. Tsimring, "Swarming ring patterns in bacterial colonies exposed to ultraviolet radiation," *Phys. Rev. Lett.* **87**, 158102 (2001).

[176] T. Neicu, A. Pradhan, D. A. Larochelle, and A. Kudrolli, "Extinction transition in bacterial colonies under forced convection," *Phys. Rev. E* **62**, 1059–62 (2000).

[177] R. G. Taylor and R. D. Welch, "Chemotaxis as an emergent property of a swarm," *J. Bacteriol.* **190**, 6811–16 (2008).

[178] D. Debois, K. Hamze, V. Guérineau, J.-P. L. Caër, I. B. Holland, P. Lopes, J. Ouazzani, S. J. Séror, A. Brunelle, and O. Laprévote, "In situ localisation and quantification of surfactins in a *Bacillus subtilis* swarming community by imaging mass spectrometry," *Proteomics* **8**, 3682–91 (2008).

[179] L. McCarter, M. Hilmen, and M. Silverman, "Flagellar dynamometer controls swarmer cell differentiation of *V. parahaemolyticus*." *Cell* **54**, 345–51 (1988).

[180] L. McCarter and M. Silverman, "Surface-induced swarmer cell differentiation of *Vibrio parahaemolyticus*," *Mol. Microbiol.* **4**, 1057–62 (1990).

[181] B. V. Jones, R. Young, E. Mahenthiralingam, and D. J. Stickler, "Ultrastructure of *Proteus mirabilis* swarmer cell rafts and role of swarming in catheter-associated urinary tract infection," *Infect. Immun.* **72**, 3941–50 (2004).

[182] A. Elfwing, Y. LeMarc, J. Baranyi, and A. Ballagi, "Observing growth and division of large numbers of individual bacteria by image analysis," *Appl. Environ. Microbiol.* **70**, 675–8 (2004).

[183] A. Métris, Y. L. Marc, A. Elfwing, A. Ballagi, and J. Baranyi, "Modelling the variability of lag times and the first generation times of single cells of *E. coli*," *Int. J. Food Microbiol.* **100**, 13–19 (2005).

[184] Z. Kutalik, M. Razaz, A. Elfwing, A. Ballagi, and J. Baranyi, "Stochastic modelling of individual cell growth using flow chamber microscopy images," *Int. J. Food Microbiol.* **105**, 177–90 (2005).

[185] J. Baranyi, S. M. George, and Z. Kutalik, "Parameter estimation for the distribution of single cell lag times," *J. Theor. Biol.* **259**, 24–30 (2009).

[186] G. W. Niven, T. Fuks, J. S. Morton, S. A. C. G. Rua, and B. M. Mackey, "A novel method for measuring lag times in division of individual bacterial cells using image analysis," *J. Microbiol. Meth.* **65**, 311–17 (2006).

[187] L. Guillier, P. Pardon, and J.-C. Augustin, "Automated image analysis of bacterial colony growth as a tool to study individual lag time distributions of immobilized cells," *J. Microbiol. Meth.* **65**, 324–34 (2006).

[188] H. C. Berg and D. A. Brown, "Chemotaxis in *Escherichia coli* analysed by three-dimensional tracking," *Nature* **239**, 500–4 (1972).

[189] S. M. Block, J. E. Segall, and H. C. Berg, "Impulse responses in bacterial chemotaxis," *Cell* **31**, 215–26 (1982).

[190] J. Yuan, K. A. Fahrner, and H. C. Berg, "Switching of the bacterial flagellar motor near zero load," *J. Mol. Biol.* **390**, 394–400 (2009).

[191] C. Patlak, "Random walk with persistence and external bias," *Bull. Math. Biol.* **15**, 311–38 (1953).

[192] F. Chalub, P. Markowich, B. Perthame, and C. Schmeiser, "Kinetic models for chemotaxis and their drift-diffusion limits," *Monatsh. Math.* **142**, 123–41 (2004).

[193] P. Lewus and R. M. Ford, "Quantification of random motility and chemotaxis bacterial transport coefficients using individual-cell and population-scale assays," *Biotechnol. Bioeng.* **75**, 292–304 (2001).

[194] M. Burkart, A. Toguchi, and R. M. Harshey, "The chemotaxis system, but not chemotaxis, is essential for swarming motility in *Escherichia coli*," *Proc. Nat. Acad. Sci. USA* **95**, 2568–73 (1998).

[195] I. Cohen, A. Czirok, and E. BenJacob, "Chemotactic-based adaptive self-organization during colonial development," *Physica A* **233**, 678–98 (1996).

[196] H. C. Berg, "Bacterial behaviour," *Nature* **254**, 389–92 (1975).

[197] N. Verstraeten, K. Braeken, B. Debkumari, M. Fauvart, J. Fransaer, J. Vermant, and J. Michiels, "Living on a surface: swarming and biofilm formation," *Trends Microbiol.* **16**, 496–506 (2008).

[198] H. C. Berg and L. Turner, "Movement of microorganisms in viscous environments," *Nature* **278**, 349–51 (1979).

[199] S. Nakamura, Y. Adachi, T. Goto, and Y. Magariyama, "Improvement in motion efficiency of the spirochete *Brachyspira pilosicoli* in viscous environments," *Biophys. J.* **90**, 3019–26 (2006).

[200] M. Kim, J. C. Bird, A. J. V. Parys, K. S. Breuer, and T. R. Powers, "A macroscopic scale model of bacterial flagellar bundling," *Proc. Nat. Acad. Sci. USA* **100**, 15481–5 (2003).

[201] T. Ishikawa and M. Hota, "Interaction of two swimming *Paramecia*," *J. Exp. Biol.* **209**, 4452–63 (2006).

[202] A. P. Berke, L. Turner, H. C. Berg, and E. Lauga, "Hydrodynamic attraction of swimming microorganisms by surfaces," *Phys. Rev. Lett.* **101**, 038102 (2008).

[203] E. Lauga, W. R. Diluzio, G. M. Whitesides, and H. A. Stone, "Swimming in circles: motion of bacteria near solid boundaries," *Biophys. J.* **90**, 400–12 (2006).

[204] E. Ben-Jacob, I. Cohen, O. Shochet, A. Tenenbaum, A. Czirók, and T. Vicsek, "Cooperative formation of chiral patterns during growth of bacterial colonies," *Phys. Rev. Lett.* **75**, 2899–902 (1995).

Index

Printed in the United States
By Bookmasters